BIOTECHNOLOGY AND SUSTAINABLE AGRICULTURE 2006 AND BEYOND

Heike Dörnenburg

Biotechnology and Sustainable Agriculture 2006 and Beyond

Proceedings of the 11th IAPTC&B Congress, August 31-18, 2006 Beijing, China

Edited by

Zhihong Xu

Peking University, China

Jiayang Li

Institute of Genetics and Developmental Biology, Chinese Academy of Sciences, Beijing, China

Yongbiao Xue

Institute of Genetics and Developmental Biology, Chinese Academy of Sciences, Beijing, China

and

Weicai Yang

Institute of Genetics and Developmental Biology, Chinese Academy of Sciences, Beijing, China

Springer

A C.I.P. Catalogue record for this book is available from the Library of Congress.

ISBN 978-1-4020-6634-4 (HB)
ISBN 978-1-4020-6635-1 (e-book)

Published by Springer,
P.O. Box 17, 3300 AA Dordrecht, The Netherlands.

www.springer.com

Printed on acid-free paper

CONTENTS

2006 Science and Technology Exhibitors...xix

Opening Speech by Zhihong Xu, President of the IAPTC&B....................xxi

PLENARY LECTURES

Plant Biotechnology and Crop Improvement in China........................3
Zhihong Xu

Facing the Future with Pharmaceuticals from Plants.........................13
Rainer Fischer, Richard M. Twyman, Stephan Hellwig, Jürgen Drossard, and Stefan Schillberg

Embryogenesis In Vitro...35
Jim M. Dunwell

Multigene Strategies for Engineering the Phytoremediation of Mercury and Arsenic...49
Richard B. Meagher, Aaron P. Smith, Melissa Pischke, Tehryung Kim, Om Prakash Dhankher, and Andrew C.P. Heaton

Molecular Control of S-RNase-based Self-Incompatibility....................63
Jian Huang, Lan Zhao, and Yongbiao Xue

The Use of Sexual Model Systems to Identify Elements of Apomixis...77
Ueli Grossniklaus, Philippa Barrell, Amal Johnston, Ramaurthy Baskar, Jacqueline Gheyselinck, Lukas Brand, and Mark Curtis

Agricultural Biotechnology in China: Past, Present and Future...81
Valerie J. Karplus and Xing Wang Deng

SYMPOSIA

Current Advances in Plant Tissue Culture Techniques

Regulation of Plant Morphogenesis In Vitro:
Role of Ethylene and Polyamines..89
Eng-Chong Pua

Lingfasu and Its Use on Plant Tissue Culture In Vitro.........................97
Xu Hong-Yuan, Xu Hong-Zhang, He Bing, and Zhou Feng-Jue

PPV-Specific Hairpin RNAs is an Effective Method
for Plum Pox Potyvirus Resistance..103
Lining Tian, Shuocheng Zhang, HJIPne SanfaHon, Antonet Svircev, Daniel C. Brown, and Rui Wen

An Efficient Novel Method of Producing Virus Free
Plants from Garlic Root Meristem..107
Md. Shahidul Haque, K. Hattori, A. Suzuki, and T. Tsuneyoshi

Site-Directed Homologous Recombination in Tobacco
Cell Cultures via Zinc Finger Nucleases..111
Charles Q. Cai, W. Michael Ainley, Trevor N. Collingwood, Robbi J. Garrison, Lisa L. Schulenberg, Andrew Worden, Philip D. Gregory, Beth C. Rubin-Wilson, and Joseph F. Petolino

Generation of Genetically Modified Strawberries
in Temporary Immersion Bioreactor System..115
Kati Hanhineva, Harri Kokko, and Sirpa Karenlampi

Current Advances in Plant Molecular Biology

The Positioning of the Division Plane Is Determined
by a Cortical Division Site at the Plasma Membrane..........................121
Danny Geelen

Repression of Plant Gene Expression via Chromosomal
Remodelling Using Histone Deacetylases..125
Lining Tian, Keqiang Wu, Brian Miki, and Dan Brown

Functional Genomics

DNA-METHYLTRANSFERASE 1 is a Member of FIS Polycomb Complex and is Involved in Seed Development in *Arabidopsis*............... 131
Ming Luo, Mingzhu Luo, Fred Berger, E.S. Dennis, Jim W. Peacock, and Abed Chaudhury

Integrating Transcriptional and Metabolic Profiling to Unravel Secondary Metabolite Biosynthesis in Plants.............................. 135
Heiko Rischer, Matej Oresic, Alain Goossens, Dirk Inzé, and Kirsi-Marja Oksman-Caldentey

Plant Transformation Technology

Application of Two New Selectable Marker Genes, *dsdA* and *dao1* in Maize Transformation....................................... 141
Fang-Ming Lai, Kangfeng Mei, Luke Mankin, and Todd Jones

Efficient *Agrobacterium*-Mediated Transformation of Various Barley (*Hordeum vulgare* L.) Genotypes............................ 143
Goetz Hensel, Vladimir Valkov, Conny Marthe, and Jochen Kumlehn

In vitro System as the Model for Discovery in Molecular Biology

Use of Alfalfa In Vitro Cultures in Studies on Regulation of Cyclin-Dependent Kinase (CDK) Functions.............................. 149
Dénes Dudits, Mátyás Cserháti, Pál Miskolczi, Attila Fehér, Ferhan Ayaydin, and Gábor V. Horváth

Phytohormone Regulation of Cotton Fiber Development In Vitro......... 153
Barbara A. Triplett, Hee Jin Kim, Doug Hinchliffe, Sing-Hoi Sze, Peggy Thaxton, David Stelly, and Z. Jeffrey Chen

Patterns of Growth and Cell Death in Diploid *Arabidopsis* Cell Cultures with Synchronised Cell Cycle.............................. 157
Ranjith Pathirana, Phillip West, and Jocelyn Eason

Molecular Control of Abiotic Resistance

Engineering of Reactive Species Detoxification Pathways for Increasing Stress Tolerance in Plants..................................163
Gábor V. Horváth, Zoltán Turóczy, Mátyás Cserháti, Petra Kis, Katalin Török, László Sass, and Dénes Dudits

Orchid Flower Fragrance Biosynthesis Pathway and Its Key Enzyme Geranyl Diphosphate Synthase..................................167
Yu-Yun Hsiao, Wen-Chieh Tsai, Chang-Sheng Kuoh, Tian-Hsiang Huang, Hei-Chia Wang, Yann-Lii Leu, Tian-Shung Wu, Wen-Huei Chen, and Hong-Hwa Chen

Cytokinin Function in Drought Stress Response and Subsequent Recovery..................................171
M. Novakova, P. Dobrev, V. Motyka, A. Gaudinova, J. Malbeck, J. Pospisilova, D. Haisel, H. Storchova, J. Dobra, M.C. Mok, D.W.S. Mok, R. Martin, and R. Vankova

Xenogenomics: Bioprospecting for Genetic Determinants of Cold and Freezing Stress Tolerance in the Cryophilic Antarctic Hair Grass *Deschampsia antarctica* E. Desv. 175
U.P. John, R.M. Polotnianka, A. Sivakumaran, L. Mackin, M.J. Kuiper, J.P. Talbot, O. Chew, E. Ribarev, D. Azria, D. Gunawardana, and G.C. Spangenberg

Plant Insect Interaction

Biotechnological Control of Plant Parasitic Nematodes..................181
P.E. Urwin

Molecular Basis of Colorado Potato Beetle Adaptation to Potato Defence Mechanisms..................................191
N. Slapar, M. Pompe-Novak, M. Buh, M. Ravnikar, B. Štrukelj, M.A. Jongsma, and K. Gruden

Molecular Mechanisms on Plant-Microbe Interactions

An Arabidopsis Nudix Hydrolase Modulates the Innate Immune Response Independent of NPR1 and Salicylate..................199
Yiji Xia, Xiaochun Ge, Guo-Jing Li, and Huifen Zhu

RNA-Mediated Resistance to Cassava Geminiviruses in Transgenic Cassava..201
Hervé Vanderschuren, Rashid Akbergenov, Mikhail M. Pooggin, Thomas Hohn, Wilhelm Gruissem, and Peng Zhang

The Molecular Basis of Cold-Induced Pollen Sterility in Rice...............205
Sandra N. Oliver, Xiaochun Zhao, Elizabeth S. Dennis, and Rudy Dolferus

Regulation of the Rice NADPH Oxidase by Binding of Small GTPase Rac and Ca^{2+} to Its N-terminal Extension...........................209
Hann Ling Wong, Reinhard Pinontoan, Kana Hasegawa, Takashi Yaeno, Koh Iba, Ryo Tabata, Kokoro Hayashi, Chojiro Kojima, Tsutomu Kawasaki, and Ko Shimamoto

Post-Transcriptional Gene Silencing of the p23 Silencing Suppressor of *Citrus tristeza Virus* Confers Resistance to the Virus in Transgenic Mexican Lime....................................211
Carmen Fagoaga, Carmelo López, Alfonso Hermoso de Mendoza, Pedro Moreno, Luis Navarro, Ricardo Flores, and Leandro Peña

Phloem Feeding Regulates the Plant Defense Pathways Responding to Both Aphid Infestation and Pathogen Infection..............215
Yinghua Huang

Metabolic Engineering

Towards Metabolic Engineering of Carotenoid Content in Sweet-orange (*Citrus sinensis* (L.) Osb.)...................................223
Marcio G.C. Costa, Amanda F.S. Mendes, Luciana C. Cidade, Walter S. Soares-Filho, Wagner C. Otoni, and Gloria A. Moore

Engineering New Crops for Safe Castor Oil Production.........................227
Grace Q. Chen, Yeh-Jin Ahn, and Louisa Vang

Biochemical and Structural Bases for (Iso)flavonoid Biosynthetic Diversity..231
Chang-Jun Liu, Wang Cheng, Joseph P. Noel, and Richard A. Dixon

Utilization of Plant Metabolic Engineering for Production of Pharmaceuticals in Tea.....235
A. Borthakur, R.K. Dutta, S. Borchetia, and Sudripta Das

Improving Nutritional Quality of Crops

Success Towards Alleviating Peanut Allergy: The Major Allergen Ara h 2 is Silenced via RNA Interference (RNAi).....261
K.N. Konan, O.M. Viquez, F.C. Chen, and H.W. Dodo

Transgenic Poplar for Phytoremediation.....265
Won-Yong Song, Young-Im Choi, Donghwan Shim, Do-Young Kim, Eun-Woon Noh, Enrico Martinoia, and Youngsook Lee

The Africa Biofortified Sorghum Project – Applying Biotechnology to Develop Nutritionally Improved Sorghum for Africa.....273
Zuo-yu Zhao

Genetic Engineering of Strawberry for Taste Improvement and Enhanced Disease Resistance by Introduction of *thauII* Gene.....279
K.A. Schestibratov, and S.V. Dolgov

Genetic Engineering of Crops for Improved Nutritional Quality.....283
Samuel S.M. Sun, Qiaoquan Liu, and Rebecca M.L. Chan

Transcriptome Profiling of Wheat Genotypes Differing in Bread Baking Quality under Different Environments.....289
Yongfang Wan, Rowan Mitchell, Tim Wheeler, Mike Gooding, Rebecca Poole, Keith Edwards, Clare Mills, Peter Skeggs, Salvador Gezan, Sue Welham, Christina Shenton, Tong Zhu, and Peter Shewry

Signal Transduction

Cotton *14-3-3L* Gene Is Preferentially Expressed in Fiber.....295
Xue-Bao Li, Hai-Yan Shi, Xiu-Lan Wang, Wen-Liang Xu, Deng-Di Li, and Hong Wang

Inositol Polyphosphate 6-/3-kinase (*AtIpk2β*), an Early Auxin-Responsive Gene, Positively Regulates Axillary Shoot Branching in *Arabidopsis thaliana*..299
Zai-Bao Zhang, Guang Yang, Zhen Chen, Yan Li, Fernando Arana, and Hui-Jun Xia

Apomixis

Genetic Analyses of Aposporous Embryo Sac Formation in Sorghum..305
John G. Carman, Michelle S. Jamison, Jayasree Pattanayak, Jeff Lacey, Jeong-Soon Kim, Estella G. Elliott, Patricia Klein, Tom Ulrich, and Krishna Dwivedi

Epigenetic Regulation of Seed Development................................309
Claudia Köhler

Biotechnology in Cereals

Optimisation of Tissue Culture, Regeneration and *Agrobacterium*-Mediated Transformation Paramaters in Winter Wheat Cultivars (*T.durum* cv. Kızıltan-91 and *T.aestivum* cv. Bezostaja-01)..315
M. Kavas; H.A. Öktem, and M. Yücel

Enhancement of the Brassinosteroid Biosynthesis Pathway Improves Grain Yield in Rice..319
Chuan-Yin Wu, Shing Kwok, Sam Harris, Anthony Trieu, Parthiban Radhakrishnan, Andres Salazar, Ke Zhang, Jiulin Wang, Jianmin Wan, Shozo Fujioka, Ken Feldmann, and Roger Pennell

Functional Genomics of Hexaploid Wheat..................................323
Yasunari Ogihara

Biotechnology in Economic Crops

Isolation and Characterization of Senescence-Associated Ethylene Genes from *Dendrobium* Orchids...............................327
Zuliana Razali, S. Chandran, Lee Ai Ling, A.N. Boyce, and Helen Nair

Biotechnological Improvement of Forage Crops..............................333
Zeng-Yu Wang, Jeremey Bell, Xiaofei Cheng, Yaxin Ge, Kun Jun Han, Xuefeng Ma, Elane Wright, Yajun Xi, Xirong Xiao, Ji-Yi Zhang, Andrew Hopkins, and Joseph Bouton

Biotechnology in Medicinal, Woody and Horticultural Plants

Lignan and Norlignan Biosynthesis and Biotechnology.....................341
Toshiaki Umezawa

Polyketide Metabolism in *Hypericum perforatum* and Related Species..345
Ludger Beerhues and Benye Liu

In Vitro Selection and Molecular Markers for Early Screening of Fusarium Head Blight Resistance Wheat.....................349
F. Eudes, Ana Badea, A. Laroche, D. Gaudet, R. Graf, and S. Sadasivaiah

The Geo-Model for Cell Wall Formation: A Design Tool for Plants..353
Carolina Cifuentes and Anne Mie C. Emons

Tissue Culture for Biodiversity Preservation

Integrating In Vitro Methods for Propagating and Preserving Endangered Plants..363
V.C. Pence, S.M. Charls, B.L. Plair, M.A. Jaskowiak, G.D. Winget, and L.L. Cleveland

Micropropagation and Germplasm Conservation of Staple Food Crops and Traditional Medicinal Plants in Micronesia..............375
Virendra Mohan Verma

Efficient In Vitro Bulblet Regeneration from Immature Embryos of Endemic and Endangered Geophyte Species in *Sternbergia*, *Muscari* and *Fritillaria* Genera............................381
Sebahattin Özcan, İskender Parmaksız, Semra Mirici, Satı Çöçü, Serkan Uranbey, Arif İpek, Cengiz Sancak, Ercüment O. Sarıhan, Bilal Gürbüz, Cafer S. Sevimay, and Neşet Arslan

In Vitro Strategies for Conservation of Plant Diversity in India..........385
Pramod Tandon

Plants as Bioreacter/Molecular Pharming

Molecular Farming in Plants: An Update..389
Elisabetta Lupotto and Maria Rosaria Stile

In Vivo and In Vitro Studies on Huperzine A Producing Ferns............397
Somika Bhatnagar, Yuan Qing Deborah Hong, Guang Yuan Wang, and Yan Hong

Production of Recombinant Antibodies in Pea Seeds and Their Oral Application in Piglets..399
Isolde Saalbach, Marcus Riehl, Martin Giersberg, Jochen Kumlehn, and Dieter Falkenburg

Molecular Breeding

Breeding for Improved and Novel Starch Characteristics in Potato........405
Farhad Nazarian Firouzabadi, Geraldine Kok-Jacon, Ji Qin, Jean-Paul Vincken, Luc Suurs, and Richard G.F. Visser

Biofortification of Cassava Using Molecular Breeding......................409
Kirsten Jørgensen, Charlotte Sørensen, Susanne Jensen, Marc Morant, Martin Fregene, and Birger Lindberg Møller

Evaluation of GM Poplars Expressing Relevant Traits for Herbicide Tolerance, Disease Resistance and Production of Pharmaceutics: Biochemical, Molecular and Microbiological Studies on Plants and Detection of Transgene Sequences in Soil.............................413
S. Zelasco, D. Carbonera, A. Giorcelli, M. Confalonieri, F. Mattivi, M. Bonadei, M. Gennaro, E. Quattrini, P. Calligari, F. Picco, G. Deandrea, and A. Balestrazzi

Molecular Breeding of Low-Phytic-Acid Grains in Rice by Using the Promoter of 18 kDa Oleosin....................................419
Mio Kuwano, Makoto Suzuki, Fumio Takaiwa, and Kaoru T. Yoshida

Regeneration, Micropropagation and Industrialization

A Highly Efficient Protocol for Micropropagation of North American Ginseng..425
Sijun Zhou and Daniel C.W. Brown

The Use of Tissue Culture for Successful and Sustainable Cane..429
Satjaporn Chantawong, Upsorn Pliansinchai, Chalermpol Kirdmanee, and Pipat Weerathaworn

Development of Subclonal Variants from Interspecific Hybrids of Sugarcane...433
S. Rajeswari, Shini Sekar, and M. Krishnamurthi

Forcing and In Vitro Establishment of Softwood Shoots from Large Stem Segments of Woody Plants...................................437
Faheem Aftab and John E. Preece

Protoplasts, Haploids and Embryo Culture in Crop Improvement

Introgression Breeding Program in *Lolium*/*Festuca* Complex Using Androgenesis...447
Toshihiko Yamada, Yang-dong Guo, Yuuko Mizukami, Ken-ichi Tamura, and Kazuhiro Tase

Recent Advances in Anther Culture of Coconut (*Cocos nucifera* L.)...451
P.I.P. Perera, V. Hocher, J.L. Verdeil, D.M.D. Yakandawala, and L.K. Weerakoon

Protoplast Isolation and Cultivation from Embryogenic Suspension Cultures and Leaves of Grapevines (*Vitis* L.)...................457
Jiang Lu, Xia Xu, and Judy Grosser

Protoplast Technology and Citrus Improvement...461
W.W. Guo, X.D. Cai, Y.J. Cheng, J.W. Grosser, and X.X. Deng

Microspore Embryogenesis in the Apiaceae.................................465
A.M.R. Ferrie and T.D. Bethune

Technology Transfer to Developing Countries

Integration of Cold Signal Transduction Pathway Related to ABA 8'-hydroxylase in Arabidopsis...471
Xin Zhou, Fei He, Fengxia Liu, Xue Zheng, Chao Di, Shaoxia Zhou, Hongya Gu, Wenying Xu, and Zhen Su

Rice Genetics and Genomics Information as the Ultimate Tool for Rice Improvement...475
Takuji Sasaki, Jianzhong Wu, and Takashi Matsumoto

The Development of Agricultural Biotechnology Capacities in Palestine through the UNESCO Biotechnology Educational and Training Center at Bethlehem University....................................487
Naim Iraki, Omar Dar-Issa, Basma Sandouka, Michael Sansour, Rami Arafeh, Amani Abu-Sa'da, and Nida' Salah

Author Index..491

From left: Arie Altman, Zhihong Xu (President), Indra K. Vasil, Atsushi Komamine, Yongbiao Xue (Chair-Local Organizing Commitee, Co-Editor of In Vitro Plant), Eng-Chong Pua (Editor of In Vitro Plant).

IAPTC&B President Zhihong Xu announcing the election of Dr. Roger Beachy of USA as President of IAPB for 2006–2010, and of St. Louis as the host city of the 12th Congress in 2010.

Indra K. Vasil on behalf of President Roger Beachy of 12th IAPB (2006–2010) receive the logo from President Zhihong Xu of 2002–2006 of IAPTC&B.

2006 Science and Technology Exhibitors

Alto Holdings Limited, New Zealand

DBN Group, China

Duchefa Biochemie bv, The Netherlands

Eppendorf China Ltd., China

Phyto Technology Laborateries, USA

Sigma-Aldrich Trading Co. Ltd., China

Springer, Germany

Sunshine Horticulture Co., Ltd., China

Yantze Deltaregion Institute of Tsinghua University, Zhejiang, China

Opening Speech by Zhihong Xu, President of IAPTC&B 2006

Good Morning, Ladies and Gentlemen!

It has been 4 years since we met in Orlando, USA in June, 2002.

On behalf of the International Association for Plant Tissue Culture and Biotechnology, I would like to express my warm welcome to all the delegates attending the 11th International Congress of Plant Tissue Culture and Biotechnology.

This Congress is certainly a world-class event with over 800 participants from 94 countries. In the following 5 days, the best and the brightest of international scientific leaders will present 14 plenary, 34 keynote and 120 symposia lectures covering the entire research field of plant biotechnology, as well as laws and policies regarding transgenic plants and technology transfer. More than 600 posters will augment the formal lecture sessions. The IAPTC&B awarded 15 fellowships to graduate students and post-doctoral associates from developing countries.

The International Association for Plant Tissue Culture and Biotechnology is one of major international professional associations in the field of life science. The mission of the Association is to promote research in all aspects of basic and applied research in plant tissue culture and biotechnology. The theme of this Congress is 'Biotechnology and Sustainable Agriculture'. It is a theme that is especially timely for plant biologists working today. I hope that this Congress will further promote global research activities in the field of plant biotechnology and thus make greater contributions to the world people.

Besides scientific activities in the Congress, I hope you will also enjoy your stay in Beijing. Beijing is one of the most attractive and cosmopolitan cities in the world. It is the political and cultural center of ever-growing China, home of 11 million citizens, seat of the Chinese government, a city highly celebrated for its beautiful surroundings and well-preserved historic sites and cultural relics. I would like to suggest that you enjoy your hot summer by visiting the Summer Palace where the emperors enjoyed their hot summer in old days.

Finally, I wish the Congress success, and all of you a pleasant stay in Beijing.

PLENARY LECTURES

Zhihong Xu

Plant Biotechnology and Crop Improvement in China

Zhihong Xu

College of Life Sciences, Peking University, Beijing, China.
Email: xuzh@pku.edu.cn

In 2005, the total food production in main land of China reached 484 millions tons. By the year of 2030, with the population increase and economic development, food production in China is supposed to increase by 30%, reaching 650 millions tons. Thus the main targets of agriculture in China are to develop modern agriculture with higher yield, better quality and higher efficiency, and to provide various agricultural products to meet continuously increasing demands of the Chinese people. The application of modern techniques of crop improvement could be one of the most important factors for sustainable agriculture.

In the 10th Five-Year Economic Plan (2001–2005), the budget for biotechnology from the government increased three to four times more than that in the past 15 years, and came to 5 billion RMB ($600 millions). Total funding for life sciences and biotechnology from government was about 15 billion RMB. Since the Chinese National High-Tech R&D Program was initiated in 1986, agricultural biotechnology has always been one of the priorities in the Program. The research projects mainly include plant tissue culture, transgenic plants and genetic engineering, genetic mapping and marker-assistant breeding, and genomics and functional genomics of main crops.

Z. Xu et al. (eds.), Biotechnology and Sustainable Agriculture 2006 and Beyond, 3–10.

Plant tissue culture techniques have been extensively applied in agriculture in China for a long time, especially in horticulture, such as clonal propagation combined with virus-free technique, somatic hybridization by protoplast fusion for producing somatic hybrids, and embryo rescue technique for obtaining hybrid plants, etc.

Since the late 1980s, the research projects on transgenic plants and genetic engineering have been initiated with different plant species, covering pest-resistance, stress tolerance, male sterility, nutrition improvement, phytoremediation, and using transgenic plants as bioreactor. Transgenic plants obtained include those main crops, such as rice, wheat, maize, potato, soybean, rapeseed, and a lot of horticultural plants. Total acreage of transgenic plants reached 3.7 millions ha in 2006. During the period of 2000–2005, 473 transgenic plant strains and lines, covering over 40 plant species, were approved for biosafety evaluation. Among them, 199 were for small scale field testing, 114 for environmental release, 102 for preproduction test, and 58 for commercialization (the data from the China Biotechnology Development Center, Ministry of Science and Technology, 2006; also see Biosafety Office of Agricultural GMOs, MOA, 2006). In this report, I had like to describe some of main progress made in China, since 2000.

Cotton and rice are the most successful species among the transgenic crops in China. Transgenic cotton and rice have experienced almost 10 years' biosafety evaluation in field test. Bt cotton is the first commercialized crop in China that has really used in agriculture so far (see Shirong et al., 2004). Fifty eight transgenic cotton varieties (Bt cotton) have been released for commercial production. The planting acreage of Bt cotton reached 3.5 millions ha in 2005, accounting for 65% of total cotton planting acreage in China. It resulted in increasing economic benefit of 40 billion RMB (about $5 billion) during 1999–2006, and remarkable decrease of the use of pesticide, 20–30 millions kg/year (it is about 7.5% of annual insecticide production in China).

Lepidopteran insects (stripped stem borer and leaffolder, etc.) caused serious damage to rice production. In China, rice yield loss caused by insects is more then 5% of the total production, i.e., 10 millions tons. Expenditure for chemical insecticide and application cost about $3 billion per year. Bt, CpTI and SCK (a modified CpTI) genes have been transferred to rice for the resistance to stripped stem borer and leaffolder. Preproduction test of two insect-resistant rice varieties in Fujian shows that the yield increases 6%, pesticide use reduces 80% (Huang et al., 2005). The Ministry of Agriculture is reviewing the results of biosafety evaluation of insect-resistant rice. We are expecting some of new transgenic rice varieties could be released for commercialization in China in the near future. Transgenic maize and poplar with Bt gene has also showed good insect resistance in preproduction test. Besides, aphid-resistant soybean line with *gna* gene from *Galanthus nivalis* (encoding agglutinin) has been approved for field testing in Jilin Academy of Agricultural Sciences.

Disease resistance is another important area in plant molecular biology and biotechnology in China. For example, Xa 21 has been extensively used for increasing the resistance to rice bacterial blight, since it is one of the most serious diseases for rice in the world, generally reducing rice output by 10–30%, sometimes up to 50%. Some of the transgenic rice varieties with rice bacterial blight resistance have been evaluated for productive testing. After years' field testing, transgenic chili pepper with CMV- and TMV-cp gene continues to show good performance of virus resistance in field (Cai et al., 2003). In South China Agricultural University, transgenic papaya with replicase gene of PRSV (papaya ringspot virus) shows high quality and virus resistance in field test, and has also been approved for productive testing.

Some new progress has been made in stress tolerance. *SKC1* gene as been cloned from a Japanese Indica rice, Nona Bokra, that is related to Na+ transport (Ren et al., 2005). High expression of stress responsive *NAC1* apparently improves drought tolerance of transgenic rice. The gene is drought-inducible, expressed in guard cells (Hu et al., 2006). Transfer of mangrove Na+/H+ antiporter gene to poplar has engineered salt-tolerant plant, that could be grown in the medium containing 150 mM NaCl. Overexpression of *Arabidopsis NHX4* gene confers enhanced tolerance to salinity in transgenic sugar beet (Zhang, 2006) and perennial ryegrass (Wu et al., 2005). For improving drought and cold tolerance, targeting for water saving and longer period of greening in winter in Beijing, dehydrin gene was cloned from *Boea crassifolia*, and *AdZFP1* (Zinc finger transcription factor) from *Artemisia desertorum* by scientists of Peking University who have been working on grass species for gene transfer, that include: tall fescue, red fescue, ryegrass, bentgrass and bluegrass. Transgenic tall fescue showed strong tolerance to water deficit, and better and longer green color in winter (Wu et al., 2004).

Since *Wx* gene was cloned from rice and a series of research work on rice *Wx* gene have been carried out in Prof. MM Hong's laboratory in Shanghai Institute of Plant Physiology and Ecology, transfer of anti-*Wx* gene has been used to regulate amylose content and then to improve the quality of rice (Liu et al., 2003). In China Agricultural University, *SB 401* gene, coding a pollen specific protein with rich lysine content, was cloned from *Solanum berthaultii.* Seed-specific expression of this gene significantly increases both lysine and total protein content in maize seeds (Yu et al., 2004). More than 20 maize inbred strains with high levels of lysine and proteins have been obtained. Among them, lysine content increases 16.1–54.8%, and total protein increases 11.6–39.0% in seeds. In Biotechnology Research Institute, CAAS, phytase gene has been cloned from *Aspergillus* sp. and transferred to maize. Its high expression (5000 U/kg seeds) improves feeding efficiency of maize seeds, resulting in reducing environmental pollution by high phosphorus content in animal manure (Fan et al., 2003). Thioredoxin h plays an important role in cereal seed germination. Transgenic wheat with antisense *thioredoxins* (*trxs*) gene (the *trxs* gene cloned from *Phalaris coerulescens*) shows remarkable pre-harvest sprouting tolerance (Liu et al., 2004; Zhou et al., 2006). A transgenic wheat line against pre-harvest germination has been approved for field testing now.

Transgenic plants are also expected to be used for phytoremediation. For example, transgenic tobacco with *merA* shows the resistance to phenyl mercuric acetate (PMA), one of very toxic pollutants. They have an increased capacity of mercury uptake, but avoid entrance of mercury to chloroplasts, that protects the chloroplasts from mercury damage (He et al., 2001; He, 2005). The strategy may be helpful for wetland phytoremediation (Czako et al., 2005). Transgenic *Arabidopsis* with *Ga LAC1* encoding a secretory laccase, cloned from *Gossypium arboreum*, have about 15 times higher laccase activity in the roots than the wild one has. The activity of laccase secreted in the medium increases 35 times. They show enhanced resistance to phenolics and trichlophenol. The results demonstrate that *in vitro* the laccase culture medium catalyzes transformation of 2,4,6-trichlorophenol (TCP, one of the most toxic and wildly spread environmental pollutants) (Wang et al., 2004). Some laboratories are working on transgenic plants as bioreactor, for example, GM tomato line with outer membrane antigen-protein gene of Hepatitis B virus has been proved for productive testing to produce Hepatitis oral vaccine, and GM rapeseed produces salmon calcitonin in seeds (3 mg/g) (Biotechnology Research Institute, CAAS). *Artemisia annua* contains artemisinis against malaria, but its content in plant normally does not exceed 0.8%. Several key branch point enzyme genes in isopronoid pathway (e.g., sesquiterpene synthase, squalene synthase) have been cloned, scientists in the Institute of Botany, CAS, are trying to increase the content of artemisinin by regulating secondary metabolism.

With rapid development of plant genetic engineering used for crop improvement, great progress has been made in plant molecular biology and genomics of both *Arabidopsis* and rice, supported by the National Basic Research Development Program (initiated in 1997), the Chinese National High-Tech R&D Program (initiated in 1987) and the National Natural Science Foundation of China (see Xue et al., 2003; Chen et al., 2006; Xu, 2006; Xu and Li, 2006). The first workshop of *Arabidopsis* research in China was held in Shanghai in 2002, organized by the Institute of Shanghai Institute of Plant Physiology and Ecology, CAS. Since then, *Arabidopsis* research community has expanded very rapidly (Xu, 2006). Since the program of rice genomics was initiated, both draft sequence of *Oryza sativa* spp. *Indica* (Yu et al., 2002) and the sequence of Chromosome 4 of *O.sativa* spp. *Japonica* (Feng et al., 2002) completed in 2002. Rice functional genomics program focuses on expression profiling and discovery of new genes of agronomic significance. Some very important genes related to rice development have been cloned in the Institute of Genetics, CAS, such as, *MOC1* gene encoding a putative GRAS family nuclear protein, that controls tillering (Li et al., 2003). Several laboratories are working on identification of genes for hybrid rice, Dr. YG Liu's laboratory in South China Agricultural

University have reported that rice male sterility with Boro II cytoplasm is caused by a cytotoxic peptide and restored by two related PPR motif genes (Wang et al., 2006). Besides rice, the functional genomics of cotton, rapeseed, soybean and tomato have been or would be initiated. The knowledge from functional genomics is expected to be used for crop breeding and genetic improvement.

In the future, the research work will focus on cloning more important genes involved in drought/salt tolerance, pest resistance, regulation of growth and development, and quality improvement that can be used for crop improvement, and more transgenic plants would be obtained and more GM crop varieties released for agriculture in China. We believe that the better plant science and biotechnology can be better for sustainable agriculture and the life of human being in the world.

References

Biosafety Office of Agricultural GMOs, MOA (2006) Research and biosafety regulation of agricultural GMOs in China.

Cai WQ, Fang RX, Shang HS, Wang X, Zhang FL, Li YR, Zhang JC, Cheng XY, Wang GL, Mang KQ (2003) Development of CMV- and TMV-resistant chili pepper: Field performance and biosafety assessment. Mol Breed, 11: 25–35.

Chen H, Karplus VJ, Ma H, Deng XW (2006) Plant biology research comes of age in China. Plant Cell 18: 2855–2864.

China Biotechnology Development Center, Ministry of Science and Technology (2006) Review report of the National Project of Transgenic Plants and Their Commercialization.

Czako M, Feng X, He Y, Marton L (2005) Genetic engineering of wetland grasses for phytoremediation: mercury contaminated soil. Environ Geochem Health 27: 397–419.

Fan YL, Chen RM, Yao B, Xue GX (2003) Patent XL 03137476.X

Feng Q, Zhang Y, Hao P, Wang S, Fu G, Huang Y, Li Y, Zhu J, Liu Y, Hu X, Jia P, Zhang Y, Zhao Q, Ying K, Yu S, Tang Y, Weng Q, Zhang L, Lu Y, Mu J, Lu Y, Zhang LS, Yu Z, Fan D, Liu X, Lu T, Li C, Wu Y, Sun T, Lei H, Li T, Hu H, Guan J, Wu M, Zhang R, Zhou B, Chen Z, Chen L, Jin Z, Wang R, Yin H, Cai Z, Ren S, Lv G, Gu W, Zhu G, Tu Y, Jia J, Zhang Y, Chen J, Kang H, Chen X, Shao C, Sun Y, Hu Q, Zhang X, Zhang W, Wang L, Ding C, Sheng H, Gu J, Chen S, Ni L, Zhu F, Chen W, Lan L, Lai Y, Cheng Z, Gu M, Jiang J, Li J, Hong G, Xue Y, Han B (2002) Sequence and analysis of rice chromosome 4. Nature. 420: 316–320.

He YK (2005) Enhanced Hg2+ uptake and reduced mercury contents in transgenic plants expressing the wild-type *merA* gene. Annual Report 2004–2005, Institute of Plant Physiology and Ecology, Shanghai Institutes for Biological Sciences, CAS, p. 34.

He YK, Sun JG, Feng XZ, Czako M, Marton L (2001) Differential mercury volatilization by tobacco organs expressing a modified bacterial *merA* gene. Cell Res 11: 231–236.

Hu HH, Dai MQ, Yao JL, Xiao BZ, Li XH, Zhang QF, Xiong LH (2006) Overexpression of aNAM, ATAF and CUC (NAC) transcription factor enhances drought resistance and salt tolerance in rice. PNAS, 103: 12987–12992.

Huang J, Hu R, Rozelle S, Pray C (2005) Insect-resistant GM rice in farmers' fields: Assessing productivity and health effects in China. Science 308: 688–690.

Li X, Qian Q, Fu Z, Wang Y, Xiong G, Zeng D, Wang X, Liu X, Teng S, Hiroshi F, Yuan M, Luok D, Han B, Li J (2003) Control of tillering in rice. Nature 422: 618–621.

Liu L, Yin J, Ren JP, Han JF (2004) Effects of antisense *trxs* on germination of transgenic wheat seeds. Acta Agron Sin 30(8): 801–805 (in Chinese).

Liu Q, Wang Z, Chen X, Cai X, Tang S, Yu H, Zhang J, Hong M, Gu M (2003) Stable inheritance of the antisense Waxy gene in transgenic rice with reduced amylose level and improved quality. Transgenic Res. 12: 71–82.

Ren ZH, Gao JP, Li LG, Cai XL, Huang W, Chao DY, Zhu MZ, Wang ZY, Luan S, Lin HX (2005) A rice quantitative trait locus for salt tolerance encodes a sodium transporter. Nat Genet 37: 1141–1146.

Shirong J, Sandui G, Daochang A, Guixian X (2004) Transgenic Cotton. Science Press, Beijing, New York.

Wang GD, Li QJ, Luo B, Chen XY (2004) *Ex planta* phytoremediation of trichlorophenol and phenolic allelochemicals via an engineered secretory laccase. Nat Biotechnol 22: 893–897.

Wang Z, Zou Y, Li X, Zhang Q, Chen L, Wu H, Su D, Chen Y, Guo J, Luo D, Long Y, Zhang Y, Liu YG (2006) Cytoplasmic male sterility of rice with Boro II cytoplasm is caused by a cytotoxic peptide and is restored by two related PPR motif genes via distinct modes of RNA silencing. Plant Cell 18: 676–687.

Wu Q, Hu Y, Ni T, Jia W, Lin Z (2004) Research situation and prospects for transgenic turf grass. Pratacultural Sci 21: 29–34.

Wu Yu-Ye, Chen Qi-Jun, Chen Min, Chen Jia, Wang Xue-Chen (2005) Salt-tolerant transgenic perennial ryegrass (*Lolium perenne* L.) obtained by *Agrobacterium tumefaciens*-mediated transformation of the vacuolar Na+/H+ antiporter gene. Plant Sci 169: 65–73.

Xu Zhi-hong (2006) Recent progress in *Arabidopsis* research in China: A preface. J Integrative Plant Biol 48: 1–4.

Xu Zhihong, Li Jiayang (2006) Plant hormones research in China: Past, present and future. Chin Bull Bot 23: 433–442.

Xue Yuebiao, Li Jiayang, Xu Zhihong (2003) Recent highlights of the China rice functional genomics progrm. Trends Genet 19: 390–394.

Yu J, Hu S, Wang J, Wong GK, Li S, Liu B, Deng Y, Dai L, Zhou Y, Zhang X, Cao M, Liu J, Sun J, Tang J, Chen Y, Huang X, Lin W, Ye C, Tong W, Cong L, Geng J, Han Y, Li L, Li W, Hu G, Huang X, Li W, Li J, Liu Z, Li L, Liu J, Qi Q, Liu J, Li L, Li T, Wang X, Lu H, Wu T, Zhu M, Ni P, Han H, Dong W, Ren X, Feng X, Cui P, Li X, Wang H, Xu X, Zhai W, Xu Z, Zhang J, He S, Zhang J, Xu J, Zhang K, Zheng X, Dong J, Zeng W, Tao L, Ye J, Tan J, Ren X, Chen X, He J, Liu D, Tian W, Tian C, Xia H, Bao Q, Li G, Gao H, Cao T, Wang J, Zhao W, Li P, Chen W, Wang X, Zhang Y, Hu J, Wang J, Liu S, Yang J, Zhang G, Xiong Y, Li Z, Mao L, Zhou C, Zhu Z, Chen R, Hao B, Zheng W, Chen S, Guo W, Li G, Liu S, Tao M, Wang J, Zhu L, Yuan L, Yang H (2002) A draft sequence of the rice genome (Oryza sativa L. ssp. indica). Science. 296: 79–92.

Yu J, Peng P, Zhang X, Zhao Q, Zhy D, Sun X, Liu J, Ao G (2004) Seed-specific expression of a lysine rich protein *sb-401* gene significantly increases both lysice and total protein content in maize seeds. Mol Breed 14: 1–7.

Zhang HX (2006) Overexpression of AtNHX4 confers enhanced salt tolerance in sugar beet. Annual Report, National Laboratory of Plant Molecular Genetics, Institute of Plant Physiology and Ecology, CAS, p. 57.

Zhou SM, Yin J, Ren JP, Zhang R (2006) Study on molecular identification and pre-harvest sprouting characteristic of the transgenic anti-*trxs*-gene wheat line 00T89. Chin J Biotechnol 22(3): 438–444 (in Chinese).

Rainer Fischer

Facing the Future with Pharmaceuticals from Plants

Rainer Fischer[1,2], Richard M. Twyman[3], Stephan Hellwig[1], Jürgen Drossard[1], and Stefan Schillberg[1]

[1]Fraunhofer Institute for Molecular Biology and Applied Ecology, Forckenbeckstrasse 6, 52074 Aachen, Germany.
[2]RWTH, Worringerweg 1, 52074 Aachen, Germany.
[3]Department of Biology, University of York, Heslington, York YO10 5DD, UK.

Abstract

Plants are the ultimate source of many of today's pharmaceutical compounds, but most of our protein drugs are derived from animal sources, and are produced either in cultured animal cells or microbes. The biopharmaceutical industry has developed with mammalian cells treated as the gold standard for production, hence the regulations governing biopharmaceutical production have been tailored for these systems. This has had a negative impact on the use as plants for biopharmaceutical production, despite their many potential advantages which include the prospect of inexpensive, large-scale biopharmaceutical production without sacrificing product quality or safety. The first plant-derived pharmaceutical products have now been approved but these represent a tiny proportion of the products in development, products which could have a profound impact on the cost and availability of medicines to those most in need. In this review, we summarize the state-of-the-art in plant-based production systems and discus the development issues which remain to be addressed before plants become an acceptable mainstream production technology.

Z. Xu et al. (eds.), Biotechnology and Sustainable Agriculture 2006 and Beyond, 13–32.

Introduction

Humans have exploited the medicinal properties of plants since the dawn of history, and even today more than one quarter of our drugs are either derived from plants or contain plant extracts (Raskin et al., 2002). The ability of plant cells to express heterologous genes was first demonstrated in 1983, but it was not until 1989 that the first recombinant human protein was expressed in plants (human serum albumin, expressed in tobacco and potato – Sijmons et al., 1990). At about the same time, the ability of plants to express recombinant antibodies was also demonstrated (Hiatt et al., 1989). Since then, over 100 therapeutic and diagnostic recombinant proteins and vaccines have been produced in a variety of plants, including tobacco, cereals, legumes, fruit and vegetable crops, fodder crops, edible foliage such as lettuce and spinach, oilseeds, and aquatic or unicellular plant species grown in bioreactors (Ma et al., 2003; Twyman et al., 2003). Plants have many advantages over traditional production systems based on microbial or mammalian cells, particularly in terms of economy, production scale, safety and practicality (Ma et al., 2003; Fischer et al., 2004; Twyman et al., 2005).

A large number of companies have now been set up to investigate and exploit the potential of plant-derived recombinant vaccines, antibodies and other therapeutic entities. Two such products have recently gained approval for use, the first a tobacco-derived vaccine against Newcastle disease, approved by the USDA in February 2006 for use in poultry. The second was a recombinant antibody against the hepatitis B virus surface antigen, approved for use in Cuba by the Center for State Control of Medication Quality (CECMED) in June 2006. The tobacco-derived antibody is not used directly as a pharmaceutical product, but as an affinity reagent to purify the viral surface antigen which is produced using conventional fermentation technology. CECMED has approved both the product and the process in which it is being used, and importantly has granted permission for the antibody to be produced in field-grown plants. Other pharmaceutical proteins from plants have progressed through clinical or veterinary trials and are approaching regulatory approval.

Lead pharmaceutical proteins produced in plants

Leading plant-derived pharmaceutical products can be divided into three major categories – recombinant antibodies, subunit vaccines and replacement human proteins (enzymes, hormones, blood products etc.). The focus on antibodies reflects the fact that monoclonal antibodies are the fastest-growing segment of the biopharmaceutical market, with 23 approved antibodies or antibody-related proteins on the market and hundreds more in preclinical and clinical development (Gottschalk, 2006). For this reason, the industry is facing a capacity crunch and antibody production in plants offers one way to alleviate this bottleneck. Furthermore, since antibodies have been expressed in plant tissue since 1989 (Hiatt et al., 1989), an impressive collection of proof-of-principle studies has accumulated with antibodies (or 'plantibodies') in many of the alternative plant-based systems listed above discussed in the section above. The possibility of using antibodies as topical therapeutic agents is also attractive, as these might attract a lower regulatory burden compared to injectable molecules. Indeed these three principles lie behind the choice of antibodies against HIV as the major 'fast-track' target molecules by the Pharma-Planta consortium (Ma et al., 2005a, b).

Much the same reasoning applies to the development of subunit vaccines expressed in plants (sometimes known as 'plantigens') since these could be developed as oral vaccines that again would need less stringent purification than those developed for parenteral administration. The development of plant-derived subunit vaccines is also buoyed by the many veterinary applications that are possible, since these could be used as a 'dry run' to determine the suitability of such vaccines for human use. Pharma-Planta is developing several potential vaccines as second-stage targets following in the wake of the anti-HIV antibodies (Ma et al., 2005a, b). The final category of replacement products is popular because plants offer an alternative, and more economical, production platform for replacement products with a low demand, such as products for rare inherited disorders. The most advanced products in these categories are discussed below (Table 1).

Table 1. PDP products that are approved for market or have reached the clinical development stage

Protein	Production system	Company involved	Target disease	Progress
Recombinant antibodies				
Avicidin, IgG against Ep-CAM	Transgenic maize	NeoRx/Monsanto Protein Technology (now dissolved)	Colorectal cancer	Phase II withdrawn
CaroRx, Secretory IgA against S. mutans adhesin	Transgenic tobacco	Planet Biotechnology	Dental caries	Phase II in progress, approved as 'medical device'
scFv antibodies against B-cell tumor	TMV vectors in tobacco	LSCB (now insolvent)	non-Hodgkin's lymphoma	Phase I completed for at least 12 patient-specific products
Recombinant subunit vaccines				
HN protein of Newcastle disease virus	Transgenic tobacco (nicotine-free)	DowAgro	Newcastle disease in poultry	USDA approved
ICAM1	Transgenic tobacco	Planet Biotechnology	Common cold	Phase I
Lt-B vaccine	Transgenic maize	Prodigene Inc.	Traveler's disease	Phase I completed
TGE vaccine	Transgenic maize	Prodigene Inc.	Gastroenteritis in piglets	Phase I completed
Lt-B	Transgenic potato	Arizona State University	E. coli	Phase I
HBsAg	Transgenic potato	Arizona State University	Hepatitis B	Phase I
NV capsid protein	Transgenic potato	Arizona State University	Norwalk virus	Phase I
	Transgenic tobacco	Arizona State University	Norwalk virus	Phase I/II
Rabies glycoprotein	Virus-infected spinach	Fraunhofer CMB	Rabies	Phase I complete

(Continued)

Replacement proteins and other therapeutic products				
Gastric lipase	Transgenic maize	Meristem Therapeutics	Cystic fibrosis	Two phase II trials (formulated and unformulated)
Lactoferrin	Transgenic maize	Meristem Therapeutics	Gastrointestinal disorders	Phase I
α-Interferon	Lemna system	Biolex Inc	Hepatitis B and C, cancer	Two phase I trials (formulated and unformulated)
Fibrinolytic	Lemna system	Biolex Inc	Blood clots	Phase I
Human intrinsic factor	Transgenic Arabidopsis	Cobento Biotech AS	Vitamin B12 deficiency	Phase II/III
Glucocerebrosidase	Carrot cells	Protalix	Gaucher's disease	Phase I
Bacterial antigen	Canola	Nexgen/Guardian Biosciences	Coccidiosis in poultry	CFIA phase 2

Antibodies

The two plant-derived antibody products that have completed Phase II trials are Avicidin and CaroRx. Avicidin is a full length IgG specific for EpCAM (a marker of colorectal cancer) produced in maize and developed jointly by NeoRx and Monsanto. Although Avicidin demonstrated some anti-cancer activity in patients with advanced colon and prostate cancers, it was withdrawn from phase II trials in 1998 because it also resulted in a high incidence of diarrhea (Gavilondo and Larrick, 2000). This was not due to the use of plants as the expression host. Indeed, the same problems occurred with antibodies produced in mammalian cells, and the conclusion of the trial was that the two antibodies were comparable in terms of physicochemical properties, serum clearance, urine clearance and dosimetry. The side effects were possibly due to cross-reaction with related epitopes on the cells lining the intestine.

CaroRx, the most advanced plant-derived antibody, is a chimeric secretory IgA/G produced in transgenic tobacco plants which has completed phase II trials sponsored by Planet Biotechnology Inc. (Ma et al., 1998). As stated earlier, secretory antibody production requires the expression of four separate components, which in this case were initially expressed in four different plant lines that were crossed over two generations to stack all the transgenes in one line. The antibody is specific for SA I/II, the major adhesin of *Streptococcus mutans*, the organism responsible for tooth decay in humans. Topical application following elimination of bacteria from the mouth helped to prevent recolonization by *S. mutans* and led to the replacement of this pathogenic organism with harmless endogenous flora.

The production of anti-idiotype antibodies recognizing malignant B-cells is a useful approach for the treatment of diseases such as non-Hodgkin's lymphoma. McCormick et al. (1999) produced a scFv antibody based on the well-characterized mouse lymphoma cell line 38C13. When administered to mice, the scFv stimulated the production of anti-idiotype antibodies capable of recognizing 38C13 cells, providing immunity against lethal challenge with the lymphoma. This was the prototype for a series of at least 12 patient-specific scFv antibodies which have been tested in phase II trials (by Large

Scale Biology Company, now insolvent) to treat patients with lymphoma. Importantly, since such antibodies can only be developed once the disease is in progress and the malignant cells identified, a long development phase is inappropriate. For this reason, the product was developed using a virus expression vector in tobacco, which allows the rapid production of milligrams of product.

Most recently, the Cuban regulatory agencies have approved the production and use of a plant-derived antibody against hepatitis B virus. Although this is not used directly as a pharmaceutical product, it is used in the purification process to isolate the cognate antigen, i.e., the HBV surface antigen. Because it is used in a biopharmaceutical manufacturing process, the antibody has to meet all the same purity and homogeneity criteria imposed on the product itself.

Vaccine candidates

Plant-derived vaccines can be divided into two categories – those designed for veterinary use and those designed for medical use. A veterinary vaccine was the first PDP to be approved, in February 2006, and there is a large body of both immunogenicity and challenge data to support the efficacy of such vaccines. In a number of reports, plant-derived recombinant subunit vaccines have protected animals against (in some cases lethal) challenges with the pathogen. Clinical trials conducted by ProdiGene Inc., a former biopharmaceutical company based in College Station, TX, have shown that an oral vaccine expressed in plants gives protection against a virulent viral pathogen in livestock. The trials were conducted on swine using an edible form of a vaccine for transmissible gastroenteritis virus (TGEV) (Lamphear et al., 2002). Another PDP veterinary vaccine has been produced by Guardian Bioscience for the prevention of coccidiosis in poultry. This subunit is produced in transgenic canola and CFIA phase II trials are ongoing.

There have been several human clinical trials involving plant-derived, recombinant oral vaccines, all of which have been successful in that they produced serum and/or secretory antibody responses against the antigen in the test subjects. Tacket et al. (1998) performed the first such trial with transgenic potatoes expressing the enterotoxigenic *E. coli* (ETEC) labile toxin B-subunit (LTB), one of the most potent known oral immunogens. The LTB content of the tubers varied between 3.7 and 15.7 μg g^{-1} fresh weight. Fourteen volunteers were given either transgenic or non-transgenic potato on days 0, 7, and 21 of the trial. Almost all of those consuming the transgenic potatoes showed at least four-fold increases in serum IgG against LTB while no such increase was seen in those consuming the non-transformed potatoes. Five of these individuals also demonstrated at least a four-fold increase in anti-LTB IgA, detected in stool samples. There were few side effects, such as nausea and diarrhea. A more recent trial using LTB expressed in processed corn seed produced similar results to the potato study (Tacket et al., 2004).

The same group also described the results of a clinical trial performed using transgenic potato tubers expressing the Norwalk virus capsid protein (NVCP) (Tacket et al., 2000). Twenty adult volunteers were given two or three 150-g doses each of raw transgenic potato tuber containing 215–750 μg NVCP. Despite the fact that only 50% of the NVCP subunits assembled into virus-like particles in the potato cells, thus reducing the effective dose of the vaccine, nearly all of the volunteers showed significant increases in the numbers of IgA-antibody forming cells (AFCs), and six of these individuals also showed increases in IgG AFCs. There were also noticeable increases in serum antibodies against NVCP and stool IgA antibodies in a few of the participants.

A clinical trial has also been carried out using orally delivered HBV surface antigen produced in lettuce (Kapusta et al., 1999). Two of three volunteers who were given two 150-g doses of transgenic lettuce containing about 2 μg of the antigen per dose, produced protective serum antibodies after the second dose, although the titers declined in a few weeks. Even so, the study confirmed that naïve subjects could be seroconverted by the oral delivery of a plant-derived viral antigen. A similar trial in the United States involved the HBV surface antigen expressed in transgenic potatoes although participants in this trial had already been seroconverted with the standard, yeast-derived vaccine (Richter et al., 2000). Of 33 participants given either two or three 1-mg doses of the antigen, about half showed increased serum IgG titers against the virus.

Finally, Yusibov et al. (2002) have carried out a trial involving 14 volunteers given spinach infected with alfalfa mosaic virus vectors expressing the rabies virus glycoprotein and nucleoprotein. Five of these individuals had previously received a conventional rabies vaccine. Three of those five and all nine of the initially naïve subjects produced antibodies against rabies virus while no such response was seen in those given normal spinach. It should be noted that the spinach used in this trial was not strictly transgenic, since the recombinant protein was produced by an engineered plant virus.

Other plant-derived biopharmaceuticals in clinical trials

Meristem Therapeutics has conducted trials of two PDPs produced in maize, gastric lipase and lactoferrin. Gastric lipase has been tested in two phase II trials in Germany and France, one with the product formulated and one with it unformulated. It will be used to treat exocrine pancreatic insufficiency, which occurs in pancreatic disorders, after surgery and as one of the symptoms of cystic fibrosis, and should be launched onto the market in 2006/7. Lactoferrin is currently in Phase I trials. Similarly, Biolex is conducting two phase I trials with α-interferon produced using the *Lemna* system, one using a formulated product and one with the protein unformulated. An Israeli biotechnology company, Protalix, is conducting phase I trials of glucocerebrosidase, for the treatment of Gaucher's disease, produced in cultured carrot cells.

Although there are many PDP products in pre-clinical development, it is perhaps worth singling out insulin produced by SemBioSys in safflower for special mention. The company has achieved a greater than 1% accumulation of the product in safflower seeds and intends to submit an IND application to the FDA in 2007. What is exciting about this application is that because insulin is a simple protein with extensive clinical safety and efficacy data, the company expects that regulatory approval of plant-derived insulin will be based on abbreviated clinical trial protocols once its purity and bioequivalence of commercial insulin (produced in bacteria) has been established. This could mean the product reaches the market as early as 2009 and may therefore be among the first PDPs to see commercial release. SemBioSys estimates that the whole world's demand for insulin could be satisfied by fewer than 16,000 acres of safflower at 30% of the current capital costs of fermenter-based production (SemBioSys, 2006).

Current challenges

The obstacles currently preventing the rapid commercialization of plant-based expression platforms and PDPs can be divided into two major groups – technical and political. Overall, it is the technical hurdles which are falling the most quickly. Key developments over the last few years include yield improvements and addressing issues concerning the presence of plant glycan groups that are not found in mammals. Advances in both these areas have been extensive – high yields are now possible (up to 36% of total soluble protein in nuclear transgenic plants, when 1% is considered the threshold for commercial viability) and the genetic modification of various plants to alter glycan synthesis has resulted in the production of humanized glycoproteins that resemble their native counterparts even more closely than the so-called authentic recombinant proteins produced in rodent cells.

Protein folding, assembly and glycosylation

The ability of plants to fold and assemble complex human proteins has been demonstrated by the successful production of functional serum antibodies using plants carrying two transgenes (encoding the heavy and light chain components, respectively). Such antibodies comprise four polypeptides – two heavy and two light chains – covalently joined by disulphide bonds. More remarkably, plants also assemble functional secretory antibodies, which have ten polypeptides representing four different polypeptide chains (the heavy and light chains, and additional joining chain and secretory component subunits) (Ma et al., 1995).

Although the protein synthesis and folding pathways are highly conserved between plants and animals, there are some differences in the capacity for post-translational modification. Plants do not, for example, hydroxylate proline residues in recombinant collagen.

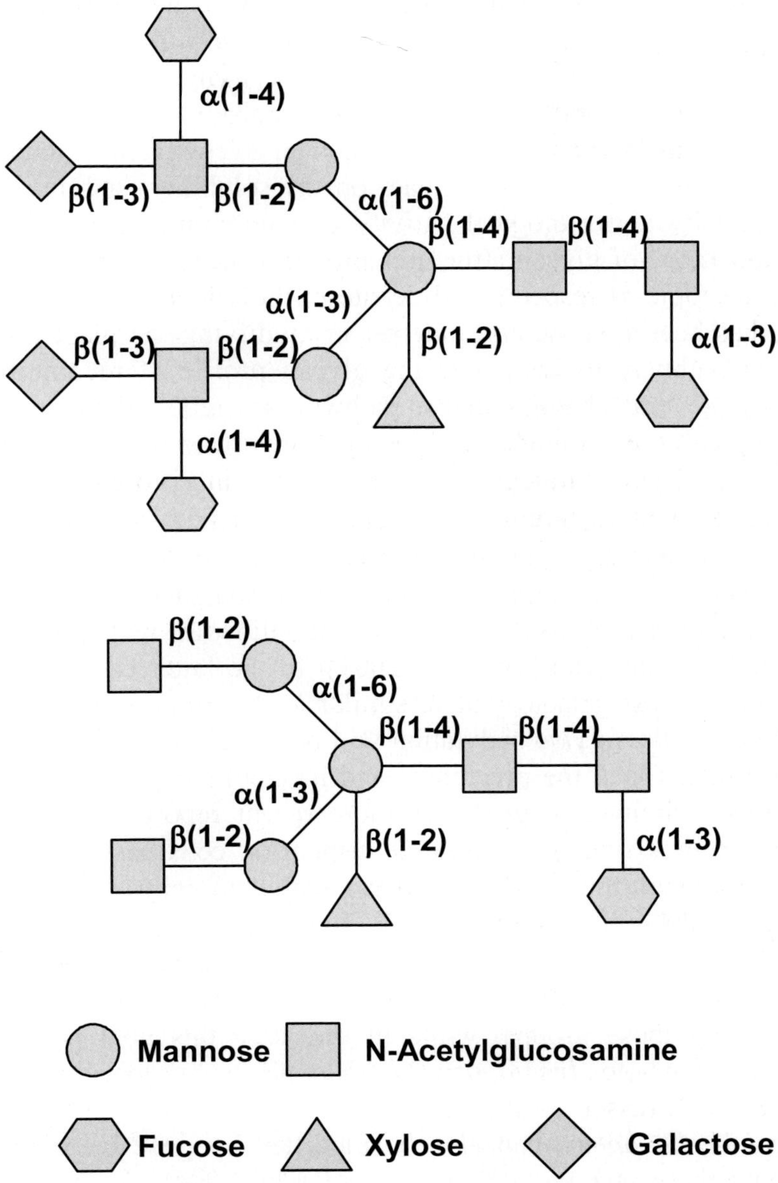

Figure 1. Two glycan structures produced in plants. The upper panel shows a galactose-extended complex glycan, while the lower panel shows a long chain complex glycan. The xylose and α(1,3)fucose residues are not found in mammals.

There are also various differences in glycan structure: plant-derived recombinant human glycoproteins tend to contain the carbohydrate groups β(1→2) xylose and α(1→3)fucose, which are absent in mammals, but generally lack the terminal galactose and sialic acid residues that are found on many native human glycoproteins (Figure 1). Since glycan structures can impact on the solubility, stability, immunogenicity and biological activity of recombinant proteins, the 'humanization' of glycan structures produced in plants has been an important topic of research and debate in the scientific community. There has been considerable interest in modifying the plant glycosylation pathway to humanize the glycan profile of recombinant proteins. Several changes in the pathway are required to produce proteins with typical human glycan structures (Gomord et al., 2005; Faye et al., 2005). Strategies used include the in vitro modification of plant-derived recombinant proteins by purified human β(1,4)-galactosyltransferase and sialyltansferase enzymes (Blixt et al., 2002) and the expression of human β(1,4)-galactosyltransferase in transgenic plants to produce recombinant antibodies with galactose-extended glycans (Bakker et al., 2001). In the latter case, 30% of the antibody was galactosylated, similar to the proportion found in hybridoma cells. *In vivo* sialylation will be more difficult to achieve because plants lack the precursors and metabolic capability to produce this carbohydrate group. A more recent report documenting sialylation in *Arabidopsis thaliana* suspension cells has been challenged, although the subject remains a matter of controversy (Shah et al., 2003, 2004; Seveno et al., 2004). To remove the non-mammalian β(1→2) xylose and α(1→3)fucose residues, some researchers have explored the possibility of inhibiting the enzymes responsible for synthesizing these groups, while in one case this goal has been achieved in whole *Arabidopsis thaliana* plants by gene knockout techniques (Strasser et al., 2004). As discussed above, the moss *Physcomitrella patens* can also be modified by gene targeting to eliminate these enzymes (Decker and Reski, 2004). Another approach is to prevent the glycoproteins passing through the Golgi, so only high-mannose glycans are added. This can be achieved simply by adding a KDEL C-terminal tag to the antibody, as demonstrated by Sriraman et al. (2004) and Triguero et al. (2005). This issue has been reviewed by Gomord et al. (2004).

Medium and process development for plant cell cultures

Since the first approved PDP product was made in cultured tobacco cells, it is interesting to note recent developments in the productivity of plant cell cultures. The productivity of plant cell cultures can vary considerably, with recombinant protein levels ranging from 0.0064% to 4% of total soluble protein (TSP) or from 0.5 μg L^{-1} to 200 mg L^{-1} based on the culture volume. However, different experiments are very difficult to compare because extraction procedures and the methods used to determine protein concentration are also variable. In most cases the expression and recovery levels obtained with plant cell cultures are approximately one or two orders of magnitude below the threshold where processes become economically feasible, DowAgro's Concert system being an important exception. One factor in this respect is whether the product is a secreted protein that can be purified readily from the fermentation supernatant. Separation of plant cells from the fermentation broth is simple compared to the same process in microbial or animal cell fermentations because the tough cell wall minimizes cell lysis under conditions that would see CHO cells largely destroyed, and can often be accomplished by straightforward filtration steps. Moreover, plant cell culture media usually contain very few proteins compared to the supernatants of microbial and mammalian cell cultures, which further facilitates recovery. Unfortunately, it has been shown that proteins such as antibodies can be very unstable in plant cell culture media (Sharp and Doran, 2001; Tsoi and Doran, 2002) which is probably why proteins targeted to the endoplasmic reticulum (ER) using a retention signal accumulate to levels (based on cultivation volume) 10–100 fold higher than identical proteins targeted for secretion (e.g., see Fischer et al., 1999; Xu et al., 2002). However, the productivities achieved by some groups are very encouraging. The use of the inducible *RAmy3D* promoter in combination with rice cell lines has provided impressive yields even though rice cell lines are inferior to BY-2 and NT-1 cell lines in terms of growth rates and growth characteristics (Huang et al., 2001, 2002). Recombinant protein concentrations >10 mg L^{-1} based on culture supernatant volume constitute promising starting points for process development (e.g., Yano et al., 2004).

The recovery of recombinant proteins from plant cell cultures can be enhanced by modifying the culture conditions or adding various agents to the growth medium. Substances that have been tested include simple inorganic compounds, amino acids, dimethylsulfoxide (DMSO), polyethylene glycol (PEG), haemin, polyvinylpyrrolidone (PVP), plant hormones such as gibberellic acid, gelatin, bovine serum albumin (BSA) and inhibitors of the secretory pathway such as Brefeldin A (reviewed by Hellwig et al., 2004).

Product yields can be improved not only by medium optimization but also through process development. All major fermentation strategies (batch, fed-batch, repeated batch or draw/fill, continuous culture and perfusion culture) can be applied to plant cell culture. Problems associated with the continuous culture of plant cells include genetic instability and non-growth-coupled production, although the latter can be circumvented using two-stage bioreactors, as described e.g., by Sahai and Shuler (1984). Even so, it remains to be seen whether simple batch processes or long-term draw-fill, semi-continuous or continuous/perfusion processes will emerge as the most suitable fermentation strategies for the production of proteins using plant cell cultures.

Scale-up, process development, integrated production, quality assurance and approval

Downstream processing, the isolation and purification of the recombinant product, is an integral part of every biomanufacturing process. Whichever production system is used, downstream processing represents up to 80% of overall production costs, although this depends on the required level of purity and is highest for clinical-grade materials (Drossard, 2003). In many cases, it is necessary to develop specific processing steps for each product, although certain classes of product can be isolated using a standardized approach (e.g., affinity chromatography to isolate recombinant antibodies; Stoger et al., 2005). Several aspects of downstream processing have to be customized specifically for plant systems, including the removal of fibers, oils and other by-products from certain crops, and process optimization for the treatment of different plant species and tissues (Menkhaus et al., 2004; Nikolov and Woodard, 2004).

For the production of clinical grade antibodies, downstream processing steps need to meet the standards that have been set for other biopharmaceutical production systems, including a strict regime of quality assurance and quality control to achieve approval of regulatory agencies (Fahrner et al., 2001). The initial stages of processing display the greatest variability and have to be optimized in a system-specific manner. Disruption of cell walls and membranes is the first post-harvesting step, but different tissue types (leaves, seeds, fruits etc.) require different forms of treatments (grinding, milling, etc.). After cell disruption, clarification of the extract is often carried out by dead-end or cross-flow filtration, sometimes preceded by bulk cell mass removal using a decanter, plate separator or centrifuge. Once a clarified feed is available, the downstream steps become more product orientated, although it may be necessary to introduce purification steps that cater for specific impurities found in plants such as phenolics and proteases. In large scale processing, the addition of protective agents such as protease inhibitors is economically unfeasible, and either the contaminant must be removed from the process stream or the product must be captured leaving such impurities in the waste stream. For the production of antibodies in plants, Protein A affinity chromatography is the most appropriate capture step although the development of 'smart' microfiltration membranes with added functionality for the trapping of specific impurities, such as proteases, is likely to become more common in the future (Mora et al., 2006).

Several liquid chromatography steps are required in a full purification protocol, although expensive early unit operations could be replaced with cheaper bulk separation methods such as aqueous two phase partitioning systems (e.g., see Platis and Labrou, 2006). In industrial processing, robust and inexpensive chromatography media are used in the initial steps, accepting that there will be some loss of selectivity and resolution (Bai and Glatz, 2003; Menkhaus and Glatz, 2005). However, important exceptions include the use of Protein A or Protein G affinity chromatography for antibody purification, and the use of affinity tags and their respective capture agents (e.g., His_6 and Ni-NTA resin) which are highly selective initial capturing methods.

One of the greatest uncertainties surrounding the use of plants for the production of pharmaceuticals is the regulatory landscape. While plants are grown in glasshouses and in enclosed bioreactors, the production of pharmaceuticals is regulated in the same way for other production systems, and comes under the authority of the FDA and equivalent agencies in other parts of the world. The switch to open-field conditions adds another layer of regulatory complexity, because the transgenic plants then come under the authority of APHIS (part of the USDA) or their counterparts in Europe and other regions. The involvement of multiple regulatory agencies makes the production process more complex because the extent of each authority's jurisdiction is not always clear, and at the current time only draft guidelines are available (CPMP 2002; FDA 2002). The impact of this is to suppress the market. It is likely that more companies will become interested in plant-derived pharmaceuticals when full guidelines become available, hopefully sometime in late 2006.

All recombinant pharmaceuticals, including those derived from plants, need to comply with the national and international GMP standards for product safety, quality, potency and efficacy. However, it is not clear at which stage GMP requirements should come into effect when plants are used as the production system, since the strict rules governing defined growth conditions are difficult to implement in the field, where variables such as the weather, differences in soil quality and the presence of other organisms needs to be considered. This is increasingly important now that European regulatory requirements regarding GMP-compliance for the manufacture of medicinal products have extended to the production of clinical trial material (Directive 2001/20/EC).

The future

The production of recombinant pharmaceuticals in plants is advantageous, theoretically offering unlimited production scales at unprecedented low manufacturing costs. We are beginning to overcome the technical limitations such as low yields, instability and non-authentic glycan structures that place hurdles along the path towards commercialization, but more needs to be done to convince industry that plants represent a true alternative to CHO cells and bacteria. Despite the further limitations of a formative and, in some cases, restrictive regulatory framework, the potential of molecular farming can be seen in the rich IP landscape and the multiple cross-licensing and collaborative ventures that are possible between companies developing production platforms, extraction and separation technologies and those with experience in the latter stages of drug development and marketing. The welcome announcement of the first approved plant-derived veterinary vaccine and process-related antibody for vaccine production may open the way for plant-derived therapeutic and diagnostic proteins to become established as a competitive and valuable source of novel medicines.

References

Bai Y, Glatz CE (2003) Bioprocess considerations for expanded-bed chromatography of crude canola extract: Sample preparation and adsorbent reuse. Biotechnol Bioeng 81: 775–782.

Bakker H, Bardor M, Molthoff JW, Gomord V, Elbers I, Stevens LH, Jordi W, Lommen A, Faye L, Lerouge P, Bosch D (2001) Galactose-extended glycans of antibodies produced by transgenic plants. Proc Natl Acad Sci USA 98: 2899–2904.

Blixt O, Allin K, Pereira L, Datta A, Paulson JC (2002) Efficient chemoenzymatic synthesis of O-linked sialyl oligosaccharides. J Am Chem Soc 124: 5739–5746.

CPMP (2002) Points to consider on quality aspects of medicinal products containing active substances produced by stable transgene expression in higher plants (CPMP/BWP/764/02). The European Agency for the Evaluation of Medicinal Products (EMEA).

Decker EL, Reski R (2004) The moss bioreactor. Curr Opin Plant Biol 7: 166–170.

Drossard J (2003) Downstream processing of plant-derived recombinant therapeutic proteins. In: Fischer R, Schillberg S (eds) Molecular Farming: Plant-made Pharmaceuticals and Technical Proteins. John Wiley & Sons Inc., NY, pp. 217–231.

Fahrner RL, Knudsen HL, Basey CD, Galan W, Feuerhelm D, Vanderlaan M, Blank GS (2001) Industrial purification of pharmaceutical antibodies: Development, operation, and validation of chromatography processes. Biotechnol Genet Eng Rev 18: 301–327.

Faye L, Boulaflous A, Benchabane M, Gomord V, Michaud D (2005) Protein modifications in the plant secretory pathway: current status and practical implications in molecular pharming. Vaccine 23: 1770–1778.

FDA (2002) Guidance for industry. Drugs, biologics, and medical devices derived from bioengineered plants for use in humans and animals. Food and Drug Administration.

Fischer R, Schumann D, Zimmermann S, Drossard J, Sack M, Schillberg S (1999) Expression and characterization of bispecific single-chain Fv fragments produced in transgenic plants. Eur J Biochem 262: 810–816.

Fischer R, Stoger E, Schillberg S, Christou P, Twyman RM (2004) Plant based production of biopharmaceuticals. Curr Opin Plant Biol 7: 152–158.

Gavilondo JV, Larrick JW (2000) Antibody production technology in the millennium. Biotechniques 29: 128–145.

Gomord V, Chamberlain P, Jefferis R, Faye L (2005) Biopharmaceutical production in plants: problems, solutions and opportunities. Trends Biotechnol 23: 559–565.

Gomord V, Sourrouille C, Fitchette AC, Bardor M, Pagny S, Lerouge P, Faye L (2004) Production and glycosylation of plant-made pharmaceuticals: The antibodies as a challenge. Plant Biotechnol J 2: 83–100.

Gottschalk U (2006) Downstream processing in biomanufacturing: Removing economic and technical bottlenecks. Bioforum Europe 10: 28–31.

Hellwig S, Drossard J, Twyman RM, Fischer R (2004) Plant cell cultures for the production of recombinant proteins. Nat Biotechnol 22: 1415–1422.

Hiatt A, Cafferkey R, Bowdish K (1989) Production of antibodies in transgenic plants. Nature 342: 76–78.

Huang J, Sutliff TD, Wu L, Nandi S, Benge K, Terashima M, Ralston AH, Drohan W, Huang N, Rodriguez RL (2001) Expression and purification of functional human α1-antitrypsin from cultured plant cells. Biotechnol Prog 17: 126–133.

Huang J, Wu L, Yalda D, Adkins Y, Kelleher SL, Crane M, Lonnerdal B, Rodriguez RL, Huang N (2002) Expression of functional recombinant human lysozmye in transgenic rice cell culture. Transgenic Res 11: 229–239.

Kapusta J, Modelska A, Figlerowicz M, Pniewski T, Letellier M, Lisowa O, Yusibov V, Koprowski H, Plucienniczak A, Legocki AB (1999) A plant-derived edible vaccine against hepatitis B virus. FASEB J 13: 1796–1799.

Lamphear BJ, Streatfield SJ, Jilka JM, Brooks CA, Barker DK, Turner DD, Delaney DE, Garcia M, Wiggins B, Woodard SL, Hood EE, Tizard IR, Lawhorn B, Howard JA (2002) Delivery of subunit vaccines in maize seed. J Control Release 85: 169–180.

Ma JKC, Barros E, Bock R, Christou P, Dale PJ, Dix PJ, Fischer R, Irwin J, Mahoney R, Pezzotti M, Schillberg S, Sparrow P, Stoger E, Twyman RM (2005a) Molecular farming for new drugs and vaccines. Current perspectives on the production of pharmaceuticals in transgenic plants. EMBO Reports 6: 593–599.

Ma JKC, Chikwamba R, Dale PJ, Fischer R, Mahoney R, Twyman RM (2005b) Plant-derived pharmaceuticals – The road forward. Trends Plant Sci 10: 580–585.

Ma JKC, Drake P, Christou P (2003) The production of recombinant pharmaceutical proteins in plants. Nat Rev Genet 4: 794–805.

Ma JK, Hiatt A, Hein M, Vine ND, Wang F, Stabila P, van Dolleweerd C, Mostov K, Lehner T (1995) Generation and assembly of secretory antibodies in plants. Science 268: 716–719.

Ma JK, Hikmat BY, Wycoff K, Vine ND, Chargelegue D, Yu L, Hein MB, Lehner T (1998) Characterization of a recombinant plant monoclonal secretory antibody and preventive immunotherapy in humans. Nat Med 4: 601–606.

McCormick AA, Kumagai MH, Hanley K, Turpen TH, Hakim I, Grill LK, Tuse D, Levy S, Levy R (1999) Rapid production of specific vaccines for lymphoma by expression of the tumor-derived single-chain Fv epitopes in tobacco plants. Proc Natl Acad Sci USA 96: 703–708.

Menkhaus TJ, Bai Y, Zhang C, Nikolov ZL, Glatz CE (2004) Considerations for the recovery of recombinant proteins from plants. Biotechnol Prog 20: 1001–1014.

Menkhaus TJ, Glatz CE (2005) Antibody capture from corn endosperm extracts by packed bed and expanded bed adsorption. Biotechnol Prog 21: 473–485.

Mora J, Sinclair A, Delmdahl N, Gottschalk U (2006) Disposable membrane chromatography. Performance analysis and economic cost model. BioProcess Int (suppl): 2–6.

Nikolov ZL, Woodard SL (2004) Downstream processing of recombinant proteins from transgenic feedstock. Curr Opin Biotechnol 15: 479–486.

Platis D, Labrou NE (2006) Development of an aqueous two-phase partitioning system for fractionating therapeutic proteins from tobacco extract. J Chromatogr A 1128: 114–124.

Raskin I, Ribnicky DM, Komarnytsky S, Ilic N, Poulev A, Borisjuk N, Brinker A, Moreno DA, Ripoll C, Yakoby N, O'Neal JM, Cornwell T, Pastor I, Fridlender B (2002) Plants and human health in the twenty-first century. Trends Biotechnol 20: 522–531.

Richter LJ, Thanavala Y, Arntzen CJ, Mason HS (2000) Production of hepatitis B surface antigen in transgenic plants for oral immunization. Nat Biotechnol 18: 1167–1171.

Sahai OP, Shuler ML (1984) Multistage continuous culture to examine secondary metabolite formation in plant cells: Phenolics from *Nicotiana tabacum*. Biotechnol Bioeng 26: 27–36.

SemBioSys (2006) SemBioSys achieves major insulin milestone. http://www.investorlook.com/adminv2/downloads/119/07.17.06%20Insulin%20Accumulation%20`final.pdf.

Seveno M, Bardor M, Paccalet T, Gomord V, Lerouge P, Faye L (2004) Glycoprotein sialylation in plants? Nat Biotechnol 22: 1351–1352.

Shah MM, Fujiyama K, Flynn CR, Joshi L (2003) Sialylated endogenous glycoconjugates in plant cells. Nat Biotechnol 21: 1470–1471.

Shah MM, Fujiyama K, Flynn CR, Joshi L (2004) Glycoprotein sialylation in plants? Reply. Nat Biotechnol 22: 1352–1353.

Sharp JM, Doran PM (2001) Strategies for enhancing monoclonal antibody accumulation in plant cell and organ cultures. Biotechnol Prog 17: 979–992.

Sijmons PC, Dekker BMM, Schrammeijer B, Verwoerd TC, Van Den Elzen PJM, Hoekema A (1990) Production of correctly processed human serum albumin in transgenic plants. Bio/Technol 8: 217–221.

Sriraman R, Bardor M, Sack M, Vaquero C, Faye L, Fischer R, Finnern R, Lerouge P (2004) Recombinant anti-hCG antibodies retained in the endoplasmic reticulum of transformed plants lack core-xylose and core-α(1,3)-fucose residues. Plant Biotechnol J 2: 279–287.

Stoger E, Ma JK, Fischer R, Christou P (2005) Sowing the seeds of success: pharmaceutical proteins from plants. Curr Opin Biotechnol 16: 167–173.

Strasser R, Altmann F, Mach L, Glossl J, Steinkellner H (2004) Generation of *Arabidopsis thaliana* plants with complex N-glycans lacking beta1,2-linked xylose and core alpha1,3-linked fucose. FEBS Lett 561: 132–136.

Tacket CO, Mason HS, Losonsky G, Clements JD, Levine MM, Arntzen CJ (1998) Immunogenicity in humans of a recombinant bacterial-antigen delivered in transgenic potato. Nat Med 4: 607–609.

Tacket CO, Mason HS, Losonsky G, Estes MK, Levine MM, Arntzen CJ (2000) Human immune responses to a novel norwalk virus vaccine delivered in transgenic potatoes. J Infect Dis 182: 302–305.

Tacket CO, Pasetti MF, Edelman R, Howard JA, Streatfield S (2004) Immunogenicity of recombinant LT-B delivered orally to humans in transgenic corn. Vaccine 22: 4385–4389.

Triguero A, Cabrera G, Cremata JA, Yuen CT, Wheeler J, Ramírez NI (2005) Plant-derived mouse IgG monoclonal antibody fused to KDEL endoplasmic reticulum-retention signal is N-glycosylated homogeneously throughout the plant with mostly high-mannose-type N-glycans. Plant Biotechnol J 3: 449–457.

Tsoi BM, Doran PM (2002) Effect of medium properties and additives on antibody stability and accumulation in suspended plant cell cultures. Biotechnol Appl Biochem 35: 171–180.

Twyman RM, Schillberg S, Fischer R (2005) Transgenic plants in the biopharmaceutical market. Expert Opin Emerg Drugs 10: 185–218.

Twyman RM, Stoger E, Schillberg S, Christou P, Fischer R (2003) Molecular farming in plants: Host systems and expression technology. Trends Biotechnol 21: 570–578.

Xu H, Montoya FU, Wang Z, Lee JM, Reeves R, Linthicum DS, Magnuson NS (2002) Combined use of regulatory elements within the cDNA to increase the production of a soluble mouse single-chain antibody, scFv, from tobacco cell suspension cultures. Protein Exp Purif 24: 384–394.

Yano A, Maeda F, Takekoshi M (2004) Transgenic tobacco cells producing the human monoclonal antibody to hepatitis B virus surface antigen. J Med Virol 73: 208–215.

Yusibov V, Hooper DC, Spitsin SV, Fleysh N, Kean RB, Mikheeva T, Deka D, Karasev A, Cox S, Randall J, Koprowski H (2002) Expression in plants and immunogenicity of plant virus-based experimental rabies vaccine. Vaccine 20: 3155–3164.

Jim M. Dunwell

Embryogenesis In Vitro

Jim M. Dunwell

School of Biological Sciences, University of Reading, Whiteknights, Reading RG6 6AS, UK.

Introduction

This review will consider several aspects of the induction and exploitation of *in vitro* embryogenesis in plants. In particular it will describe recent molecular data that has supplemented the earlier empirical information obtained largely by trial and error. It will also describe how the process of induced embryogenesis is commercially exploited.

Zygotic embryogenesis

The model for embryogenesis *in vitro* is the corresponding process that takes place during early seed development after fertilisation. Space considerations do not allow a complete description of this process but as summarised elsewhere (Ding et al., 2006), embryogenesis in Angiosperms is a highly orchestrated process of cell division, differentiation, growth, and pattern formation, with a minimal set of ~750 nonredundant *Arabidopsis thaliana* genes thought to be required to coordinate this development. Of these, an initial set of 250 embryo-defective genes (*EMBs*) essential for normal embryo development has been reported. A summary of the phenotypes of *emb* mutants can be found at www.seedgenes.org.

Z. Xu et al. (eds.), Biotechnology and Sustainable Agriculture 2006 and Beyond, 35–46.

There are many recent studies in *Arabidopsis* on the role of particular genes in embryogenesis (Leibfried et al., 2005; Long et al., 2006). These include a study of GLUTAMINE-RICH PROTEIN23 (GRP23), a pentatricopeptide repeat (PPR) protein that functions as a potential regulator of gene expression during early embryogenesis (Ding et al., 2006). Similar studies on loss of function mutants in METHYLTRANSFERASE1 (*Met1*) and CHROMOMETHYLASE3 (*Cmt3*) confirm that DNA methylation (5-methylcytosine) is critical for the regulation of embryogenesis and for seed viability (Xiao et al., 2006), as are the embryo essential gene EMB506 (Latvala-Kilby and Kilby, 2006) which plays a crucial role in the transition of the embryo from radial symmetry to bilateral symmetry, and an AMP deaminase (Xu et al., 2005). Similarly, a mutation in the *TILTED1* locus, which encodes the catalytic subunit of DNA polymerase, causes a lengthening of the cell cycle throughout embryo development and alters cell type patterning in the root (Jenik et al., 2005). This result suggests an interaction between the cell cycle, cell fate and pattern formation during embryogenesis. Plasmodesmata also have a critical role (Kim and Zambryski 2005).

Recent investigations on crops (Le et al., 2005; Ning et al., 2006) include one (Okamoto et al., 2005) on maize in which genes up-regulated in the apical or basal cell were shown to be already expressed in the early maize zygote, suggesting that the transcripts from these genes are localized to the putative apical or basal region of the zygote, or that the transcripts are rapidly degraded in one of the daughter cells after zygotic cell division.

Somatic embryogenesis

Many studies link information from zygotic and somatic embryogenesis. For example, a simple and effective system to induce somatic embryos at high frequency in *Arabidopsis* via ovule culture has been developed (Wei et al., 2006). This provides a useful alternative to create sufficient numbers of somatic embryos for the study of biochemical and molecular mechanisms of embryogenesis, especially to recover early defective embryos in some mutations. Amongst the family of genes studied in most detail in *Arabidopsis* is the *leafy cotyledon* (*lec*) genes that encode B3 domain proteins. For example it was found that contrary to the response of wild-type cultures, which produced somatic embryos via an efficient, direct process, *lec* mutants were strongly impaired in their embryogenic response. These results provide evidence that, besides their key role in controlling many different aspects of zygotic embryogenesis, LEC genes are also essential for *in vitro* somatic embryogenesis induction (Gaj et al., 2005). Most recently it has been shown that LEC2 directly controls a transcriptional programme involved in the maturation phase of seed development. Clustering analyses identified other LEC2-induced RNAs not previously shown to be involved in the maturation phase (Braybrook et al., 2006).

Among several studies comparing the protein profile of zygotic and somatic embryos is one on *Cyclamen persicum* in which high resolution mass spectrometry (nano-LC-MS/MS) was used to demonstrate the presence of 11S globulin proteins in high levels in somatic embryos, zygotic embryos and endosperm, whereas 7S globulins were detected mainly in endosperm and zygotic embryos (Winkelmann et al., 2006).

For nearly a century, the plant hormone auxin has been recognized for its effects on post-embryonic plant growth. Recent genetic and biochemical studies have revealed that much of this regulation involves the SCF(TIR1/AFB)-mediated proteolysis of the Aux/IAA family of transcriptional regulators. With the finding that the TRANSPORT INHIBITOR RESPONSE1 (TIR1)/AUXIN SIGNALING F-BOX (AFB) proteins also function as auxin receptors, a potentially complete, and surprisingly simple, signaling pathway is suggested (Quint and Gray 2006). Jenik and Barton (2005) propose that localized surges in auxin within the embryo occur through a sophisticated transcellular transport pathway and cause the proteolysis of key transcriptional repressors. The resulting downstream gene activation establishes much of the basic body plan of the embryo.

Successful embryonic development in plants, as in animals, requires a strict coordination of cell proliferation, cell differentiation, and programmed cell-death (PCD). The role of cell death is especially critical for the establishment of polarity at early stages of plant embryogenesis, when the differentiation of the temporary structure, the suspensor, is followed by its programmed elimination. The emerging knowledge of PCD, and the role of metacaspases during plant embryogenesis, as revealed by developmental analyses of embryo-specific mutants and gymnosperm model embryonic systems has been recently reviewed (Bozhkov et al., 2005a, b).

Although the greatest emphasis in molecular studies of embryogenesis has been on *Arabidopsis*, some information is available from crop species (Nolan et al., 2006). For example, expression patterns of ca. 12,000 genes were profiled during somatic embryogenesis in a regeneration-proficient maize hybrid line, Hi II, in an effort to identify genes that might be used as developmental markers or targets to optimize regeneration steps for recovering maize plants from tissue culture (Che et al., 2006).

Microspore embryogenesis

During the 1960s a major breakthrough was achieved by the discovery that immature pollen grains (microspores) could develop into haploid embryos if cultured under specific conditions *in vitro* (Guha and Maheshwari, 1964). This discovery has since been extended to many other species, principally to members of the Brassicaceae, Poaceae, and Solanaceae (Palmer and Keller, 2005). The various developmental, physiological, and environmental factors that influence this diversion of the normal pattern of gametophytic development toward a sporophytic one (Segui-Simarro et al., 2006) were the subject of many early studies (Sunderland and Dunwell 1974; Dunwell 1976, 1978, 1985, 1992). These investigations have now been complemented by a range of molecular investigations (Tang et al., 2006). These approaches include the use of gene expression profiling in *Brassica napus*, a technique that has revealed the expression of several embryogenesis-related genes like the BABY BOOM ERF/AP2 transcription factor (Boutilier et al., 2002), LEC1 and LEC2 (see above) as early as 48–72 h of microspore culture (Malik et al., 2007). Indeed ectopic expression studies have shown that BBM expression is sufficient to induce spontaneous somatic embryo formation in both *Arabidopsis* and *B. napus*. Other related studies on

Nicotiana tabacum have identified the important role of the *ntsm10* gene in microspore embryo induction (Hosp et al., 2007).

Facultative apogamy

Within the lower plant kingdom there is a phenomenon directly analogous to the induction of sporophytes from microspores; it is the apogamous production of sporophytes from gametophytes of ferns and mosses. More than a century ago Farlow (1874) described the outgrowth of a shoot from a prothallus of *Pteris cretica.* Such facultative apogamy has been described subsequently in several other fern species including *Pteridium aquilinum* gametophytes where Whittier and Steeves (1960) found that high concentrations of sucrose, glucose, fructose and maltose (a common component of the culture media for microspores) led to an increased frequency of apogamous outgrowths. Hirsch (1975) subsequently showed a definite nutritional role of sucrose in determining the balance of gametophytic to sporophytic development in Microgramma vaccinifolia. She concluded that increased carbohydrate supply led to increased production of sporophytes and that starvation conditions favoured gametophyte development. This concept of starvation is now a common theme in theories of microspore embryo development. As discussed (Dunwell, 1992), one would expect the control of this alternation of generations to become more rigorous during evolution of higher plants but this seems not to be the case. Production of microspore-derived embryos has been described in several plant families with very different advancement indices, and although it is possible that these examples only represent isolated remnants of an ability widespread amongst lower forms of plants, I consider this view to be unnecessarily pessimistic.

Ab-initio zygotic-like embryogenesis from microspores

Probably the most interesting recent advance is the work of Jan Custers and colleagues (PRI, Wageningen) on the direct induction of zygotic-like embryogenesis in microspores of *B. napus*. Although regeneration from microspores of this species has been known for many years, recent modifications in culture conditions have provided a process most analogous to zygotic embryogenesis. Using the cultivar 'Lisandra', microspores at the late unicellular stage (buds 3.3–3.4 mm in length) are subject to a mild heat stress (32°C ± 0.2°C) for 8 h. This treatment induces transverse divisions in the microspore to form a filamentous structure, of which the distal tip cell gives rise to formation of the embryo proper. The early division pattern of these embryos mimics exactly that observed during zygotic embryo production, with the lower end of the filamentous structure resembling the zygotic suspensor. These findings represent a major breakthrough and will greatly facilitate the study of plant embryogenesis in an isolated system. Other recent improvements in methodology include the development of a simple and efficient isolated microspore culture system for producing doubled haploid wheat plants in a wide range of genotypes, in which embryogenic microspores and embryos are formed without any apparent stress treatment (Shariatpanahi et al., 2006). The regeneration frequency and percentage of green plants using this protocol are significantly higher than is found with the shed microspore culture method. However, despite this continuous range of improvements, there is still no method that can be universally recommended with a new species of interest and much progress still depends on long and tedious comparisons of media and environmental conditions.

Commercial exploitation

There are many commercial applications of embryogenesis *in vitro*.

Somatic embryos

Probably the most commercially valuable application of somatic embryogenesis is in the propagation of conifers, where the method is well established on a large scale by several companies in North America and elsewhere (Nehra et al., 2005). Other proprietary multiplication techniques developed recently for other species include one for sugar cane (Lakshmanan et al., 2006). Other more specialist applications include the production of high value compounds by somatic embryos grown in bioreactors. Probably the most impressive example of this is the production of the B subunit of *Escherichia coli* heat-labile toxin (LTB), a potent mucosal immunogen and immunoadjuvant for co-administered antigens. In order to produce a large scale of LTB for the development of an edible vaccine, transgenic somatic embryos of Siberian ginseng were cultured in a 130 L airlift type bioreactor. The mature somatic embryos contained approximately 0.36% LTB of total soluble protein (Kang et al., 2006).

Microspore-derived embryos

Foremost amongst the exploitation of this material is the use of microspore-derived plants as a source of homozygous doubled haploids in plant breeding programmes. Probably the most complete list of haploid-derived varieties is that provided by the COST 851 programme, an EU funded network (http://scri.ac.uk/assoc/COST851/Default.htm). This shows a total of almost 300 such varieties from 12 species, though not all of these varieties have been derived from microspore material; some originated from other sources of haploids.

Patents

In preparing this review, extensive use has been made of the freely available patent databases (Dunwell, 2005) in the United States (http://www.uspto.gov/patft/index.html), European Union (http://ep.espacenet.com/), World International Patent Organization (http://pctgazette.wipo.int/) and other sites (e.g., http://www.surfip.gov.sg/sip/site/sip_home.htm; http://www.patentlens.net/daisy/patentlens/patentlens.html). The most recently published US patents and patent applications in this area are listed below. These include improved methods for inducing somatic and microspore embryogenesis and the application of these processes in various transformation and other protocols.

Number	Date	Inventor: Title
US Patents		
7057089	June 2006	Pioneer: Methods for transforming immature maize embryos
6995016	Feb 2006	AgCanada: Process for inducing direct somatic embryogenesis in immature scutella cells of Pooideae, and rapidly regenerating fertile plants
6964870	Nov 2005	MeadWestvaco: Enhanced selection of genetically modified pine embryogenic tissue
6946295	Sept 2005	CellFor: Process for ex vitro sowing and germination of plant somatic embryos
6906244	June 2005	University of California: Compositions and methods for modulating plant development
6897065	May 2005	Woody Plant Biotech: Method for maturation of conifer somatic embryos
6893873	May 2005	Georgia Tech Research Corp: Methods for improving conifer embryogenesis
6812028	Nov 2004	Kasha et al. Embryogenesis and plant regeneration from microspores
6764854	July 2004	Konzak et al. Methods for generating doubled haploid plants
6362393	March 2002	Konzak et al. Methods for generating doubled haploid plants
US Patent Applications		
20060143731	June 2006	Weyerhauser: Methods for processing image and/or spectral data for enhanced embryo classification
20060117407	June 2006	Monsanto: Method of producing transgenic maize using direct transformation of commercially important genotypes
20060088935	April 2006	Fan et al.: Method of ex vitro sowing, germination, growth and conversion of plant somatic embryos or germinants, and nutrient medium used therefore
20060051868	March 2006	Pullman et al.: Methods for increasing conifer somatic embryo initiation, capture, and multi plication
20060032121	Feb 2006	Weyerhauser: Method and system for producing manufactured seeds
20060031964	Feb 2006	Daniell: Plastid genetic engineering via somatic embryogenesis
20050289673	Dec 2005	Armstrong al: A novel method for Agrobacterium transformation for dihaploid corn plants

20050287660	Dec 2005	Aubry et al.: Scalable bioreactor culture process and system for the maturation of conifer somatic embryos
20050262595	Nov 2005	Niu et al.: Promotion of somatic embryogenesis in plants by PGA37 gene expression
20050071898	Mar 2005	Zuo et al.: Promotion of somatic embryogenesis in plants by Wuschel gene expression

Conclusion

In summary, although rapid progress is being made to understand embryogenesis at the molecular level, we are still a long way from being able to induce selected cells into the embryogenic pathway when and as desired.

References

Boutilier K, Offringa R, Sharma VK, Kieft H, Ouellet T, Zhang L, Hattori J, Liu CM, Van Lammeren AA, Miki BL, Custers JB, Van Lookeren Campagne MM (2002) Ectopic expression of *BABY BOOM* triggers a conversion from vegetative to embryonic growth. Plant Cell 14: 1737–1749.

Bozhkov PV, Filonova LH, Suarez MF (2005a) Programmed cell death in plant embryogenesis. Curr Top Dev Biol 67: 135–179.

Bozhkov PV, Suarez MF, Filonova LH, Daniel G, Zamyatnin AA, Rodriguez-Nieto S, Zhivotovsky B, Smertenko A (2005b) Cysteine protease mcII-Pa executes programmed cell death during plant embryogenesis. Proc Natl Acad Sci USA 102: 14463–14468.

Braybrook SA, Stone SL, Park S, Bui AQ, Le BH, Fischer RL, Goldberg RB, Harada JJ (2006) Genes directly regulated by LEAFY COTYLEDON2 provide insight into the control of embryo maturation and somatic embryogenesis. Proc Natl Acad Sci USA 103: 3468–3473.

Che P, Love TM, Frame BR, Wang K, Carriquiry AL, Howell SH (2006) Gene expression patterns during somatic embryo development and germination in maize Hi II callus cultures. Plant Mol Biol 62: 1–14.

Ding YH, Liu NY, Tang ZS, Liu J, Yang WC (2006) Arabidopsis GLUTAMINE-RICH PROTEIN23 is essential for early embryogenesis and encodes a novel nuclear PPR motif protein that interacts with RNA polymerase II subunit III. Plant Cell 18: 815–830.

Dunwell JM (1976) A comparative study of environmental and developmental factors which influence embryo induction and growth in cultured anthers of *Nicotiana tabacum*. Environ Exp Bot 16: 109–118.

Dunwell JM (1978) Division and differentiation in cultured pollen. In: Thorpe TA (ed) Frontiers of Plant Tissue Culture 1978. Int Assoc Plant Tissue Cult. Calgary, pp. 103–112.
Dunwell JM (1985) Embryogenesis from pollen *in vitro*. In: Zaitlin P, Day P, Hollaender A (eds) Biotechnology in Plant Science. Academic Press, Orlando, pp. 49–76.
Dunwell JM (1992) Mechanisms of microspore embryogenesis. In: Dattée Y, Dumas C, Gallais A (eds) Reproductive Biology and Plant Breeding. Springer-Verlag, Berlin, pp. 121–130.
Dunwell JM (2005) Review: Intellectual property aspects of plant transformation. Plant Biotech J 3: 371–384.
Farlow WG (1874) An asexual growth from the prothallus of *Pteris cretica.* Q J Microsc Sci 14: 266–272.
Gaj MD, Zhang S, Harada JJ, Lemaux PG (2005) Leafy cotyledon genes are essential for induction of somatic embryogenesis of *Arabidopsis*. Planta 222: 977–988.
Guha S, Maheshwari SC (1964). *In vitro* production of embryos from anthers of *Datura*. Nature 204: 497.
Hirsch AM (1975) The effect of sucrose on the differentiation of excised fern leaf tissue into either gametophytes or sporophytes. Plant Physiol 56: 390–393.
Hosp J, Tashpulatov A, Roessner U, Barsova E, Katholnigg H, Steinborn R, Melikant B, Lukyanov S, Heberle-Bors E, Touraev A (2007) Transcriptional and metabolic profiles of stress-induced, embryogenic tobacco microspores. Plant Mol Biol 63: 137–149.
Jenik PD, Barton MK (2005) Surge and destroy: the role of auxin in plant embryogenesis. Development 132: 3577–3585.
Jenik PD, Jurkuta RE, Barton MK (2005) Interactions between the cell cycle and embryonic patterning in *Arabidopsis* uncovered by a mutation in DNA polymerase epsilon. Plant Cell 17: 3362–3377.
Kang TJ, Lee WS, Choi EG, Kim JW, Kim BG, Yang MS (2006) Mass production of somatic embryos expressing *Escherichia coli* heat-labile enterotoxin B subunit in Siberian ginseng. J Biotechnol 121: 124–133.
Kim I, Zambryski PC (2005) Cell-to-cell communication via plasmodesmata during *Arabidopsis* embryogenesis. Curr Opin Plant Biol 8: 593–599.
Lakshmanan P, Geijskes RJ, Wang L, Elliott A, Grof CP, Berding N, Smith GR (2006) Developmental and hormonal regulation of direct shoot organogenesis and somatic embryogenesis in sugarcane (*Saccharum* spp. interspecific hybrids) leaf culture. Plant Cell Rep 25: 1007–1015.
Latvala-Kilby SM, Kilby NJ (2006) Uncovering the post-embryonic role of embryo essential genes in Arabidopsis using the controlled induction of visibly marked genetic mosaics: EMB506, an illustration. Plant Mol Biol 61:179–184.
Le Q, Gutierrez-Marcos JF, Costa LM, Meyer S, Dickinson HG, Lorz H, Kranz E, Scholten S (2005) Construction and screening of subtracted cDNA libraries from limited populations of plant cells: A comparative analysis of gene expression between maize egg cells and central cells. Plant J 44: 167–178.

Leibfried A, To JP, Busch W, Stehling S, Kehle A, Demar M, Kieber JJ, Lohmann JU (2005) WUSCHEL controls meristem function by direct regulation of cytokinin-inducible response regulators. Nature 438: 1172–1175.

Long JA, Ohno C, Smith ZR, Meyerowitz EM (2006) TOPLESS regulates apical embryonic fate in *Arabidopsis*. Science 312: 1520–1523.

Malik MR, Wang F, Dirpaul JM, Zhou N, Polowick PL, Ferrie AM, Krochko JE (2007) Transcript profiling and identification of molecular markers for early microspore embryogenesis in Brassica napus. Plant Physiol 144: 134–154.

Nehra N, Becwar MR, Rottmann WH, Pearson L, Chowdhury K, Chang S, Wilde HD, Kodrzycki RJ, Zhang C, Gause KC, Parks DW, Hinchee MA (2005) Forest biotechnology: Innovative methods, emerging opportunities. In Vitro Cell Dev Biol – Plant 41: 701–717.

Ning J, Peng XB, Qu LH, Xin HP, Yan TT, Sun M. (2006) Differential gene expression in egg cells and zygotes suggests that the transcriptome is restricted before the first zygotic division in tobacco. FEBS Lett 580: 1747–1752.

Nolan KE, Saeed NA, Rose RJ. (2006) The stress kinase gene MtSK1 in *Medicago truncatula* with particular reference to somatic embryogenesis. Plant Cell Rep 25: 711–722.

Okamoto T, Scholten S, Lorz H, Kranz E. (2005) Identification of genes that are up- or down-regulated in the apical or basal cell of maize two-celled embryos and monitoring their expression during zygote development by a cell manipulation- and PCR-based approach. Plant Cell Physiol 46: 332–338.

Palmer CED, Keller WA (2005) Challenges and limitations to the use of haploidy in crop improvement. In: Palmer CED, Keller WA, Kasha KJ (eds) Haploids in Crop Improvement II. Springer, Berlin Heidelberg. pp. 295–303.

Quint M, Gray WM (2006) Auxin signaling. Curr Opin Plant Biol 9: 448–453.

Segui-Simarro JM, Barany I, Suarez R, Fadon B, Testillano PS, Risueno MC. (2006) Nuclear bodies domain changes with microspore reprogramming to embryogenesis. Eur J Histochem 50: 35–44.

Shariatpanahi ME, Belogradova K, Hessamvaziri L, Heberle-Bors E, Touraev A (2006) Efficient embryogenesis and regeneration in freshly isolated and cultured wheat (Triticum aestivum L.) microspores without stress pretreatment. Plant Cell Rep 25: 1294–1299.

Sunderland N, Dunwell JM (1974) Pathways in pollen embryogenesis. In: Street HE (ed) Tissue Culture and Plant Science 1974. Academic Press, London, pp. 141–167.

Tang XC, He YQ, Wang Y, Sun MX (2006) The role of arabinogalactan proteins binding to Yariv reagents in the initiation, cell developmental fate, and maintenance of microspore embryogenesis in *Brassica napus* L. cv. Topas. J Exp Bot 57: 2639–2650.

Wei J, Li XR, Sun MX (2006) Establishment of a simple and efficient system for somatic embryo induction via ovule culture in *Arabidopsis thaliana*. Plant Cell Rep 25: 1275–1280.

Whittier DP, Steeves TA (1960) The induction of apogamy in the bracken fern. Can J Bot 38: 925–930.

Winkelmann T, Heintz D, Van Dorsselaer A, Serek M, Braun HP (2006) Proteomic analyses of somatic and zygotic embryos of *Cyclamen persicum* Mill. reveal new insights into seed and germination physiology. Planta 224: 508–519.

Xiao W, Custard KD, Brown RC, Lemmon BE, Harada JJ, Goldberg RB, Fischer RL (2006) DNA methylation is critical for *Arabidopsis* embryogenesis and seed viability. Plant Cell 18: 805–814.

Xu J, Zhang HY, Xie CH, Xue HW, Dijkhuis P, Liu CM (2005) *EMBRYONIC FACTOR 1* encodes an AMP deaminase and is essential for the zygote to embryo transition in *Arabidopsis*. Plant J 42: 743–756.

Richard B. Meagher

Multigene Strategies for Engineering the Phytoremediation of Mercury and Arsenic

Richard B. Meagher[1,*], Aaron P. Smith[1], Melissa Pischke[1], Tehryung Kim[3], Om Prakash Dhankher[2], and Andrew C.P. Heaton[1]

[1]Department of Genetics, University of Georgia, Life Sciences Building, Athens, GA 30602, USA.
[2]Department of Plant, Soil, and Insect Sciences, University of Massachusetts, Amherst, MA 01002, USA.
[3]Phygen, Inc.,1309 Bld., 5F, 1309 Dunsan-Dong, Seo-Gu, Taejeon, South Korea 302-120.
[*]Tel: (706) 542-1444, Fax: (706) 542-1387, Email:meagher@uga.edu, URL: www.genetics.uga.edu/rbmlab/

Abstract

Hundreds of millions of people worldwide suffer the consequences of being exposed to toxic levels of mercury and arsenic. Our laboratory is focused on developing simple genetic engineering strategies for the phytoremediation of these two toxic elements. The phytoremediation of elemental pollutants relies on plants to extract toxicants from soil and water, manage their toxicity, and concentrate them aboveground for later harvest. Plants possess many natural properties, such as using photosynthetic energy and pervasive root systems to uptake and manage 15 essential elemental nutrients from their environment, which make them suitable to process related toxic elemental pollutants. These properties may be enhanced via genetic engineering. The long-term goal of our work is to develop vigorous, field-adapted plant species that can extract mercury and arsenic from soil or water, process them in high concentrations, and prevent them

Z. Xu et al. (eds.), Biotechnology and Sustainable Agriculture 2006 and Beyond, 49–60.

from entering the food-chain. Our initial research efforts have focused on enhancing several of these properties including: 1) increasing plant tolerance to mercury and arsenic; 2) transforming these toxic elements into the chemical species that promote their short- and long-distance transport from roots to shoots; 3) transforming these toxic elements to the best chemical species for storage in leaves, stems, and vascular tissues above ground; 4) enhancing plant chemical 'sinks' that can trap these toxicants aboveground; and 5) enhancing transporters for uptake and storage. It is likely that the next decade will see unprecedented progress in developing specialized plants that can cleanup mercury and arsenic and restore contaminated land and water resources.

Introduction

Many dangerous, elemental pollutants are present at unacceptably high levels in the environment because of industrial, defense, agricultural, and municipal processes, and they are adversely affecting the health of millions of people worldwide (Nriagu, 1994; Wolz et al., 2003). Elemental pollutants include heavy metals, metalloids, and radionuclides such as mercury, lead, cadmium, arsenic, technetium, tritium, and deuterium. Unlike organic pollutants that can be degraded to harmless small molecules, elemental pollutants are immutable by all biochemical reactions, making them particularly difficult to remediate (Meagher, 2000; Kramer and Chardonnens, 2001). The phytoremediation of some elemental pollution may require genetically engineering several plant properties to enhance plant extraction, tolerance, and hyperaccumulation of these toxicants. Research on engineered phytoremediation is still in its infancy, and a better understanding of natural hyperaccumulators will be a tremendous aid in our understanding of which genes and which cellular and organismal processes can be most effectively manipulated (Baker, 2000; Wang, 2002; Kerkeb and Kramer, 2003). The genes of soil bacteria, also have much to contribute to phytoremediation schemes, because they have evolved many properties that direct the aggressive transformation or management of elemental pollutants (Meagher et al., 2000). Another focal point for research on phytoremediating elemental toxicants is plant biomass. Although some exotic native plants have exceptional elemental pollution phytoremediation capabilities (Ebbs et al., 1997), most are small and slow growing or exotic in their habitats, and therefore, their potential for large-scale commercial remediation of polluted sites is limited (Baker et al., 1994; Tordoff et al., 2000). Hence, engineering high biomass plant species for effective phytoremediation should greatly increase the efficiency of large scale cleanup operations over currently practiced physical clean up methods such as excavation and reburial (Rugh et al., 1998; Merkle and Dean, 2000; Che et al., 2003, 2006).

Enhancing various plant mechanisms in order to engineer the ideal plant species for the phytoremediation of mercury and arsenic

Our long-term goal is to develop and test highly productive, field-adapted plant species that clean up toxic elements like mercury and arsenic from polluted sites. We will, eventually, need to engineer fast-growing grasses, shrubs, and trees that control the uptake, chemical speciation, electrochemical state, and aboveground binding of toxic elements. Our experimental approach is to examine individual genes and mechanisms that accelerate phytoremediation processes in model plants like *Arabidopsis*. Once we have well-characterized phenotypes for individual transgenes and combinations of appropriate genes in model plants, these well-tested genetic systems, affecting particular processes, can be moved into more appropriate macrophytes for testing in the field. We have defined eight mechanisms and processes on which we have focused our research efforts to engineer plants for the phytoremediation of several elemental pollutants, and these are summarized in the diagram in Figure 1.

We have made experimental progress examing four of these mechanisms, while engineering plants for the phytoremediation of the toxic elements mercury and arsenic. These include increasing plant tolerance to mercury and arsenic; controlling the electrochemical state of these elements to increase mobility and to enhance storage; and increasing chemical sinks to safely sequester toxicanty in plant tissues. In the following paragraphs we discuss these four processes and summarize some of our related experimental data on mercury and arsenic. This research involves the engineered over expression of bacterial and plant genes or in some cases suppression of native plant genes in model plants like *Arabidopsis* and tobacco. As a result of this work we have engineered the first generation of high biomass plants for the field with single transgenes assisting with the phytoremediation of these two elemental pollutants. It should be noted that this research on phytoremediation of environmental pollutants is in its infancy and only a few percent of all the easily assessable genes and mechanisms have been examined so far, and these only to a minimal extent (Cobbett and Meagher, 2002; Meagher, 2002).

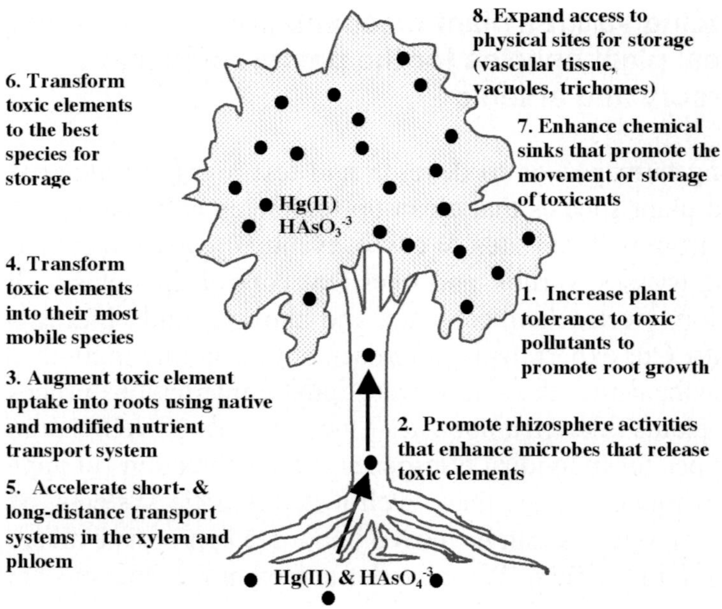

Figure 1. Suggested foci for basic research and technology development that are likely to enhance the phytoremediation activity of plants toward toxic elements such as mercury and arsenic.

Increasing plant tolerance to toxic elements enhances plant growth is essential to all aspects of a phytoremediation strategy. Plants must be able to grow vigorously on sites polluted with elemental toxins (Salt et al., 1995). Tolerance to these toxins should prevent roots from avoiding 'hot spots' in soils or sediments containing the highest levels of these pollutants (Figure 1, Mechanism 1) (Heaton et al.,

2003, 2005). Roots are highly and positively chemotactic toward water and nutrients, and this response can be even stronger than the gravitropic response (Takahashi et al., 1992). Conversely, we find that without high-level resistance, wild-type plant roots will grow away from toxins like the mineral HgS (Heaton et al., 2003, 2005), in spite of the expected low bioavailability of toxic Hg(II). In addition, leaves must be able to function with their full photosynthetic potential even when loaded with toxic pollutants, or one of the primary benefits of plants to the remediation process, their photosynthetic autotrophy, is lost. Our initial efforts at engineering plants to remediate mercury and arsenic have focused heavily on cellular resistance mechanisms, because only healthy plants will efficiently extract these toxicants and process them appropriately.

Arabidopsis and some larger plant species, such as tobacco, canola, cottonwood, rice, and yellow poplar that were engineered to use particular tolerance genes can take up and process several times higher levels of mercury and/or arsenic than would kill most plant species. For example, modified plants expressing the bacterial *merB* gene encoding organomercury lyase cleave the most toxic and biomagnified form of mercury, methylmercury (CH_3Hg^+) to less toxic ionic mercury (Hg(II)) and methane (Bizily et al., 1999, 2000; Che et al., 2003, 2006). These *merB* transgenic plants grow on levels of methylmercury or phenylmercuryacetate (PMA) (0.1–2 μM in agar media) that kill wild-type plants. This result has been repeated in transgenic *Arabidopsis*, tobacco, and cottonwood. Plants expressing the bacterial *merA* gene encoding mercuric ion reductase detoxify ionic mercury Hg(II) by electrochemically reducing it to elemental mercury (Hg(0)) (Rugh et al., 1996, 1998; Meagher, 2000; Che et al., 2003; Heaton et al., 2003). For example, when *merA* cottonwood trees and wild type cottonwood are transplanted to contaminated soil with 400 ppm of mercury the wild type trees die quickly, while the transgenic trees thrive (Che et al., 2003). These plants are resistant to levels of ionic mercury that kill wild-type plants (25–250 μM in sterile media or 100 ppm to even higher concentrations in soil). Trangenic *merA* expression has been shown to confer similar levels of resistance to *Arabidopsis*, tobacco, cottonwood, yellow poplar, *Brassica napus*, and rice suggesting this property can be engineered into most plant species. By combining the transgenic expression of both *merA* and *merB*, plants process organic mercury even more efficiently through a coupled reaction and are resistant to 2–10 μM PMA (Bizily et al., 1999, 2003), 100 times higher levels than kill most wild-type plants.

We have made similar progress in engineering resistance and processing of arsenic. *Arabidopsis* plants expressing the *E. coli* arsenic reductase gene (*ArsC*) can convert arsenate to arsenite, but they become hypersensitive to arsenate due to the greater toxicity of arsenite over arsenate, once electrochemically reduced in side cells (Dhankher et al., 2002). Expression of ArsC enzyme constitutively throughout the plant or restricting its expression only to green shoots resulted in similar levels of arsenic hypersensitivity relative to wild-type plants. Sensitivity from ArsC expression in shoots suggests that arsenite might be particularly toxic to the photosynthetic machinery. Plants expressing the *E. coli* γ-glutamylcyseine synthetase gene (*ECS*) were moderately resistant to arsenic, apparently due to the ability of the thiol-peptide product of the ECS catalyzed reaction, γ-glutamylcyseine, to bind and sequester arsenite in a less toxic state (Dhankher et al., 2002; Li et al., 2005). Plants expressing both ECS and GS can accumulate significantly higher levels of mercury (Li et al., 2006b). *Arabidopsis* plants expressing a combination of the two bacterial genes, *ECS* and *ArsC*, can couple these reactions, converting arsenate to arsenite and trapping it in thiol-peptide complexes (Dhankher et al., 2002). The transgenic *Arabidopsis* plants expressing these two genes are resistant to several times more arsenic in their media than wild-type plants or plants expressing *ECS* alone. The growth of these various plant lines cannot be distinguished on media lacking arsenic. This mechanism of engineered resistance to arsenic is currently being explored in other plant species.

Maximal plant tolerance to some toxicants may require targeting the products of transgenes to particular organs, tissues, cells or subcellular compartments. For example, when aluminum resistant and sensitive genotypes of bean are compared, resistance appears to come from activities in root border cells and acts at the root organ level (Miyasaka and Hawes, 2001). This result suggests that both understanding and engineering phytoremediation may require control of various processes at the level of tissues, cells, and subcellular organelles. Root border cells interact intimately with the soil and are particularly active in exo- and endo-cytosis and in secretion and uptake of small molecules. As a result these cells are rich in reticuloendothelial and vesicles membranes. Considering that methylmercury is moderately hydrophobic, and as such, partitions efficiently into membranes, one cause for its high level of neurotoxicity and phytotoxicity might be poisoning of these essential root epidermal cells. Targeting the bacterial methylmercury lyase protein MerB for expression to the endoplasmic reticulum of *Arabidopsis* provided the most efficient organic mercury processing, significantly more than did expression in the cytoplasm (Bizily et al., 2003), and supporting this view of the mechanism of organic mercury toxicity. Targeting the *merB* gene and MerB protein to tobacco chloroplasts is also reported to provide moderate levels of methylmercury resistance (Ruiz et al., 2003).

Transforming elements to their most mobile species. Most phytoremediation strategies for elemental pollutants rely on mobilizing the toxicant and concentrating it in aboveground tissue for later harvest (Figure 1, Mechanism 4). A few plants like the Chinese break fern, which hyperaccumulates arsenic, may already have adopted this strategy by mobilizing arsenate in its vascular system and concentrating arsenite in its fronds (Ma et al., 2001). Large numbers of nickel, cadmium, and zinc hyperaccumulators mobilize these elements from soil to aboveground organs (Baker, 2000). However, most native plants trap reactive elemental pollutants in their roots presumably protecting valuable photosynthetic machinery and reproductive organs from their toxic effects. For example, arsenate that is inadvertently taken up by plants is reduced in roots to arsenite, and this highly thiol-reactive species stays bound in roots (Pickering et al., 2000; Dhankher et al., 2002). Similarly, most mercury that is taken up as Hg(II) remains bound to root tissues, and most metallic Hg(0) taken up by leaves or roots is re-oxidized to Hg(II) and remains tightly bound to plant tissues (Waldron and Terry, 1975; Heaton et al., 1998, 2005).

Engineered phytoremediation strategies for elemental pollutants generally require that the pollutants be concentrated aboveground where they can be harvested, and hence, may have to counter these natural processes that evolved to benefit plant survival. For example, we knocked down endogenous arsenic reductase activities in roots allowing 16-20 times more arsenate to move to aboveground tissues (Dhankher et al., 2006). Similarly, one proposed strategy for phytoremediating mercury was to transform Hg(II) to Hg(0) only in the roots, thus allowing Hg(0) gas to move up the transpiration stream in these plants (Heaton et al., 1998, 2005). Alternatively, Hg(II) is an analog of Zn(II) and may be best mobilized as a zinc analog.

Transforming toxic elements into the best species for aboveground storage. The goal of many phytoremediation technologies is to store elemental pollutants at high enough concentrations aboveground. This would make it economically worthwhile to harvest and store the contaminated plant material itself, or alternatively, to further concentrate elements from harvested plant material. Once a toxic element has been transported above ground, its continued concentration and storage may require transformation into a more or less reactive chemical species that favors its accumulation (Figure 1, Mechanism 6). For example, among the first steps tested as part of an engineered arsenic phytoremediation strategy, we used the natural transport of arsenate, a phosphate analog, to aboveground tissues. But as described above, we enabled the plants to electrochemically transform arsenate to arsenite with the bacterial enzyme ArsC expressed in shoots from light-induced regulatory sequences (Dhankher et al., 2002). Arsenite is a much a more chemically reactive species than arsenate, can bond with relative stability to various thiol-peptides, and thus is stored at higher levels in this reduced electrochemical state than arsenate. As a complementary step in this study, we overexpressed the bacterial enzyme ECS to generate a thiol-peptide sink for arsenite. The resulting plants, expressing both ArsC and ECS, accumulated three-times more arsenic aboveground. In parallel to this strategy used with arsenic, we are testing strategies in which ionic mercury (Hg(II); Abdel-Mageed and Agrawal, 1997) is transformed into less toxic Hg(0) in roots, transported above ground as Hg(0), and then retransformed into Hg(II) by high levels of native peroxidases and catylases (Heaton et al., 1998). Hg(II) is highly reactive and forms particularly stable chemical products with reduced thiols, and perhaps these products can be stored at higher levels until harvested. Most toxic elemental pollutants have a diversity of chemical species that vary widely in their reactivity and solubility that can be exploited in different phytoremediation schemes. Bacteria and some eukaryotes utilize a number of enzyme catalyzed elemental transformations to manage toxic elements (Rosen, 2002), and expression of these enzymes in aboveground plant organs and tissues reactions may contribute substantially to strategies for engineered storage of pollutants.

Enhancing chemical sinks that promote movement and/or accumulation of toxic elements. Low molecular-weight chemical chelators can act as chemical 'sinks' in which to sequester large quantities of toxic elemental pollutants aboveground (Figure 1, Mechanisms 7). With the exception of exotic plant hyperaccumulators, it is unlikely that many fast-growing non-engineered plants will naturally produce sufficient quantities of the appropriate low molecular-weight chemical sinks to store large quantities of toxic elemental pollutants. The goal of most element hyperaccumulator strategies is to find or create fast growing plants with deep root systems that concentrate elemental pollutants to 0.1–2% of the dry weight of the aboveground plant material. These high levels are needed to make harvesting and processing plant material economically feasible. Indeed, intensive efforts have focused on identifying and understanding the chemical sinks that natural hyperaccumulators use to store large quantities of zinc, cadmium, and nickel. Organic acids, as mentioned above, and amino acids have been implicated in the chelation of toxic metal ions for several hyperaccumulators. For example, hyperaccumulation of nickel in *Alyssum lesbiacum* is associated with order of magnitude increases in free histidine (Kramer et al., 1996; Kerkeb and Kramer, 2003), but histidine does not necessarily play a role in other nickel hyperaccumulators (Persans et al., 1999). Zinc in plants has been found in phosphate, citrate, and malate complexes (Sarret et al., 2002). We and others are examining engineered increases in the levels of cysteine, γ-glutamylcyseine, glutathione, phytochelatins, and metalothioneins to act as sinks for Hg(II), Cd(II) and arsenite (Heiss et al., 1999; Zhu et al., 1999; Lee et al., 2003; Li et al., 2004, 2005; Zimeri et al., 2005). However, the fact remains that only few of these experiments have lead to truly significant increases in the levels of toxic element accumulation and a great deal more needs to be learned about developing chemical sinks for these toxicants. For example, we recently found that both the location of gene expression in shoots or roots and the combinations of genes expressed each effected arsenic and mercury accumulation (Kim et al., 2005; Zimeri et al., 2005; Li et al., 2006a, b).

Conclusion

Strategies for using genetically engineered plants in phytoremediation of toxic elemental pollutants focus on detoxification, extraction from soil, mobilization aboveground, and storage in shoots for later harvest. Phytoremediation takes advantage of many natural properties of plants to carry out these processes including: use of a photosynthetic energy source, nutrient uptake and transport systems, and extensive root systems that naturally extract nutrients and toxins. Significant progress has been made in the engineering of model plants to remediate mercury and arsenic and a few mechanisms have been tested in field adapted species. Application of this technology is ready for commercial development. It is likely that the next decade will see unprecedented progress in developing specialized genetically engineered plants that can clean mercury and arsenic from chemical waste sites and restore marginal lands and water resources.

Acknowledgements

We would like to thank Gay Gragson for editorial comments. This work was supported by grants from U.S. Department of Energy's Environmental Management Sciences program (DEG0796ER20257) and Biological and Environmental Research program (DEFG0203ER63620).

References

Abdel-Mageed A, Agrawal KC (1997) Cancer Gene Ther 4: 199–207.
Baker AJ (2000) In: Terry N, Baneulos GS (eds) Phytoremediation of Contaminated Soil and Water. CRC Press LLC, Boca Raton, FL, pp. 85–107.
Baker AJM, McGrath SP, Sidoli CMD, Reeves RD (1994) Res Conser Recycling 11: 41–49.
Bizily S, Kim T, Kandasamy MK, Meagher RB (2003) Plant Physiol 131: 463–471.
Bizily S, Rugh CL, Meagher RB (2000) Nat Biotechnol 18: 213–217
Bizily S, Rugh CL, Summers AO, Meagher RB (1999) Proc Natl Acad Sci USA 96: 6808–6813.
Che DS, Meagher RB, Heaton ACP, Lima A, Rugh CL, Merkle SA (2003) Plant Biotechnol 1: 311–319.
Che DS, Meagher RB, Rugh CL, Kim R, Heaton ACP, Merkle SA (2006) In Vitro Cell Dev Biol – Plant 42: 228.
Cobbett C, Meagher R (2002) In: Somerville C (ed) Arabidopsis. Cold Spring Harbor Laboratory Press, Cold Spring Harbor, NY.
Dhankher OP, Li Y, Rosen BP, Shi J, Salt D, Senecoff JF, Sashti NA, Meagher RB (2002) Nat Biotechnol 20: 1140–1145.
Dhankher OP, Rosen BP, McKinney EC, Meagher RB (2006) Proc Natl Acad Sci USA 103: 5413–5418.
Ebbs SD, Lasat MM, Brady DJ, Cornish J, Gordon R, Kochian LV (1997) J Environ Qual 26: 1424–1430.
Heaton A, Rugh C, Wang NJ, Meagher R (2005) Water, Air, Soil Pollut 161: 127–155.
Heaton ACP, Rugh CL, Kim T, Wang NJ, Meagher RB (2003) Environ Toxicol Chem 22: 2940–2947.
Heaton ACP, Rugh CL, Wang N-J, Meagher RB (1998) J Soil Contamination 7: 497–509.
Heiss S, Schafer HJ, Haag-Kerwer A, Rausch T (1999) Plant Mol Biol 39: 847–857.
Kerkeb L, Kramer U (2003) Plant Physiol 131: 716–724.
Kim T, Balish RS, Heaton ACP, McKinney EC, Meagher RB (2005) Plant Biotechnol 3: 571–582.
Kramer U, Chardonnens AN (2001) Appl Microbiol Biotechnol 55: 661–672.
Kramer U, CotterHowells JD, Charnock JM, Baker AJM, Smith JAC (1996) Nature 379: 635–638.
Lee S, Moon JS, Ko TS, Petros D, Goldsbrough PB, Korban SS (2003) Plant Physiol 131: 656–663.
Li Y, Dhankher O, Carreira L, Balish R, Meagher R (2005) Environ Toxicol Chem 24: 1376–1386.
Li Y, Dhankher O, Carreira L, Cobbett S, Meagher R (2006a) Plant Physiol 141: 288–298.
Li Y, Dhankher O, Carreira L, Lee D, Chen A, Schroeder J, Balish R, Meagher R (2004) Plant Cell Physiol 45: 1787–1797.

Li Y, Heaton ACP, Carreira L, Meagher RB (2006b) Physiol Plantarum, 128: 48–57.
Ma LQ, Komar KM, Tu C, Zhang W, Cai Y, Kennelley ED (2001) Nature 409: 579.
Meagher RB (2000) Curr Opin Plant Biol 3(2): 153–162.
Meagher RB (2002) Curr Opin Plant Biol 5(2): 135–140.
Meagher RB, Rugh CL, Kandasamy MK, Gragson G, Wang NJ (2000) In: Terry N, Banuelos G (eds) Phytoremediation of Contaminated Soil and Water. Lewis Publishers, Boca Raton, FL, pp. 203–221.
Merkle SA, Dean JF (2000) Curr Opin Biotechnol 11: 298–302.
Miyasaka SC, Hawes MC (2001) Plant Physiol 125: 1978–1987.
Nriagu E (1994) Arsenic in the Environment. Wiley, New York, NY, p. 430.
Persans MW, Yan X, Patnoe JM, Kramer U, Salt DE (1999) Plant Physiol 121: 1117–1126.
Pickering IJ, Prince RC, George MJ, Smith RD, George GN, Salt DE (2000) Plant Physiol 122: 1171–1177.
Rosen BP (2002) Comp Biochem Physiol A Mol Integrative Physiol 133: 689–693.
Rugh CL, Senecoff JF, Meagher RB, Merkle SA (1998) Nat Biotechnol 16: 925–928.
Rugh CL, Wilde D, Stack NM, Thompson DM, Summers AO, Meagher RB (1996) Proc Natl Acad Sci USA 93: 3182–3187.
Ruiz ON, Hussein HS, Terry N, Daniell H (2003) Plant Physiol 132: 1344–1352.
Salt DE, Blaylock M, Kumar NPBA, Dushenkov V, Ensley BD, Chet I, Raskin I (1995) Biotechnology 13: 468–474.
Sarret G, Saumitou-Laprade P, Bert V, Proux O, Hazemann JL, Traverse A, Marcus MA, Manceau A (2002) Plant Physiol 130: 1815–1826.
Takahashi H, Brown CS, Dreschel TW, Scott TK (1992) HortScience 27: 430–432.
Tordoff GM, Baker AJ, Willis AJ (2000) Chemosphere 41: 219–228.
Waldron LJ, Terry N (1975) J Environ Qual 4: 58–60.
Wang J, Zhao FJ, Meharg AA, Raab A, Feldmann J, McGrath SP (2002) Plant Physiol 130: 1552–1561.
Wolz S, Fenske RA, Simcox NJ, Palcisko G, Kissel JC (2003) Environ Res 93: 293–300.
Zhu YL, Pilon-Smits EA, Jouanin L, Terry N (1999) Plant Physiol 119: 73–80.
Zimeri AM, Dhankher OP, McCaig B, Meagher RB (2005) Plant Mol Biol 58: 839–855.

Yongbiao Xue

Molecular Control of S-RNase-based Self-Incompatibility

Jian Huang, Lan Zhao, and Yongbiao Xue

Laboratory of Molecular and Developmental Biology, Institute of Genetics and Developmental Biology, Chinese Academy of Sciences, Datum Road, Chaoyang District, Beijing 100101, China.

Self-incompatibility (SI) is a genetically controlled system used by many flowering plant species to prevent inbreeding and thus promote out-crossing. Generally, self-incompatibility response involves the molecular interactions between pollen and pistil. As a result of the interactions, the self- or genetically related pollen is unable to germinate or grow in the style to complete the fertilization (de Nettancourt, 2001).

SI responses can be classified into various types based on the underlying molecular mechanisms. In most SI species, it is controlled by a single multiallelic locus known as the *S*-locus (de Nettancourt, 2001). Because recent molecular studies have revealed that more than one gene reside in the *S*-locus, they are often described as a haplotype. Since SI signaling in some species has been extensively studied at molecular level, we can define these types of self-SI as Brassicaceae-type, Solanaceae-type and Papaveraceae-type SI. In this review, we will discuss the current progress on the Solanaceae-type SI (Kao and Tsukamoto, 2004; McClure, 2004).

The Solanaceae-type SI is also often referred to as S-RNase-based self-incompatibility, since the three families (the Solanaceae, Rosaceae and Scrophulariaceae) of this type all share the similar

Z. Xu et al. (eds.), Biotechnology and Sustainable Agriculture 2006 and Beyond, 63–73.

female *S*-determinants, *S*-ribonucleases (S-RNases). These pistil-specific proteins are expressed with different molecular masses and isoelectric points that co-segregated with *S*-hapotypes, and accumulated to very high levels in the extracellular matrix. These features enabled people to identify the *S-RNase* genes through peptide sequencing and cDNA cloning, from several species in the Solanaceae and Rosaceae (Anderson et al., 1986; Ai et al., 1990; Clark et al., 1990; Sassa et al., 1993 and Ishimizu et al., 1996). Sequence alignments revealed conserved regions of *S-RNase* genes; then more *S-RNase* genes were cloned by a homology-based PCR approach, especially from *Antirrhinum hispanicum*, (Xue et al., 1996), demonstrating that the Scrophulariaceae could share the same SI mechanism with the other two families.

S-RNase was confirmed to be the sole female-specificity determinant for the SI response by gain- and loss-of-function experiments (Lee et al., 1994; Murfett et al., 1994). Other transgenic experiments revealed that the RNase activity per se of S-RNases is essential for their function in SI response, by site-directed mutagenesis of the *S-RNase* genes (Huang et al., 1994).

The determinants for SI should have haplotype diversity to ensure the specific recognition of self or non-self pollen. As one would expect, S-RNases in the Solanaceae, Rosaceae and Scrophulariaceae are all highly divergent, often with ~70% sequence identity. However, all S-RNases contained the similar conserved regions and hypervariable regions. These regions are believed to be the essential elements for ribonuclease activity or S-specificity recognition and interaction. It is notable that S-RNases in the Rosaceae appeared to have a different structure, especially in hypervariable regions, from those in the Solanaceae (Matsuura et al., 2001). This may implicate the divergence of the SI mechanisms between the Solanaceae and Rosaceae.

It is now widely accepted that S-RNases could enter the pollen cytoplasm in an *S*-haplotype-independent manner (Luu et al., 2000). Thus, whether the pollen can grow or not appears to be dependent on whether the ribonuclease activity can be inhibited in the pollen tube, since this activity had been demonstrated by McClure et al. (1990) to be cytotoxic for the pollen tube growth.

Identification of the pollen *S*-determinant in S-RNase-based SI

Genetic studies had suggested that the pollen and pistil function could be controlled by separate genes (de Nettancourt, 2001). Thus, besides S-RNases, there should be a player as the *S*-specificity determinant in pollen. It can be expected that the pollen *S*-determinant could be specifically expressed in pollen, diversified in different *S*-haplotypes, genetically linked to *S-RNase* genes and with a potential to physically associate with S-RNase proteins.

It was a real challenge to identify the pollen *S*-determinant since this gene appeared not to be highly expressed as *S-RNase* genes. After many attempts such as RNA differential display and subtractive hybridization (Dowd et al., 2000; Li et al., 2000; McCubbin et al., 2000), the first promising candidate for the pollen *S* was identified by Lai et al. (2002), who sequenced a 63.7 kb region containing the S_2-*RNase* of *A. hispanicum*. This gene was predicted to encode a protein that contains an F-box domain at its N-terminus (Bai et al., 1996), and then named as *AhSLF* (*S*-Locus F-box)-S_2. More extensive searches of the adjacent regions of *S-RNase* revealed the existence of multiple genes that encode F-box proteins (FBP). However, recombination analyses confirmed that *AhSLF*-S_2 is the most adjacent haplotype-specific gene among those F-box-encoding genes linked to the *S*-locus (Zhou et al., 2003).

Similar experiments have been conducted by several groups to identify pollen determinant. In the Rosaceae, several FBP-encoding genes were also found to be linked to the *S-RNase* genes, referred to as *SLF* in *P. mume* by Entani et al. (2003) and as *SFB* in *P. dulcis* by Ushijima et al. (2003) respectively. In *Petunia inflata*, a Solanaceae species, *PiSLFs* were also identified through an exhaustive search of a 328-kb BAC contig of S_2-haplotype (Sijacic et al., 2004).

The SLF genes were confirmed to be the pollen determinant soon by the transgenic experiments. As it had been observed in the Solanaceae plants, the pollen cannot be rejected by genetically related style if it is diploid (produce by tetraploid plants). This phenomenon was termed as ‘competitive interaction’ (de Nettancourt, 2001), since it was believed that one diploid pollen grain contains two different pollen *S*-factors and the interaction between the two factors leads to the SI breakdown. Two groups introduced an extra copy of SLF gene (*AhSLF-*S_2 or *PiSLF*$_2$, respectively) into self-incompatible *Petunia* lines and found the SI breakdown of transgenic pollen (Qiao et al., 2004b; Sijacic et al., 2004). Recently, Tsukamoto et al. (2005) reported a naturally occurring mutant of *P. axillaris* in which duplication of the *SLF* gene caused the breakdown of the pollen function in SI response.

How do SLF and S-RNase work together to elicit SI response?

Generally, it is widely accepted not to associate the SI breakdown with a dominant-negative effect, which maybe caused by the introduction of another *S* gene in the single pollen grain. Instead, the S-RNase inhibitor model was proposed to explain the S-RNase-based SI (Kao and Tsukamoto, 2004; McClure, 2004). Since the S-RNases of any *S*-haplotype can enter the pollen tubes without difference, the pollen *S* product should inhibit all non-self-S-RNases, but leave the self S-RNase intact. Thus, the two different pollen *S* products in one pollen grain could inhibit all S-RNases. It can be also concluded that the pollen *S* deletion will be lethal due to the inability of inhibiting any S-RNases, according to the extensive studies of irradiation generated pollen-part mutants by Golz et al. (2001), who suggested that the loss-of-function mutants of pollen *S* might be absent in the Solanaceae.

Intriguingly, the pollen *S* proteins in the Solanaceae and Scrophulariacea, which are not highly variable (~90% identity) among various *S*-haplotypes, while the Rosaceae SLF/SFB proteins display a higher level of allelic sequence diversity (70–80%). Furthermore, *Prunus* pollen-part mutants examined by two groups (Sonneveld et al., 2005; Ushijima et al., 2004) were self-compatible because of the loss or significant alteration of *SFB* genes. In addition, the heteroallelic pollen of tetraploid *Prunus* is self-incompatible, contrary to the 'competitive interaction' phenomena observed in the Solanaceae and Scrophulariaceae. These tetraploid plants can also be self-compatible, when the deletion or mutation of the *S* gene, whether it is the pollen *S* or pistil *S*, is accumulated to allow the presence of two or more non-functional *S*-haplotypes within an individual (Hauck et al., 2006). A modified S-RNase inhibitor model was used to explain the SI mechanism in the Rosaceae. This model allows for the deletion of pollen *S*, because it argues that SLF/SFB prevents the self S-RNase from the inactivation by a general S-RNase inhibitor, through the *S*-haplotype-specific interaction between SLF/SFB and its cognate S-RNase (Sonneveld et al., 2005).

The conventional S-RNase inhibitor model would be consistent with a potential role of SLF in the ubiquitin/26S proteasome pathway (Lai et al., 2002), since an inhibitor for S-RNase is required and F-box proteins are often recruited as the substrate receptor of SCF (SKP1-Cullin1-F-box) ubiquitin ligase (Bai et al., 1996). The SCF complex is widely distributed throughout eukaryotes and absolutely essential for a wide range of cellular processes (reviewed by Cardozo and Pagano, 2004; Petroski and Deshaies, 2005). As one type of ubiquitin ligase (E3), the SCF complex recognizes the substrate and tags it with a polyubiquitin chain. Thus, the substrate is undergoing a process called ubiquitination, and then is delivered to 26S proteasome for its degradation (reviewed by Hershko and Ciechanover, 1998).

However, it is not determined that every FBP has a SKP1 homolog as the interacting partner and then form SCF complex. For example, Rcy1 (Galan et al., 2001) and Ctf13 (Russell et al., 1999), which have irregular motifs compared with the other six FBPs in budding yeast, do not assemble into SCF complexes. Interestingly, compared with other plant FBPs with known function (Huang et al., 2006; Smalle and Vierstra, 2004), SLF/SFB are quite exceptional because it contains no conventional protein–protein interaction domain and do not interact with FAPs (Ingram et al., 1997), the housekeeping SKP1 homologs in *Antirrhinum* (Huang et al., 2006).

Through a yeast two-hybrid screening against a pollen cDNA library, Huang et al. (2006) identified AhSLF-interacting SKP1-like1 (AhSSK1) protein. Subsequent yeast two-hybrid assays demonstrated that AhSSK1, but not other SKP1 homologs, can interact with the orthologous AhSLF, but not with the homologous FBPs linked to *S*-locus. Particularly, AhSSK1 is specifically expressed in pollen like its interacting partner SLF. Moreover, AhSSK1 is very unique in its amino acid sequence and phylogenetic relation, compared with most identified plant SKP1-like proteins. Thus, it appeared that SSK1 protein does not occur in those plants without S-RNase-base SI, suggesting that SSK1 protein could be exclusively recruited in the self-incompatibility response.

Owing to the quite different features in the structures of SSK1 and SLF proteins, the SSK1-SLF interaction may have some structural elements and functional implications more than a canonical SKP1-FBP interface (Schulman et al., 2000; Zheng et al., 2002). SSK1 could be used as an adaptor protein that connects SLF to CUL1 protein to form an SCF complex, as suggested by a pull-down assay (Huang et al., 2006). Thus, the existence of SCF^{SLF} complex in SI response can be confirmed by the identification of SSK1 protein and this complex have the homologous function with other identified SCF ubiquitin ligases in plants. However, it will be not surprising that the SCF^{SLF} complex may also provide novel insights for those who study plant ubiquitin-proteasome system, and the SSK1-SLF interface could define a new pattern of SKP1-F-box interaction for numerous 'orphan' F-box proteins (FBPs) and SKP1 homologs in plants (Gagne et al., 2002; Risseeuw et al., 2003).

As genetic evidence indicated that the SLF/SFB proteins in the Solanaceae, Scrophulariaceae or Rosaceae could play different roles in SI response, though they share quite extensive similarities in structure. Thus, it is intriguing whether the AhSSK1 counterpart in the Rosaceae could exist. It would be required that the SSK1 orthologs should be the adaptor between SLF/SFB and CUL1, specifically expressed in pollen and phylogenetically related.

No loss-of-function mutant for the pollen *S* gene had ever been observed in the Solanaceae and Scrophulariaceae, although it was predicted that this mutation could be lethal if S-RNase is not absent. Thus, studying the SLF null mutant is extremely important to establish the real relation between the male and female components: if S-RNase is toxic to fertilization, is SLF a rescuer or an accomplice? The two different models discussed above offer two answers. While even both models may not cover the truth, this question is still fundamental and needed to be clarified. Functional dissection of SSK1, which is physically associated to SLF in SI signaling, may provide an important clue to this question.

A new clue to SI response

A recent study at cytological level provides a new vision for the restriction of the S-RNase cytotoxicity. Goldraij et al. (2006) performed immunolocalization experiments to demonstrate that S-RNase is transported to pollen vacuole after its uptake from extracellular matrix. Thus, the compartmentalization of S-RNases could be a mechanism for protecting pollen tube from S-RNase cytotoxic effect. While most compartments always remained intact to sequestrate the S-RNases in compatible pollen tubes, the vacuole-like endomembrane system for S-RNases appeared to break down in the latter stages of pollen rejection in incompatible pollination. Thus, in the incompatible pollen tubes, S-RNase is unleashed and then its toxic activity causes the pollen tube to stop growing.

Goldraij et al. (2006) also used the immunolocalization approaches to investigate some plant materials with style-side defects (McClure et al., 1999, 2000; Hancock et al., 2005). In these self-compatible plants of which the pistil did not express HT-B or 4936-factor, S-RNase remains sequestered during all stages of self-pollination. This prompted them to examine the HT-B level in the pollen tubes growing in wild-type styles. As expected, large-scale HT-B degradation occurred in compatible pollen tubes, but not in incompatible pollen tubes.

Apparently, the results about S-RNase compartmentalization are quite revolutionary, compared with the popular S-RNase inhibitor models. At this point, it would be interesting to examine the protein degradation of HB-B protein in SI response of several other S-RNase-based SI species. It is also the first time to relate the repression of S-RNase cytotoxicity to compartmentalization within which S-RNase is trapped, rather than to the physical elimination of S-RNase. This new work reported an endomembrane system, which were demonstrated to be the hot zone for several style factors, and could be also the place where the pollen components interact with S-RNase.

These results can also be described as a new phenotype of SI response. People have never applied a multi-antibody-labeling technique to study the pollen tube in SI response before and thus have missed such observations for years. It will be extremely helpful to examine this proposed new model in other S-RNase-based SI species. Nevertheless, one may argue that the break down of compartment could be a consequence, but not the cause, of pollen rejection in SI response. More importantly, since some key SI factors in both pollen and style have been identified thanks to the great efforts for so many years, it will be very interesting to investigate where the pollen factors are precisely localized and how SLFs interact with S-RNases in compatible or incompatible pollen tubes. We can also expect more novel and important data from the work using molecular genetics and biochemistry approaches, for better understanding of SI, such a delicate and complicated system widespread in flowering plants including many important economic crops.

References

Ai Y, Singh A, Coleman CE, Ioerger TR, Kheyr-Pour A, Kao TH (1990) Self-incompatibility in *Petunia inflata*: Isolation and characterization of cDNAs encoding three S-allele-associated proteins. Sex Plant Reprod 3: 130–138.

Anderson MA, Cornish EC, Mau SL, Williams EG, Hoggart R, Atkinson A, Bonig I, Greg B, Simpson R, Roche PJ, Haley JD, Penschow JD, Niall HD, Tregear GW, Coghlan JP, Crawford RJ, Clarke AE (1986) Cloning of cDNA for a stylar glycoprotein associated with expression of self-incompatibility in *Nicotiana alata*. Nature 321: 38–44.

Bai C, Sen P, Hofman K, Ma L, Goebel M, Harper W, Elledge S (1996) *SKP1* connects cell cycle regulation to the ubiquitin proteolysis machinery through a novel motif, the F-box. Cell 86: 263–274.

Cardozo T, Pagano M (2004) The SCF ubiquitin ligase: Insights into a molecular machine. Nat Rev Mol Cell Biol 5: 739–753.

Clark KR, Okuley JJ, Collins PD, Sims TL (1990) Sequence variability and developmental expression of *S*-alleles in self-incompatible and pseudo-self-compatible Petunia, Plant Cell 2: 815–826.

de Nettancourt D (2001) Incompatibility and Incongruity in Wild and Cultivated Plants. Springer, Berlin.

Dowd PE, McCubbin AG, Wang X, Verica JA, Tsukamoto T, Ando T, Kao TH (2000) Use of *Petunia* as a model for the study of solanaceous type self-incompatibility. Ann Bot Suppl A 85: 87–93.

Entani T, Iwano M, Shiba H, Che FS, Isogai A, Takayama SFS, Isogai A, Takayama S (2003) Comparative analysis of the self-incompatibility (*S*-) locus region of *Prunus mume*: Identification of a pollen-expressed F-box gene with allelic diversity. Genes Cells 8: 203–213.

Gagne JM, Downes BP, Shiu SH, Durski AM, Vierstra RD (2002) The F-box subunit of the SCF E3 complex is encoded by a diverse superfamily of genes in *Arabidopsis*. Proc Natl Acad Sci USA 99: 11519–11524.

Galan JM, Wiederkehr A, Seol JH, Haguenauer-Tsapis R, Deshaies RJ, Riezman H, Peter M (2001) Skp1p and the F-box protein Rcy1p form a non-SCF complex involved in recycling of the SNARE Snc1p in yeast. Mol Cell Biol 21: 3105–3117.

Goldraij A, Kondo K, Lee CB, Hancock CN, Sivaguru M, Vazquez-Santana S, Kim S, Phillips TE, Cruz-Garcia F, McClure B (2006) Compartmentalization of S-RNase and HT-B degradation in self-incompatible *Nicotiana*. Nature 439: 805–810.

Golz JF, Oh HY, Su V, Kusaba M, Newbigin E (2001) Genetic analysis of *Nicotiana* pollen-part mutants is consistent with the presence of an *S*-ribonuclease inhibitor at the *S* locus. Proc Natl Acad Sci USA 98: 15372–15376.

Hancock CN, Kent L, McClure B (2005) The 120 kDa glycoprotein is required for *S*-specific pollen rejection in *Nicotiana*. Plant J 43: 716–723.

Hauck NR, Yamane H, Tao R, Iezzoni AF (2006) Accumulation of non-functional S-haplotypes results in the breakdown of gametophytic self-incompatibility in tetraploid *Prunus*. Genetics 172: 1191–1198.

Hershko A, Ciechanover A (1998) The ubiquitin system. Annu Rev Biochem 67: 425–479.

Huang J, Zhao L, Yang Q, Xue Y (2006) AhSSK1, a novel SKP1-like protein that interacts with the S-locus F-box protein SLF. Plant J 46: 780–793.

Huang S, Lee HS, Karunanandaa B. Kao TH (1994) Ribonuclease activity of *Petunia inflata* S proteins is essential for rejection of self-pollen. Plant Cell 6: 1021–1028.

Ingram GC, Doyle S, Carpenter R, Schultz EA, Simon R, Coen ES (1997) Dual role for *fimbriata* in regulating floral homeotic genes and cell division in *Antirrhinum*. EMBO J 16: 6521–6534.

Ishimizu T, Sato Y, Saito T, Yoshimura Y, Norioka S, Nakanishi T, Sakiyama F (1996) Identification and partial amino acid sequences of seven S-RNases associated with self-incompatibility of the Japanese pear, *Pyrus pyriforia* Nakai. J Biochem 120: 326–334.

Kao TH, Tsukamoto T (2004) The molecular and genetic bases of S-RNase-based self-incompatibility. Plant Cell 16(suppl): S72–S83.

Lai Z, Ma W, Han B, Liang L, Zhang Y, Hong G, Xue Y (2002) An F-box gene linked to the self-incompatibility (*S*) locus of *Antirrhinum* is expressed specifically in pollen and tapetum. Plant Mol Biol 50: 29–42.

Lee HS, Huang S, Kao TH (1994) *S* proteins control rejection of incompatible pollen in *Petunia inflata*. Nature 367: 560–563.

Li JH, Nass N, Kusaba M, Dodds PN, Treloar N, Clarke AE, Newbigin E (2000) A genetic map of the *Nicotiana alata S* locus that includes three pollen-expressed genes. Theor Appl Genet 100: 956–964.

Luu DT, Qin X, Morse D, Cappadocia M (2000) *S*-RNase uptake by compatible pollen tubes in gametophytic self-incompatibility. Nature 407: 649–651.

Matsuura T, Sakai H, Unno M, Ida K, Sato M, Sakiyama F, Norioka S (2001) Crystal structure at 1.5 Å resolution of Pyrus pyrifolia pistil ribonuclease responsible for gametophytic self-incompatibility. J Biol Chem 276: 45261–45269.

McClure BA (2004) S-RNase and SLF determine S-haplotype-specific pollen recognition and rejection. Plant Cell 16: 2840–2847.

McClure BA, Cruz-Garcia F, Beecher BS, Sulaman W (2000) Factors affecting inter- and intra-specific pollen rejection in *Nicotiana*. Anal Bot 85: 113–123.

McClure BA, Gray JE, Anderson MA, Clarke AE (1990) Self-incompatibility in *Nicotiana alata* involves degradation of pollen rRNA. Nature 347: 757–760.

McClure BA, Mou B, Canevascini S, Bernatzky R (1999) A small asparagine-rich protein required for *S*-allele-specific pollen rejection in *Nicotiana*. Proc Nat Acad Sci USA 96: 13548–13553.

McCubbin AG, Wang X, Kao TH (2000) Identification of self-incompatibility (*S*-) locus linked pollen cDNA markers in *Petunia inflate*. Genome 43: 619–627.

Murfett J, Atherton TL, Mou B, Gasser CS, McClure BA (1994) S-RNase expressed in transgenic *Nicotiana* causes S-allele-specific pollen rejection. Nature 367: 563–566.

Petroski MD, Deshaies RJ (2005) Function and regulation of cullin-RING ubiquitin ligases. Nat Rev Mol Cell Biol 6: 9–20.

Qiao H, Wang F, Zhao L, Zhou J, Lai Z, Zhang Y, Robbins TP, Xue, Y (2004b) The F-box protein AhSLF-S_2 controls the pollen function of *S*-RNase-based self-incompatibility. Plant Cell 16: 2307–2322.

Risseeuw EP, Daskalchuk TE, Banks TW, Liu E, Cotelesage J, Hellmann H, Estelle M, Somers DE, Crosby WL (2003) Protein interaction analysis of SCF ubiquitin E3 ligase subunits from Arabidopsis. Plant J 34: 753–767.

Russell ID, Grancell AS, Sorger PK (1999) The unstable F-box protein p58-Ctf13 forms the structural core of the CBF3 kinetochore complex. J Cell Biol 145: 933–950.

Sassa H, Hirano H, Ikehashi H (1993) Identification and characterization of stylar glycoproteins associated with self-incompatibility genes of Japanese pear, *Pyrus serotina* Rehd. Mol Gen Genet 241: 17–25.

Schulman BA, Carrano AC, Jeffrey PD, Bowen Z, Kinnucan ERE, Finnin MS, Elledge SJ, Harper JW, Pagano M, Pavletich NP (2000) Insights into SCF ubiquitin ligases from the structure of the Skp1-Skp2 complex. Nature 408: 381–386.

Sijacic P, Wang X, Skirpan AL, Wang Y, Dowd PE, McCubbin AG, Huang S, Kao TH (2004) Identification of the pollen determinant of S-RNase-mediated self-incompatibility. Nature 429: 302–305.

Smalle J, Vierstra RD (2004) The ubiquitin 26S proteasome proteolytic pathway. Annu Rev Plant Physiol Plant Mol Biol 55: 555–590.

Sonneveld T, Tobutt KR, Vaughan SP, Robbins TP (2005) Loss of pollen-S function in two self-compatible selections of *Prunus avium* is associated with deletion/mutation of an S haplotype-specific F-Box gene. Plant Cell 17: 37–51.

Tsukamoto T, Ando T, Watanabe H, Marchesi E, Kao TH (2005) Duplication of the *S*-locus F-box gene is associated with breakdown of pollen function in an *S*-haplotype identified in a natural population of self-incompatible *Petunia axillaris*. Plant Mol Biol 57: 141–153.

Ushijima K, Sassa H, Dandekar AM, Gradziel TM, Tao R, Hirano H (2003) Structural and transcriptional analysis of the self-incompatibility locus of almond: Identification of a pollen-expressed F-box gene with haplotype-specific polymorphism. Plant Cell 15: 771–781.

Ushijima K, Yamane H, Watari A, Kakehi E, Ikeda K, Hauck NR, Iezzoni AF, Tao R (2004) The *S* haplotype-specific F-box protein gene, *SFB*, is defective in self-compatible haplotypes of *Prunus avium* and *P. mume*. Plant J 39: 573–586.

Xue Y, Carpenter R, Dickinson HG, Coen ES (1996) Origin of allelic diversity in *Antirrhinum S* locus RNases. Plant Cell 8: 805–814.

Zheng N, Schulman BA, Song L, Miller JJ, Jeffrey PD, Wang P, Chu C, Koepp DM, Elledge SJ, Pagano M, Conaway RC, Conaway JW, Harper JW, Pavletich NP (2002) Structure of the Cul1-Rbx1-Skp1-F box Skp2 SCF ubiquitin ligase complex. Nature 416: 703–709.

Zhou J, Wang F, Ma W, Zhang Y, Han B, Xue Y (2003) Structural and transcriptional analysis of S-locus F-box genes in *Antirrhinum*. Sex Plant Reprod 16: 165–177.

Ueli Grossniklaus

The Use of Sexual Model Systems to Identify Elements of Apomixis

Ueli Grossniklaus*, Philippa Barrell, Amal Johnston, Ramaurthy Baskar, Jacqueline Gheyselinck, Lukas Brand, and Mark Curtis

Institute of Plant Biology and Zürich-Basel Plant Science Center, University of Zürich, CH-8008 Zürich, Switzerland.
*Email: grossnik@botinst.unizh.ch

Most flowering plants reproduce sexually. Viable seed formation depends on the fertilization of the egg and central cell and the coordinated development of the embryo, the endosperm, and the maternal seed coat. Apomictic plants employ a closely related, yet asexual strategy, in which progeny of maternal origin are produced in the absence of reduction and fertilisation. This alternative approach to seed production is thought to have arisen from the spatial and temporal deregulation of ancestral sexual processes. Thus, apomixis and sexual reproduction are closely interrelated. Therefore, a better understanding of the genetic and molecular basis of sexual reproduction will provide new tools towards the engineering of apomixis in sexual crops. We have used Arabidopsis and maize as model systems to identify genes that play a role in key steps of sexual reproduction, which may also function during apomixis. We concentrate on three developmental processes that are highly relevant to the engineering of apomixis: (1) megasporogenesis (apomeiosis), (2) fertilization (parthenogenesis), and (3) endosperm initiation (functional endosperm). By screening through a few thousand enhancer detector lines we identified genes expressed in specific cell types

Z. Xu et al. (eds.), Biotechnology and Sustainable Agriculture 2006 and Beyond, 77–78.

such as the nucellar region, where aposporic initials form in apomictic species, or the egg cell and central cell, which can initiate development in the absence of fertilization in apomicts. We will present a strategy in which we will use these isolated promoters can be used in mis-expression experiments to induce elements of apomixis in Arabidopis. Furthermore, screens in maize have identified genes that, if mutated, cause changes in reproductive development towards apomixis. Our progress in the characterization of these maize mutants will be reported.

Xingwang Deng

Agricultural Biotechnology in China: Past, Present and Future

Valerie J. Karplus and Xing Wang Deng

National Institute of Biological Sciences, No. 7 Science Park Road, Zhongguancun Life Science Park, Beijing 102206, China.

China's scientists have developed a large number of transgenic (also known as genetically modified or GM) crops in laboratories, but only one, insect-resistant Bt cotton, is planted widely in farmers' fields. Understanding the origins of this apparent paradox requires a closer look at how history, politics, and China's agricultural system have influenced the technology's development. This paper summarizes a forthcoming book that explores the rationale and prospects for China's large public investment in transgenic crops, which rank among biotechnology's most controversial applications. It is our hope that this book will interest scientists, policymakers, students of the biological and social sciences, and anyone interested the unfolding story of China's bold pursuit of agricultural biotechnology.

Agriculture's importance to China's economy has long provided impetus for technology development. For thousands of years, China's political stability and prosperity depended on the productivity of its agricultural sector, which employed the vast majority of the population. Today, just under half of China's population earns a living from agriculture or related activities, which account for a substantial fraction of the country's economic output. Now as before, technology plays an important role in protecting or expanding harvests. In the early to mid-20th century, post-industrial advances in biology and chemistry led to the development of technologies (such as high-yielding crop varieties, pesticides, and fertilizers) that, when adopted

Z. Xu et al. (eds.), Biotechnology and Sustainable Agriculture 2006 and Beyond, 81–86.

widely in China and other parts of the world, ushered in a period of expanding harvests known as the Green Revolution. At the end of the 1970s, as China began to reform and open to international markets, Green Revolution gains had already demonstrated how modern biology could help to boost agricultural productivity. In the early 1980s, when advances in molecular biology enabled specific and powerful alterations in crop traits at the single-gene level, China's policymakers paid close attention. The new technology, which enabled the creation of transgenic crops, looked poised to offset major challenges facing the agricultural sector, including population growth, loss of valuable farmland to development, and land degradation.

Reformers were also keenly aware that a newly industrializing China's future growth would depend on more than just the economic boost that accompanied reform and opening. In many advanced industrialized countries, knowledge creation had long surpassed agricultural expansion as a principle source of sustained economic growth. China's reformers were likewise keen to reap the economic benefits of technological innovation. Several ambitious publicly-funded initiatives to revitalize the country's science and technology accompanied the early years of labor and market reforms. These initiatives tied basic science and technology development objectives to the practical needs of agriculture, energy, and medicine, among other sectors. Biotechnology, as a relatively young field, was viewed as a promising research area in which China might quickly gain a seat at the frontier. This view, coupled with concerns about the entry of foreign competition and pressing challenges facing China's agricultural sector, inspired many initiatives that strongly supported biotechnology research with a significant focus on agricultural applications.

Starting from nearly zero in the mid-1980s, China's public investment in agricultural biotechnology has since dramatically increased. Newly established grant programs began to provide generous funding to top scientists pursuing biotechnology research with a wide range of applications. Among the first of several high profile initiatives was the 863 Program, which channeled generous funding to agricultural biotechnology research. Today, China's growing investment in agricultural biotechnology has enabled it to develop the largest plant biotechnology program outside North America, with annual expenditures surpassing $120 million in 2003 (Huang et al., 2005). This rise is especially significant when considering that both personnel and non-personnel research budgets stretch farther in China than in the United States or Western Europe. In contrast to private sector dominance in most developed countries, nearly all plant biotechnology research in China is funded by government sources. By 2001, over 150 individual laboratories were working on agricultural biotechnology, and 30 National Key Laboratories for biotechnology research had been funded, fifteen of which work on agricultural applications. The quality of equipment, total research staff, and funding per researcher likewise increased remarkably since 1986.

The number and type of transgenic crops produced by China's laboratories has since blossomed. Scientists have developed transgenic varieties of staple crops including rice, corn, soybean, wheat and potato as well as vegetables, fruits, and nuts. Most of these crops are endowed with traits such as pest or disease resistance that would primarily benefit farmers, but a few crops that are nutrient enriched or better suited for cooking are designed to appeal to consumers. Many of the crops developed are already in or have completed field trials – several varieties of transgenic rice fall into the latter category – and are awaiting final commercial approval. Research priorities roughly reflect the planted area and economic importance of each crop. However, despite rapid progress in laboratories, only four transgenic crops (virus-resistant tomato, and late-ripening tomato, virus-resistant sweet pepper, and insect-resistant Bt cotton) have received approval for commercial planting. Only one crop, insect-resistant Bt cotton, is planted on a large scale.

First planted in 1996, insect-resistant Bt cotton made its debut prior to the controversy over transgenic crops, and was considered an attractive investment for several reasons. First, and most importantly, pesticides were once again losing their effectiveness against pests that fed on conventional (non-transgenic) cotton. Farmers had to spray more often with increasingly toxic and costly pesticide cocktails, compromising their earnings and health. Second, as cotton is China's most important cash crop, economic losses due to bollworm outbreaks were compromising the country's competitive position in domestic and export markets. Third, since cotton is not consumed by humans, adoption was less controversial than it might have been for a food crop. Studies by the Chinese Center for Agricultural Policy have shown that the shift to Bt cotton has indeed helped to reduce yield losses and pesticide requirements (thereby raising farmer incomes) as well as mitigate adverse health effects.

For several years, China has been on the verge of comercializing several other crops. Most prominent among them are strains of blight resistant and insect-resistant (Bt) rice. Impact studies by the Chinese Center for Agricultural Policy for the varieties of transgenic rice now in pre-production trials show that aggregate economic and health benefits due to reductions in pesticide sprayings could be significant. For the time being, the absence of an imminent threat to rice harvests (as in the case of cotton) coupled with calls for additional health and environmental impact studies – both from within and outside of China – has favored a standstill in the approval process.

Despite demonstrated benefits, several obstacles remain to establishing an integrated system for developing and distributing transgenic crops, which may further help to explain why commercialization has proceeded slowly. Vigorous investment in agricultural biotechnology has not been accompanied by a corresponding investment in the agricultural system as a whole. In the mid-1980s and 1990s, public expenditures on the agricultural research and extension system have not grown nearly as fast as expenditures on plant biotechnology, and have fallen in some years (Huang et al., 2004). As a result, provincial- or prefecture-level institutes that focus on lower tech but vitally important research such as agronomics or breeding may not have the support or incentives to shoulder the development and regulatory costs for new transgenic varieties, especially if they are not able to benefit from their commercialization.

Reforms are also reshaping channels for delivering seed to farmers in ways that may hinder the commercialization of transgenic crops. Starting in the late 1990s, the seed industry has been undergoing privatization and consolidation, with competition in local markets varying significantly by locality and crop. With growing numbers and rapid turnover of seed sellers, farmers are left to guess seed origin and quality, especially as officials often cannot prevent lawbreakers from reinventing themselves and reentering the market. When Bt cotton was first introduced into China's market, fake seed spread quickly and undermined the integrity of the seed supply as well as developers' profit margins. Only recently has enforcement of developers' and breeders' rights reportedly improved, but remains weak and patchy. However, poor enforcement of these rights may discourage domestic as well as international companies from marketing transgenic seed in China.

The ongoing reforms in many parts of China's agricultural system may affect the strength of biosafety oversight as well. The biosafety system relies in part on local authorities to prevent unapproved varieties from reaching the market, while ensuring that approved varieties are monitored carefully to detect adverse effects, such as the emergence of resistant pests. It is not yet clear if China will continue to require separation of transgenic from non-transgenic harvests, but this would further add to the responsibilities of local companies and bureaucracies. Although no harm from transgenic crops has been reported so far, a few unapproved varieties have reached fields. In the case of Bt cotton, unapproved varieties were disseminated, prompting the government to strengthen biosafety regulations in 2004. Additional reports of farmers planting unapproved transgenic rice seed and transgenic corn showing up in export shipments suggest that the biosafety system is far from airtight. Since breaches of biosafety can undermine trust – both at home and abroad – in China's ability to manage introduction of transgenic crops, strengthening oversight may be an important prerequisite for further approvals.

If approvals resume, the strength of intellectual property protection, biosafety oversight, and China's seed distribution channels will all influence the success of China's homegrown transgenic crops, as well as the entry and impact of foreign competition. As China's seed industry grows more market oriented, property rights are strengthened, and provincial level biosafety monitoring capacity is enhanced, the conditions for both domestic and international companies to market their products will improve. Given that persistent weaknesses still remain, it may be no accident that laboratory activity has been allowed to proceed while commercial approvals remain on hold. This approach is further consistent with China's acceptance of imported, labeled transgenic soy and corn products, while foreign seed companies are restricted to minority ownership in joint partnerships, and are not permitted to engage in biotechnology research.

China's public investment in agricultural biotechnology has two main goals: producing homegrown transgenic varieties that farmers and consumers will buy and helping China to gain a leading position in what China's policymakers view as a promising new area of research. The number and variety of transgenic crops developed in China's laboratories suggest the first goal is in sight, but further commercial approvals and continued acceptance are likely to depend on the strength of the agricultural research and seed delivery system, biosafety monitoring, and intellectual property management at the local level. The acuteness of a particular agronomic problem may also influence decisions to commercialize transgenic crops on a case-by-case basis, as in the case of insect-resistant Bt cotton. Greater acceptance of the technology at home and abroad could also further hasten approvals. However, stability in rural areas in the midst of rapid economic growth as well as the safety and security of the food supply remains paramount, and further commercialization is unlike to proceed if either is jeopardized. Meanwhile, pursuit of the second goal – leadership in the laboratory – is in some ways more straightforward, as it does not encounter the inevitable complexities and uncertainties of commercialization. If more transgenic crops are eventually approved, research advances and reform progress may leave domestic industry better prepared to manage their introduction in the long term. For now, China is emerging as an important center of research in agricultural biotechnology, even if farmers have to wait.

References

Huang J, Hu R, Rozelle S (2004) China's agricultural research system and reforms: Challenges and implications for developing countries. Center for Chinese Agricultural Policy Working Paper 04-E5, 1–18.

Huang J, Hu R, Rozelle S, Pray C (2005) Development, policy, and impacts of genetically modified crops in China: A comprehensive review of China's agricultural biotechnology sector. Paper for the Belfer Center for Science and International Affairs, Kennedy School of Government at Harvard University, 1–58.

SYMPOSIA

Current Advances in Plant Tissue Culture Techniques

Regulation of Plant Morphogenesis In Vitro: Role of Ethylene and Polyamines

Eng-Chong Pua

School of Arts and Sciences, Monash University Sunway Campus, 46150 Bandar Sunway, Selangor, Malaysia

Introduction

Transgenic plants can be efficiently produced using the floral dip method to circumvent tedious tissue culture process for *de novo* plant regeneration. However, this method has been confined mostly to *Arabidopsis thaliana*. The ability to regenerate plants from cultured cells and tissues at high frequency remains crucial for the success of genetic engineering. This is particularly true when genetic engineering is used for the crop improvement program, through which novel transgenes encoding desirable traits can be transferred to a given crop species.

Plant regeneration from cultured cell and tissues has been reported for a wide range of species. This has been achieved mostly based on empirical studies through manipulation of culture medium, especially growth regulators, and selection of genotypes and explant source. However, many plant species and cultivars remain recalcitrant in regeneration. Although results from several lines of study indicate that shoot regeneration may be associated with cascade of genes that are controlled transcriptionally (Chee et al., 2006), the underlying mechanism that regulates plant regeneration *in vitro* is not clear.

Z. Xu et al. (eds.), Biotechnology and Sustainable Agriculture 2006 and Beyond, 89–95.

Shoot regeneration can be enhanced by ethylene inhibition

Ethylene is a gaseous plant hormone that has been involved in the regulation of various plant physiological processes during growth and development. Plant cells and tissues grown *in vitro* have been known to emanate ethylene since the 1960s but the physiological role of ethylene in plant tissue culture was not clear until recently. In our earlier work, we discovered that poor regeneration of mustard and Chinese cabbage in vitro was associated with accumulation of ethylene, whose level increased rapidly after 3 day of culture and reached a maximum after 10–14 days. Addition of ethylene synthesis or action inhibitor, aminoethoxyvinylglycine (AVG) or $AgNO_3$, in culture medium greatly enhanced shoot regeneration from cultured explants. Similar regeneration enhancement using ethylene inhibitors has also been observed in other plant species, including monocots and dicots (Pua, 1999; Pua and Gong, 2004). Results of ethylene inhibitor studies have been supported by transgenic studies, in which ethylene synthesis was inhibited by expression of an antisense cDNA encoding an enzyme 1-aminocyclopropane-1-carboxylate (ACC) oxidase (Pua and Lee 1995), which catalyzes the conversion of ACC to ethylene. Explants from these transgenic plants were highly regenerative in culture. Similar results were also observed in transgenic plants expressing an antisense ACC synthase cDNA (Cheng, 2002). ACC synthase catalyzes the conversion of *S*-adenosylmethionine (SAM) to ACC, which is the rate-limiting step of ethylene biosynthesis. For transgenic plants overexpressing ACC synthase cDNA, cultured explants were shown to produce higher levels of ethylene and possessed lower shoot regenerability compared to untransformed control explants. These results are in line with the findings of ethylene inhibitor studies, suggesting that ethylene plays a regulatory role in shoot regeneration *in vitro*. However, we know very little about the mechanism whereby decreased ethylene production or action upregulates shoot regeneration.

Interactive role of ethylene and polyamines (PAs) in shoot regeneration

The synthesis of ethylene is metabolically linked to that of PAs as both utilize SAM as a common intermediate (Figure 1). It has been hypothesized that changes in ethylene production may lead to altered PA synthesis and *vice versa*. The implication of PAs in morphogenic response of cultured tissues has been derived mostly from studies showing that shoot regeneration and somatic embryogenesis can be enhanced by exogenous applications of PAs such as putrescine (Put), spermidine (Spd) and spermine (Spm) (Pua, 1999). We had previously shown that enhancement of shoot regeneration by ethylene inhibitors might be attributed to accumulation of PAs in cultured explants. This was demonstrated in Chinese kale (*Brassica alboglabra*), where enhanced shoot regeneration from cultured hypocotyls in the presence of AVG or $AgNO_3$ could be suppressed by application of difluoromethylarginine (DFMA) (Pua et al., 1996), which inhibits Put synthesis by inhibiting the enzyme arginine decarboxylase. The inhibitory effect of DFMA was nullified by exogenous application of Put. Furthermore, characterization of transgenic plants expressing antisense ACC synthase cDNA also showed the implication of PAs in shoot regeneration, as explants accumulated Spd and Spm in culture (Cheng, 2002). Inhibition of Spd and Spm synthesis by methylglyoxal(bis-guanylhydrazone) (MGBG), which is a potent inhibitor of SAM decarboxylase (SAMDC) that catalyzes decarboxylation of SAM, was shown to inhibit shoot regeneration and stimulate ethylene production. The inhibitory effect of MGBG on regeneration could be nullified by exogenous PAs but ethylene production was not affected (Cheng, 2002). These results suggest that high frequency shoot regeneration may be associated with accumulation of PAs, especially Spd and Spm, regardless of the presence of the ethylene level. Similar culture response has also been observed in transgenic *A. thaliana* by modulation of SAMDC expression. Inhibition of SAMDC expression by RNA interference has resulted in higher Spd+Spm/Put ratios and a marked increase in shoot regeneration from cultured explants compared to explants overexpressing SAMDC cDNA, but the pattern of ethylene production in both types of explant was comparable (Hu et al., 2006).

To understand the mechanism as to how shoot regeneration was upregulated by PAs, attempts were made to investigate the PA metabolism in cultured explants. This was achieved by the cloning and characterization of genes encoding diamine oxidase (DAO) and polyamine oxidase (PAO). While DAO catalyzes Put oxidation, PAO is responsible for Spd and Spm oxidation. Both reactions can result in the release of hydrogen peroxide (H_2O_2), which has been associated with somatic embryogenesis (Pua and Gong, 2004). Transgenic plants expressing sense, antisense and doubled stranded cDNA of DAO or PAO were generated and analyzed in terms of the endogenous PA content, ethylene production and shoot regeneration capacity in culture. In general, results supported the conclusions of the above study (data not shown).

Identification of genes associated with shoot regeneration

Evidence from several lines of study indicates that plant regeneration from cultured cells and tissues, via either shoot organogenesis or somatic embryogenesis, may be controlled genetically and/or epigenetically (Chee et al., 2006). We had previously attempted identifying genes associated with shoot regeneration as part of effort to elucidate the regulatory mechanism of this morphogenic event *in vitro*. Using the mRNA differential display method, we had isolated 87 expressed sequence tags (ESTs) from highly and poorly regenerative mustard explants grown on the medium in the presence and absence of 10 μM $AgNO_3$, respectively (Gong and Pua, 2004). Sequence analysis revealed that 30% of the ESTs were associated with ethylene- and/or stress-induced responses. Expression of several ESTs homologous to pectate lyase, glutathione *S*-transferase (GST), peroxidase, superoxide dismutase and catalyase was shown to be upregulated in response to H_2O_2 (Gong and Pua, 2004).

One H_2O_2-responsive EST homologous to GST was selected for further study. GSTs are encoded by large gene families. The enzymes have been known to catalyze a wide variety of hydrophobic, electrophilic, and cytotoxic substrates, and serve as glutathione-dependent peroxidases by catalyzing the reduction of organic hydroperoxide to the less toxic monohydroxy alcohols. Expression characterization of the GST gene in transgenic *A. thalina* showed that it expressed highly in the roots, cotyledons, anthers and both ends of the silique of but expression was low or barely detectable in the leaves, seeds, petals and stamens (Gong et al., 2005a). Furthermore, expression of the β-glucuronidase gene under the control of GST promoter in transgenic plants was upregulated by exogenous application of H_2O_2, confirming the regulatory role of H_2O_2 in GST expression. Results of promoter analysis revealed that the promoter region between positions of −2623 and −1521-bp upstream of the translation start site might be responsible for transcriptional activation of GST induced by H_2O_2 and salicylic acid (Gong et al., 2005a).

Role of GST in shoot regeneration

Information regarding the role of GST in plant morphogenesis in vitro has been limited, although GST has been implicated in early stages of microspore embryogenesis in barley and somatic embryogenesis of chicory (see Gong et al., 2005b). To investigate whether shoot regeneration was associated with GST expression, transgenic plants expressing sense (GST-S6), antisense (GST-A4) and double-stranded (GST-DS1) GST cDNAs were generated (Gong et al., 2005b). RNA gel blot analysis showed that GST transcripts were barely detectable in GST-DS plants, while it present at lower levels in GST-A plants compared to wild type and GST-S plants. These results suggest that RNAi is more effective than the antisense approach in suppressing GST expression.

Explants derived from different transgenic plants also exhibited differential shoot regeneration in culture. GST-S6 was most regenerative (70.3%), followed by wild type (53.6%), GST-A4 (44.3%) and GST-DS1 (6.7%). The regeneration capacity appears to be associated with the level of ethylene production, which was highest in

GST-DS, followed by GST-A4, wild type and GST-S6 (Gong et al., 2005b). This inverse relationship between shoot regenerability and ethylene production was also observed in transgenic plants that expressed antisense ACC oxidase (Pua and Lee, 1995) or ACC synthase cDNA (Cheng, 2002), indicating that PAs may be also involved. However, as the PA content in GST-S, GST-A and GST-DS plants was not determined, it is not certain as to whether PA metabolism is also affected by GST modulation. Nevertheless, the highest level of ethylene production by explants of different transgenic lines after 6 days of culture was coincided with the level of accumulated GST transcripts, suggesting that ethylene biosynthesis may be regulated, at least in part, by GST at the transcriptional level (Gong et al., 2005b).

References

Chee P, Lall S, Nettleton D, Howell SH (2006) Gene expression programs during shoot, root, and callus development in Arabidopsis tissue culture. Plant Physiol 141: 620–637.

Cheng W (2002) Molecular characterization and regulation of 1-aminocyclo-propane-1-carboxylate (ACC) synthase in shoot morphogenesis of mustard *in vitro*. PhD Thesis, National University of Singapore, Singapore.

Galland R, Randoux B, Vasseur J, Hilbert JL (2001) A glutathione-*S*-glutathione cDNA identified by mRNA differential display is upregulated during somatic embryogenesis in *Cichorum*. Biochim Biophys Acta 1522: 212–216.

Gong H, Hu W, Jiao Y, Pua EC (2005a) Molecular characterization of a phi class mustard (*Brassica juncea*) glutathione *S*-transferase gene in *Arabidopsis thaliana* by 5' deletion analysis of its promoter. Plant Cell Rep 24: 439–447.

Gong H, Jiao Y, Hu WW, Pua EC (2005b) Expression of glutathione *S*-transferase and its role in plant growth and development *in vivo* and shoot morphogenesis *in vitro*. Plant Mol Biol 57: 53–66.

Gong H, Pua EC (2004) Identification and expression of genes associated with shoot regeneration from leaf disc explants of mustard (*Brassica juncea*) *in vitro*. Plant Sci 167: 1191–1201.

Hu WW, Gong H, Pua EC (2006) Modulation of SAMDC expression in *Arabidopsis thaliana* alters *in vitro* shoot organogenesis. Physiol Plant 128: 740–750.

Pischke MS, Huttlin EL, Hegeman AD, Sussman MR (2006) A transcriptome-based characterization of habituation in plant tissue culture. Plant Physiol 140: 1255–1278.

Pua EC (1999) Morphogenesis in cell and tissue culture: Role of ethylene and polyamines. In: Soh WY, Bhojwani SS (eds) Morphogenesis in Plant Tissue Cultures. Kluwer Academic Publications, Dordrecht, pp. 255–303.

Pua EC, Gong H (2004) Regulation of plant morphogenesis *in vitro*. In: Pua EC, Douglas CJ (eds) Biotechnology in Agriculture and Forestry, Vol 54 *Brassica*. Springer-Verlag, Berlin Heidelberg, pp. 83–102.

Pua EC, Lee JEE (1995) Enhanced *de novo* shoot morphogenesis *in vitro* by expression of antisense 1-aminocyclopropane-1-carboxylase oxidase in transgenic mustard plants. Planta 196: 69–76.

Pua EC, Teo SH, Loh CS (1996) Interactive role of ethylene and polyamines on shoot regenerability of Chinese kale (*Brassica alboglabra* Bailey) *in vitro*. J Plant Physiol 149: 138–148.

Vrinten PL, Nakamura T, Kasha KJ (1999) Characterization of cDNAs expressed in the early stages of microspore embryogenesis in barley (*Hordeum vulgare* L.). Plant Mol Biol 41: 455–463.

Zeng F, Zhang X, Zhu L, Tu L, Guo X, Nie Y (2006) Isolation and characterization of genes associated to cotton somatic embryogenesis by suppression subtractive hybridization macroarray. Plant Mol Biol 60: 167–183.

Lingfasu and Its Use on Plant Tissue Culture In Vitro

Xu Hong-Yuan[1,*], Xu Hong-Zhang[2], He Bing[1], and Zhou Feng-Jue[1]

[1]Agricultural College, Guangxi University, Nanning 530005, China.
[*]Email: xuhy08@126.com
[2]Institute of Medicinal Biotechnology of Chinese Academy of Medical Sciences, Beijing 100050, China.

'Lingfasu'(LFS), code name PGR-08 (Plant growth regulator-08) was a new kind of CTK discovered by the authors (Xu et al., 2003). It was a white crystal substance isolated and purified from the ferment liquor of a soil microorganism (S. *sp.*) Its mp. was 180°C – 181°C; $[\alpha]_D^{21}$ was + 43° (H_2O). Molecular weight 279 (MS). It dissolved easily in H_2O (≥60°C), HCl, NaOH, $(CH_3)_2CO$, CH_3OH and C_2H_5OH, but difficultly in $NaHCO_3$. Its molecular formula was $C_{11}H_{13}O_4N_5$, the chemical structure was identified by MS, UV, IR, NMR and X-ray diffraction. Results showed that LFS was a nucleoside, N^9-angustose Adenine (Xie et al., 2004) (Figure 1).

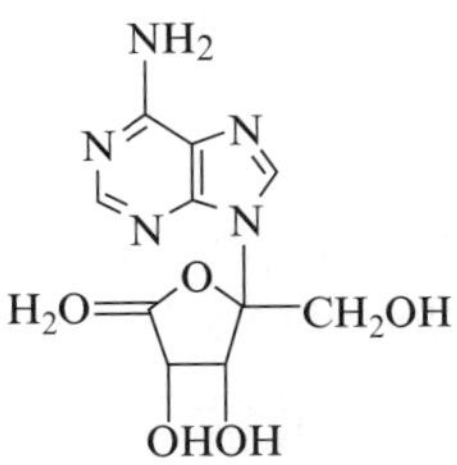

Figure 1. Structural formula of LFS.

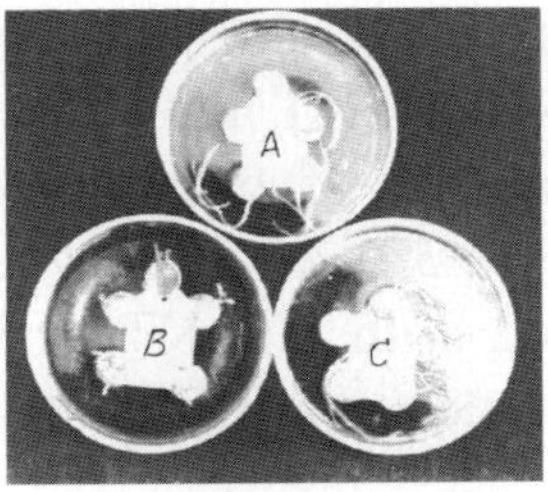

Figure 2. LFS induced the adventitious root (A,C) or bud(B) differentiation.

Z. Xu et al. (eds.), Biotechnology and Sustainable Agriculture 2006 and Beyond, 97–101.

Bioassays showed that LFS possessing some activities of CTK and Auxin. For example, 10~100 mg·L^{-1} could prolong the live of the round slice of Dantelion's leaf more one to two times than CK (H_2O), and induced the adventitious root or bud differentiation (Figure 2), even to grow up a complete regenerate plantlet on the round slice of Dantelion's leaf (Xu et al., 2003) (Figure 3). LFS had been used on some plant tissue culture , and got good results:

Figure 3. LFS induced regenerate plantlet set up.

1. *Panax notoginseng* F.H. Chen was called as 'tian-qi' or 'san-qi' in china, a famous and precious traditional Chinese medicine, produced mainly in guang-xi and yun-nan province of china. Its major medical functions were to stop bleeding; ease pain; invigorate the circulation of blood; nutriment and strong body. So, it was always used on tradition to treat coughing blood; spitting blood; excessive menstruation; injuries from falls, fractures, contusions and strains or knife wound. But it was acknowledged a difficulty plant to tissue culture. However, when LFS 2 mg·L^{-1} was added to the basic medium (MS + 2,4-D 2 mg·L^{-1}, + sugar 3% + agar 0.5%, pH 6.0, CK) the callus formation from stem cutting was promoted earlier 1–2 weeks than CK, and the rate of callus inducing as high as 81%, higher more 30% than added ZT, KT and N^6-BA at the same concentrations. The regenerating ability of callus cultured on the medium with LFS could keep more 3 years (at last, stopped by people). But on other mediums added with ZT, KT or N^6-BA, the callus hardening and aging quickly, and very difficult to regenerate (Figure 4).

Out of the ordinary , when the callus cultured continuously on the medium with LFS, there were a large number of asexual embryos formed (Figure 5), the inducing ratio could reach to more 85%, and

30% of asexual embryos could grow and develop to be robust plantlets (Figure 6).

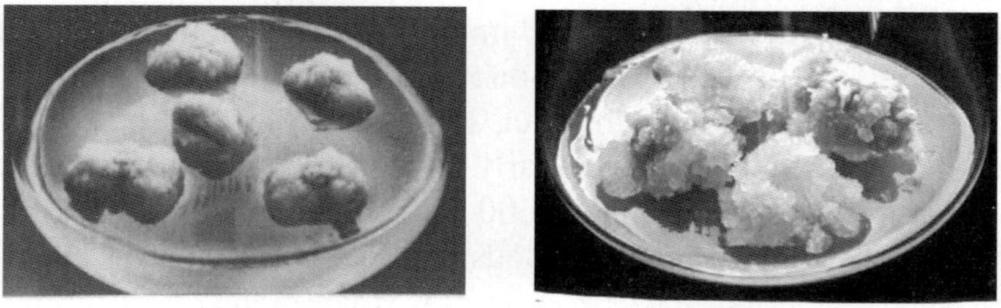

(a) (b)

Figure 4. Effects of different CTK on the growth of callus of *Panax notoginseng* (a) Basic medium added with 6-BA, KT or ZT, (b) Basic medium added with LFS.

Figure 5. LFS induced a large number of asexual embryos on the callus of *Panax notoginseng*.

Figure 6. LFS induced embryos grow to be plantlets.

Toke the leaf blade or leafstalk as explants, there were similar results could be got, but the concentration of LFS only needed 0.2 $mg{\cdot}L^{-1}$.

2. *Siraitia grosvenori* (Swingle) C. Jeffrey was another famous and precious traditional Chinese medicine produced mainly in the north area of guang-xi province of china. Its fruit called by Chinese as 'luo-han-guo' (meaning fruit of immortals) had medical functions that to get rid of fever and moisten the respiratory organs, stop cough and eliminate phlegm, etc., It was used to treat pharyngitis, tonsillitis, gastritis, etc., Particularly its fruits contented sweet glucoside that sweetness was higher 300 times than sugar, so be used to process many kinds of healthy foods.

The stem cutting with a bud was used generally as explants for tissue culture of *Siraitia grosvenori* (Swingle) C. Jeffrey. But for a long time, there was a outstanding problem that the callus formatting and regenerating were too fast to the sprouting of bud, when the explants cultured on the medium (MS + NAA 0.1 $mg{\cdot}L^{-1}$) added N^6-BA, KT, ZT or CPPU 0.2 $mg{\cdot}L^{-1}$, the bud growing was inhibited obviously, even could not sprouting and died, cultured 25~30 d the sprouting rate always less than 30%, so, some people said, 'no callus, no tissue culture of *Siraitia grosvenori* (Swingle) C. Jeffrey'.

If LFS was used to replay N^6-BA, KT, ZT and CPPU, the results would be very different:

1. The bud of explants sprouting and growing quickly, cultured only 10~15 d the sprouting rate could reach 80~100%, however, the callus formatting and growing very slow at the same time (Figure 7).

Figure 7. Effects of different CTK on the bud sprouting and callus growing of explants of *Siraitia grosvenori*. (Cultured 10 d).

2. When the regenerated plantlets were proliferated on the MS basic medium, we had not to add any kind of auxin (NAA ,IAA or IBA), only added LFS 0.2 $mg{\cdot}L^{-1}$, the plantlet could normally grow and

the adventitious roots formed easily, the rooting rate could reach more 90%, but the callus were very little. However, if added N^6-BA, KT, ZT or CPPU 0.2 mg·L^{-1} to the MS basic medium, the callus grew rapidly and seldom or never happened rooting in the 20~30 d cultured period (Figure 8).

People considered that to inhibit callus developing in plant tissue culture to get regenerated plantlets was advantageous to the stable of hereditary features.

The applied studies of LFS on other plant tissue culture *in vitro* had been in progress.

MS + LFS 0.2 MS + BA 0.2 MS + KT 0.2 MS + CPPU 0.2

Figure 8. Effects of LFS, BA KT and CPPU on the adventitious roots and callus of regenerate plantlets of *Siraitia grosvenori*. (Cultured 20 d).

References

Xie YY, Jiang R, Xu HZ (2004) Studies on the spectral analyses of angustmycins obtained from two *Streptomycetes hygroscopicus* strains. Chin J Antibiot 29: 645–647.

Xu HY, Xu HZ, Yang MC, Zhou W (2003) Studies on the physical and chemical characteristics and bioactivities of PGR-08. Guihaia 23: 461–463.

PPV-Specific Hairpin RNAs Is an Effective Method for Plum Pox Potyvirus Resistance

Lining Tian[1,*], Shuocheng Zhang[2], HJIPne SanfaHon[2], Antonet Svircev[1], Daniel C. Brown[1], and Rui Wen[2]

[1]Southern Crop and Food Research Centre, Agriculture and Agri-Food Canada, London, Ontario, Canada N5V 4T3.
[2]Pacific Agri-Food Research Centre, Agriculture and Agri-Food Canada, Summerland, B.C., Canada V0H 1Z0.
[*]Email: tianl@agr.gc.ca

Plum pox virus (PPV) is a viral disease that affects *Prunus* fruits, including plum, peach, nectarine, cherry, and apricot. The disease was first reported in Eastern Europe and has progressively spread to many other countries. PPV is the most serious viral disease of *Prunus* fruits and causes significant damage to *Prunus* fruit trees. Few naturally occurring resistance sources have been found which can be used to develop highly resistant fruit varieties. Alternatively, genetic resistance to plum pox virus may be developed via introducing molecularly-designed resistance into *Prunus* species via genetic engineering (Ravelonandro et al., 1998). In an effort to develop resistance to PPV in *Prunus* species, we introduced PPV-specific hairpin RNAs into *Nicotiana benthamiana* plants and evaluated PPV resistance. We showed that expression of PPV-specific hairpin RNAs is an effective method to induce post-transcriptional gene silencing directed against PPV.

Z. Xu et al. (eds.), Biotechnology and Sustainable Agriculture 2006 and Beyond, 103–106.

Materials and methods

Two conserved regions in P1 and HC-Pro of the PPV genome were selected for the study using RNAi approach. The first region shares nucleotide sequence similarity at 96% among different PPV isolates. The second region has a level of nucleotide sequence identity of 98% among PPV isolates. cDNA fragments corresponding to these two regions of PPV were inserted in an inverted repeat orientation in a transformation vector. The intron 1 S13 of sweet cherry S-allele was introduced at the junction between the two inverted repeats. A double 35S promoter was used to drive the gene. The expression cassette was transferred to a binary vector (pBin19). Standard molecular techniques were used for gene cloning and vector construction.

Transgenic *Nicotiana benthamiana* lines were generated using *Agrobacterium*-mediated transformation. The presence of the insert was verified by PCR using primers specific for the *nptII* gene and for the S13 cherry intron. A large number of plants were generated from R0 transgenic leaves via *in vitro* approach and used for the resistance assays. Plants at the eight to ten leaf stage were inoculated with Canadian PPV isolate. Four to 5 weeks post-inoculation, the presence of the virus in the plants was tested on leaf samples. ELISA was conducted using a commercially available PPV-detection kit for infection assay. Plants with an ELISA value three times above background were considered positive.

Northern blot was used for siRNA detection. Total RNA was isolated from transgenic *N. benthamiana* leaves. Two separate probes P1 and HC-Pro were prepared to detect siRNAs specific for the P1 and HC-Pro regions. Northern assay was conducted via standard approach.

Results and discussion

Transgenic lines containing hair-pin DNA were inoculated with PPV and resistance was evaluated. Transgenic plants which containing the empty pBin19 vector were used as the control. PPV-specific symptoms such as chlorosis and striation of the leaves appeared 21–28 days post-inoculation (dpi) in young systemic leaves of the control plants. ELISA was also used to confirm PPV infection in the control plants. In 14 lines containing hair-pin design, all inoculated plants remained negative for PPV using ELISA assay. In a separate test, only 5–10% of the inoculated plants became positive with ELISA. These results suggest that expression of PPV-specific hairpin RNA induced resistance to PPV in the transgenic lines.

Four selected transgenic lines were further tested using RT-PCR method which is more sensitive for viral detection. For each time point tested, ten plants from each transgenic line and ten plants from a control line were inoculated with a PPV isolate. All inoculated control plants were positive for PPV in the RT-PCR assay 14 and 21 dpi. In contrast, virus was not detected at 7, 14 or 21 dpi in any of the transgenic plants tested. Two transgenic lines and two control lines were also tested at 50 dpi. Virus was readily detected in the control lines but not in the transgenic lines. The results indicate that PPV replication was basically restricted in the inoculated leaf of transgenic lines, if occurring at all. Taken together, the results presented suggest that PPV resistance in the selected transgenic lines is highly effective, not only in the systemic leaves but also in the inoculated leaf.

The presence of transgene-specific siRNAs was tested in all six selected transgenic lines using a Northern blot assay. Two species of siRNAs of approximately 21–22 nts and 25–27 nts in length were detected in all transgenic lines tested. The size of these RNAs was the same to the two classes of siRNAs which are associated with the induction of PTGS in transgenic plants (Hamilton et al., 2002). These results confirmed that PPV-specific PTGS was active transgenic lines.

Summary

In this study, we showed that PPV resistance can be efficiently induced in transgenic plants using PPV-specific hairpin RNA approach. We have introduced the above strategies into *Prunus* and will evaluate PPV resistance in *Prunus* plants.

References

Hamilton A, Voinnet O, Chappell L, Baulcombe D (2002) Two classes of short interfering RNA in RNA silencing. EMBO J 21: 4671–4679.

Ravelonandro M, Scorza R, Dunez J (1998) Characterization of phenotype resistance to plum pox of transgenic plums expressing plum pox virus capsid gene. Acta Virol 42: 270–272.

An Efficient Novel Method of Producing Virus Free Plants from Garlic Root Meristem

Md. Shahidul Haque[1], K. Hattori, A. Suzuki, and T. Tsuneyoshi

Laboratory of Plant Genetics and Breeding, Graduate School of Bioagricultural Sciences, Nagoya University, Chikusa, Nagoya 464-8601, Japan.
[1]Professor and Head, Department of Biotechnology, Bangladesh Agricultural University, Mymensingh 2202, Bangladesh.
Email: haquems@yahoo.com

Introduction

Garlic (*Allium sativum* L.) is an important crop with anti-AIDS, anti-cancer, and anti-vascular diseases and anti-infectious properties. Commercially cultivated clones have been shown to be infected with a complex of viruses that include leek yellow stripe virus (LYSV), onion yellow dwarf virus (OYDV), shallot (garlic) latent virus (SLV, GLV), garlic common latent virus (GCLV), and garlic viruses (GarVs) (Walkey, 1990; Ayabe and Sumi, 2001) causing severe yield and quality losses. Plant tissue culture is an efficient tool for the elimination of viruses from infected plants. Like other species, garlic shoot meristem culture has long been used for the production of virus-free plants (Walkey, 1990). However, the production and propagation rate of virus-free plantlets is very low (Ayabe and Sumi, 2001). Moreover, excision of meristems containing only two primordial leaves under microscope is a laborious, time-consuming procedure. Root tips of plants infected with viruses were found to be free from detectable virus (Appiano and D'Agostino, 1983). Unfortunately, production of virus free plant from root meristem was not possible.

Z. Xu et al. (eds.), Biotechnology and Sustainable Agriculture 2006 and Beyond, 107–110.

We found direct shoot regeneration from the root meristem (Haque et al., 1997, 1999). Here we report successful elimination of viruses and production of virus free garlic plantlets by root meristem culture.

Methodology

Root-tips (2–3 mm including meristems) were excised from aseptically sprouted cloves and micropropagated plantlets. Explants were cultured on MS medium containing 1 μM NAA and 10 μM BA (Haque et al., 1997). After 35–40 days when the shoots became ca. 5 mm, they were cut off from the mother explant and transferred to growth regulator free MS medium for their further development and virus testing. RT-PCR was used to detect garlic viruses according to Tsuneyoshi and Sumi (1996). Reverse transcription (RT) was performed using a first-strand cDNA Synthesis Kit (Pharmacia) with approximately 1–2 μl of the RT mix was added to 100 μg of a polymerase reaction mixture containing 20 mM Tris-HCl (pH 8.4), 50 mM KCl, 1.5 mM $MgCl_2$, 200 μM each of dNTP, 2.5 U Taq polymerase (GIbco BRL), and 100 ng each of the upstream and downstream primers. Thirty reaction cycles were used, with periods of 30 s for annealing at 50°C, 45 s for synthesis at 72°C and 45 s for melting at 94°C. Following PCR, 10 μl portions of the reaction mixtures were analyzed on 1.5% agarose gels (Tsuneyoshi and Sumi, 1996).

Results and discussion

Production of virus free propagation materials by meristem culture requires the use of highly specific, sensitive and rapid detection methods to screen the regenerants. RT-PCR has been used for detection of viruses in many plants including garlic. In our study, we followed RT-PCR to detect viruses from the regenerated shoots from garlic root meristems. The shoots regenerated from garlic root meristems and the mother plantlets showed the presence of viruses (Figure 1). Among the four different types of viruses, the mother

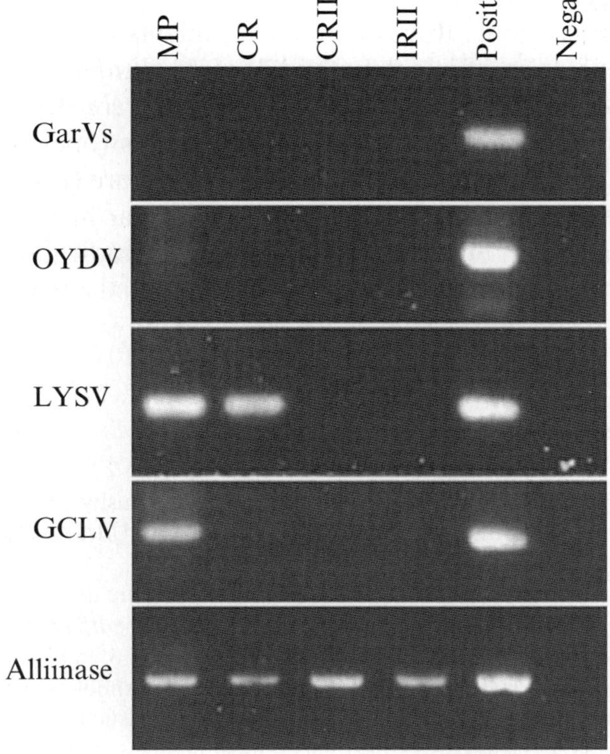

Figure 1. RT-PCR amplification of viral cDNA. MP: Mother plant; CR: Shoots from sprouted clove derived root tips; CRII: Shoots from sprouted clove derived root tips (two-step culture); IRII: Shoots regenerated from root tips derived from *in vitro* shoots (two-step culture). Allinase was used as internal control.

plants showed the presence of three viruses (OYDV, LYSV and GCLV); while the regenerants (one step culture) indicated the presence of LYSV, while OYDV and GCLV were eliminated (Figure 1). Since elimination of two viruses and presence of one virus in the regenerated shoots were confirmed by RT-PCR, success of total elimination was expected through additional precautions. In the next step, the individual shoot primordia were separated from the explants before they became ca. 5 mm and transferred to same medium without growth regulator. The shoots grew further, developed roots within 3 weeks and formed rooted plantlets on growth regulator free medium. The shoots from the two-step culture were tested using RT-PCR for the same viruses and they were found to be free from all viruses (Figure 1). Therefore, root tip culture followed by separation of the regenerated tiny shoots from the mother explants (two-step process) was confirmed as an efficient method of virus elimination using root meristem. Four different types of viruses that were detected in the mother garlic plant and in the regenerated shoots (one-step culture) were absent in the regenerants from two-step culture (Figure 1). This result proves for the first time by using molecular method that root tip culture can be an efficient technique for the eradication of multiple viruses from garlic. This method has potential for use in other plants also.

References

Appiano A, D'Agostino G (1983) Distribution of tomato bushy stunt virus in root tips systematically infected *Gomphrena globosa*. J Ultrastructure Res 85: 239–248.

Ayabe M, Sumi S (2001) A novel and efficient tissue culture method – 'stem-disc dome culture' – for producing virus-free garlic (*Allium sativum* L.). Plant Cell Rep 20: 503–507.

Haque MS, Wada T, Hattori K (1997) High frequency shoot regeneration and plantlet formation from root tip of garlic. Plant Cell Tissue Organ 50: 83–89.

Haque MS, Wada T, Hattori K (1999) Anatomical changes during *in vitro* direct formation of shoot bud from root tips in garlic (*Allium sativum* L.). Plant Prod Sci 2(2): 146–153.

Tsuneyoshi T, Sumi S (1996) Differentiation among garlic viruses in mixed infections based on RT-PCR procedures and direct tissue blotting immunoassay. Phytopathology 86: 253–259.

Walkey DAG (1990) Virus diseases. In: Rabinowitch HD, Brewster JL (eds) Onion and Allied Crops, Vol 2. CRC Press, Boca Raton. pp. 191–211.

Site-Directed Homologous Recombination in Tobacco Cell Cultures via Zinc Finger Nucleases

Charles Q. Cai[1,*], W. Michael Ainley[1], Trevor N. Collingwood[2], Robbi J. Garrison[1], Lisa L. Schulenberg[1], Andrew Worden[1], Philip D. Gregory[2], Beth C. Rubin-Wilson[1], and Joseph F. Petolino[1]

[1] Dow AgroSciences, LLC, 9330 Zionsville Rd., Indianapolis, IN 46268, USA.
[*]Email: ccai2@dow.com
[2] Sangamo BioSciences, Inc., 501 Canal Blvd, Suite A100, Richmond, CA 94804, USA.

Targeted transgene integration via homologous recombination occurs at a very low frequency in plant cells compared to random integration, even when the incoming DNA comprises large stretches of sequence homologous to host DNA. As such, gene targeting in plants via homologous recombination has not been practical. Most recently, substantial increases in the frequency of homologous recombination have been observed following the induction of DNA double stranded breaks in host cells and apparent stimulation of cellular repair mechanisms. Restriction enzymes whose recognition sites are rare in the plant genome have been shown to stimulate homologous recombination following the formation and repair of DNA double stranded breaks in the host DNA. Strategies to achieve targeted DNA double stranded breaks have been developed by fusing zinc finger DNA binding proteins with sequence-independent nuclease domains derived from Type II restriction endonucleases.

In the present study, engineered zinc finger proteins fused to nuclease domains, the so-called zinc finger nucleases, were used to

Z. Xu et al. (eds.), Biotechnology and Sustainable Agriculture 2006 and Beyon, 111–113.

facilitate site-specific transgene integration via homologous recombination in tobacco cell cultures. A target DNA sequence was designed to address both inter- and intra-chromosomal homologous recombination and stably integrated into tobacco cell cultures using *Agrobacterium*. To test inter-chromosomal homologous recombination, two strategies were designed using two types of zinc finger nucleases (ZFN). In first strategy, the binding site for one of the ZFN (ZFN-1) was included in the middle of the target construct, which was flanked by ~3 kb of non-homologous sequences on both sides followed by left- and right-side homologous sequences. In the second strategy, two binding sites of the second ZFN (ZFN-2) were included in the target vector with one directly downstream of the left-side homologous sequence and the other directly upstream of the right-side homologous sequence. In between the two binding sites were ~6 kb of sequence to be replaced upon homologous recombination. To monitor the outcome of inter-chromosomal homologous recombination, a non-functional test gene to be corrected was included in the target. In addition, a full expression cassette of another functional reporter gene was also included in the target to serve as an indicator of the accessibility of the target.

A few transgenic events containing a single integrated copy of the target sequence were selected for re-transformation using *Agrobacterium* strains harboring two different T-DNAs. One *Agrobacterium* strain contained a donor DNA sequence comprising the bases necessary to correct the non-functional test gene, flanked by sequences homologous to the pre-integrated target DNA. The second *Agrobacterium* strain contained a gene encoding either ZFN-1 or ZFN-2 that specifically recognized binding sites in the integrated target sequence. Gene correction via site-directed inter-chromosomal homologous recombination was achieved with both strategies tested as evidenced by the re-constitution of a functional test gene. Homologous recombination-facilitated correction of the test gene was verified by PCR analysis using one primer in the target sequence and the other in the donor sequence and confirmed by sequencing the entire amplicon. The recombinants with a corrected test gene were also confirmed by Southern blot as well as biochemical analyses.

To test ZFN-facilitated intra-chromosomal homologous recombination, two nonfunctional reporter gene fragments with 540 bp overlap sequences were included in the target sequence. In between these two fragments was the functional reporter gene expression cassette used as an indicator of target site accessibility. The binding site of ZFN-1 was fused with the reporter gene coding sequence at its N-terminal. A few transgenic events containing a single integrated copy of the target sequence were selected for re-transformation using an *Agrobacertium* strain harboring the gene encoding ZFN-1. Cells were plated onto non-selection medium after transformation. The expression of the reconstituted functional reporter gene was observed around 5–8 days after transformation. No appreciable expression beyond slight background was observed in the negative controls. The tissue expressing the reporter gene was enriched and subjected to molecular analysis. PCR and Southern analysis confirmed that reporter gene expression resulted from reconstitution of a functional gene cassette. Thus ZFN-facilitated intra-chromosomal homologous recombination was demonstrated.

Generation of Genetically Modified Strawberries in Temporary Immersion Bioreactor System

Kati Hanhineva*, Harri Kokko, and Sirpa Karenlampi

Institute of Applied Biotechnology, University of Kuopio,
P. O. Box 1627, 70211 Kuopio, Finland.
*Email: kati.hanhineva@uku.fi

Abstract

A temporary immersion bioreactor system (TIB system) provides a convenient and efficient way to propagate plant material *in vitro* in liquid medium with significantly lower labour input than the conventional methods on semi-solid medium. The TIB system is used in our laboratory for the micropropagation of several berry species and has been now adapted for the production of genetically modified strawberries (*Fragaria x ananassa*). The process is started with a standard *Agrobacterium tumefaciens*-mediated gene transfer method with co-cultivation on semi-solid medium. For the regeneration and selection phases the plant material is moved to TIB containers and the subculturing is carried out by renewing the medium in the containers. After the regenerating shoots start to emerge, compatible antibiotic is applied to select the transgenic progeny.

Introduction

The octoploid genome composition (56 chromosomes) of the cultivated strawberry *Fragaria x ananassa* Duch. is the most complicated one among the Rosaceae family, which is a potential difficulty

Z. Xu et al. (eds.), Biotechnology and Sustainable Agriculture 2006 and Beyon, 115–118.

in the successful genetic modification. During the past few years several reports on fine-tuning the transformation process for certain strawberry cultivars have been published (e.g., Folta et al., 2006), but widely applicable techniques are still lacking.

Temporary immersion bioreactor (TIB) systems (Etienne and Berthouly, 2002) are presently used for *in vitro* propagation of several agricultural and forestry species. In the TIB system used in our laboratory the plant material is placed into a double-barreled container with liquid medium. When overpressure of sterile air is applied in the lower container, the medium is pushed into the upper container flushing the plants and simultaneously renewing the air inside the vessel. When the pressure is released, the liquid medium returns to the bottom of the vessel and the plants remain covered by a film of medium by capillary action.

This method has been proven useful for the *in vitro* cultivation of strawberry (Hanhineva et al., 2005) and has now been adapted for the regeneration of genetically modified strawberries (Figure 1).

Figure 1. (**A**) TIB container flushing the leaf pieces in regeneration medium. (**B**) Vigorously regenerating leaf piece 6 weeks after the onset of the study.

Materials and methods

A. tumefaciens strain LBA4404 containing the gene transfer construct with hygromycin resistance as the selection marker was used for the inoculation of young folded leaves of the strawberry cultivar Jonsok. The plant material was surface-sterilized, chopped and after inoculation (30 min in *Agrobacterium*/MS suspension) the leaf pieces were co-cultivated on MS agar plate overnight (22°C, duration of light period 19 h at 45 $\mu Em^{-2}s^{-1}$). For the regeneration, the pieces were placed into TIB containers with the regeneration medium (MS, TDZ, 2.0 mg L^{-1}; IBA, 0.5 mg L^{-1}). For the first 2 weeks the antibiotic cefotaxime (400 mg L^{-1}) was added in order to eliminate *Agrobacterium* growth. As the leaf pieces started to regenerate (3–8 weeks), they were picked into another container with the regeneration medium containing hygromycin (10 mg L^{-1}). The shoots that remained viable and continued developing in the presence of the antibiotic were moved to another container with MS medium for rooting.

Results and discussion

The shoots remaining viable in the antibiotic selection were planted in soil and subsequently analysed by molecular biology methods for the presence and expression of the transferred gene of interest, and also tested for the possible *Agrobacterium* contamination by PCR with *virG*-specific primers (Figure 2).

The main advantage of the TIB system in strawberry gene transfer is in the ease of handling the plant material, as the subculturing is achieved by chancing the medium in the container rather than treating the leaf pieces one by one. This is especially beneficial for the transformation of recalcitrant varieties which have low transformation frequencies and require considerable amount of starting material.

In the TIB system the liquid medium rinses the leaf explants thoroughly throughout the subcultures and the constituents in the medium get into direct contact with the whole explant. A major advantage of the direct contact is especially in the antibiotic selection, when the selective agent gets to interact directly with all the cells in the regenerating explant. This enables the 'post-regeneration'-selection of the emerging shoots, and shortens the overall time needed for the regeneration of transgenic strawberries, as the regeneration initiates faster in the absence of the selective antibiotic. When the selection is applied the transgenic individuals are distinguishable within 2–3 weeks. In the elimination of *Agrobacterium* effective contact with the medium is inevitable so that all the bacteria on the explant can be removed completely, and this can be achieved by keeping the antibiotic in the medium for the first 2 weeks. Furthermore, the ability to remove cefotaxime from the culture medium as early as after 2 weeks is beneficial, as cefotaxime has inhibitory effect on the regeneration of strawberry.

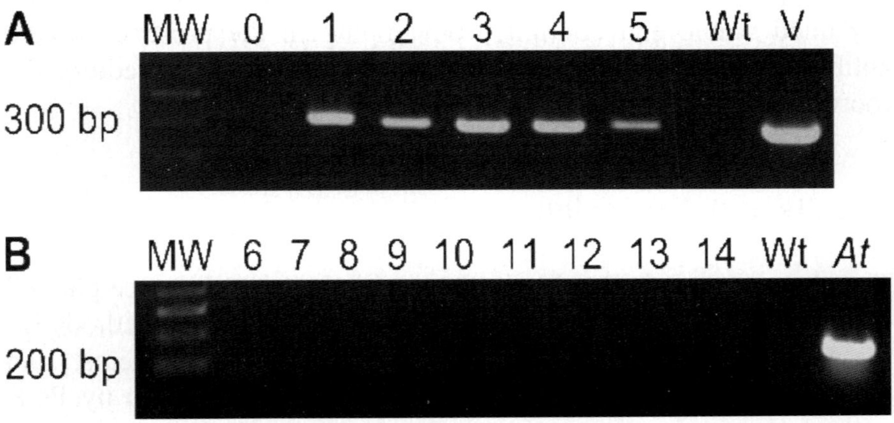

Figure 2. (**A**) PCR by *nptII* primers. (**B**) PCR by *virG* primers. Abbreviations are 0: PCR blank, 1–14: transgenic strawberries, Wt: wild type strawberry, V: vector control, *A.t*: *Agrobacterium* control.

References

Etienne H, Berthouly M (2002) Temporary immersion systems in plant micropropagation. Plant Cell Tissue Organ Cult 69: 215–231.

Folta KM, Dhingra A, Howard L, Stewart PJ, Chandler CK (2006) Characterization of LF9, an octoploid strawberry genotype selected for rapid regeneration and transformation. Planta 224: 1058–1067.

Hanhineva K, Kokko H, Karenlampi S (2005) Shoot regeneration from leaf explants of five strawberry (Fragaria x ananassa) cultivars in temporary immersion bioreactor system. In Vitro Cell Dev Biol – Plant 41: 826–831.

Current Advances in Plant Molecular Biology

The Positioning of the Division Plane Is Determined by a Cortical Division Site at the Plasma Membrane

Danny Geelen

Department of Plant Production, University Ghent, Coupure links 653, B-9000 Ghent, Belgium.

Positioning of the new cell wall after mitosis is critical for the establishment of highly structured plant tissues and the overall morphology of a plant. To secure correct orientation of the separating cell walls, plants have developed cytoskeletal structures that guide the newly formed cell plate toward a predefined position. The preprophase band of microtubules (PPB) that surround the usually central-localized nucleus is one of the first structural indications that a cell is preparing to divide. The PPB is removed before chromosome segregation and cytokinesis and therefore cannot contribute directly to the cell plate guidance process. A bundle of two sets of microtubules of opposite polarity called the phragmoplast, appearing at the central zone of the anaphase spindle, provides the structural scaffold for the coordination of cell plate expansion and positioning. The microtubules of the phragmoplast spread centrifugally into the cortical cytoplasm as the cell plate expands, eventually completing division by joining with parental walls at sites previously occupied by the PPB. The function of actin in determining the division plane and plate guidance is more complex. Actin filaments are present in the PPB and phragmoplast structures and may in fact contribute to the formation and organization of the microtubules that are the major components of both structures. When the PPB microtubules degrade, cortical actin disappears at the corresponding former position of the

Z. Xu et al. (eds.), Biotechnology and Sustainable Agriculture 2006 and Beyond, 121–124.

PPB, leaving behind an actin depleted zone (Cleary et al., 1992). The dense actin mesh flanking the cortical division site observed in BY-2 suspension cells described as twin peaks (Sano et al., 2005), is principally the same as the actin depleted zone originally described for *Tradescantia.* Actin drug treatments slow down mitotic progression and lead to aberrant cell plate formation with subtle changes in position and morphological aspects of the cell plate (Hoshino et al., 2003). In fact, depolymerisation of actin does not prevent cell plate formation and it does not disturb the overall positioning and orientation of the division plane (Vanstraelen et al., 2006). Incubation of actin drugs during preprophase leads to broadend PPBs which may have difficulties in the precision control or the exact positioning of the future cell wall insertion site (Eleftheriou and Palevitz, 1992; Li et al., 2006).

The relative insensitivity of division plane determination to actin drugs suggests that the signal transmitted by the PPB is not actin dependent and that actin is not directly responsible for maintaining the signal that positions the cortical division site. Other cellular structures like the Golgi apparatus and the endoplasmic reticulum (ER) have been found to accumulate after PPB degradation (Nebenführ et al., 2000; Zachariadis et al., 2001; Gupton et al., 2006). Whether organelle redistribution is of significance is not clear as membrane trafficking inhibitor Brefeldin A (BFA) does not interfere with division plane establishment (Dixit and Cyr, 2002). Perhaps there exist BFA insensitive vesicle-trafficking processes that support the modficiations of the cortical cell periphery closest to the PPB as to ascertain transmission of the positional cue(s) throughout mitosis. Recently, a novel protein marker was identified that delineates a cortical plasma membrane domain at the division site predicted by the PPB (Vanstraelen et al., 2006). The marker, a GFP-tagged kinesin, accumulates at the plasma membrane during preprophase but is restrained from concentrating at the position corresponding to that of the PPB and the actin-depleted zone. The domain depleted of the KCA kinesin was hence called the KCA-depleted zone (KDZ). The KDZ persists throughout mitosis until the new cell plate inserts into the mother wall. The position and size of the KDZ perfectly match with the accumulation of another marker identified in a GFP-based screen for cytokinesis related proteins (Van Damme et al., 2006). The protein TPLATE accumulates at the cortical division site prior to cell plate maturation, when the expanding rim of the plate makes contact with the mother wall (Van Damme et al., 2006). TPLATE contains sequences that show similarity to domains present in ß-COP proteins, suggesting that it is implicated in vesicle targeting. Taken together, the localization patterns of KCA and TPLATE favor the viewpoint that the PPB is involved in an unknown modification of the plasma membrane that leads to the establishment of a fixed cortical division site that later, during division, is used by the phragmoplast as a beacon for cell plate attachment.

References

Cleary AL, Gunning BES, Wasteneys GO, Hepler PK (1992) Microtubule and F-actin dynamics at the division site in living *Tradescantia* stamen hair cells. J Cell Sci 103: 977–988.

Dixit R, Cyr R (2002) Golgi secretion is not required for marking the preprophase band site in cultured tobacco cells. Plant J 29: 99–108.

Eleftheriou EP, Palevitz BA (1992) The effect of cytochalasin D on preprophase band organization in root tip cells of *Allium*. J Cell Sci 103: 989–998.

Gupton SL, Collings DA, Allen NS (2006) Endoplasmic reticulum targeted GFP reveals ER organization in tobacco NT-1 cells during cell division. Plant Physiol Biochem 44: 95–105.

Hoshino H, Yoneda A, Kumagai F, Hasezawa S (2003) Roles of actin-depleted zone and preprophase band in determining the division site of higher-plant cells, a tobacco BY-2 cell line expressing GFP-tubulin. Protoplasma 222: 157–165.

Li CL, Chen ZL, Yuan M (2006) Actomyosin is involved in the organization of the microtubule preprophase band in arabidopsis suspension cultured cells. J Integrative Plant Biol 48: 53–61.

Nebenführ A, Frohlick JA, Staehelin LA (2000) Redistribution of Golgi stacks and other organelles during mitosis and cytokinesis in plant cells. Plant Physiol 124: 135–151.

Sano T, Higaki T, Oda Y, Hayashi T, Hasezawa S (2005) Appearance of actin microfilament 'twin peaks' in mitosis and their function in cell plate formation, as visualized in tobacco BY-2 cells expressing GFP-fimbrin. Plant J 44: 595–605.

Van Damme D, Coutuer S, De Rycke R, Bouget F-Y, Inzé D, Geelen D (2006). Somatic cytokinesis and pollen maturation depend on TPLATE, a novel protein with domains similar to coat proteins. Plant Cell 18: 3502–3518.

Vanstraelen M, Van Damme D, De Rycke R, Mylle E, Inze D, Geelen D (2006) Cell cycle-dependent targeting of a kinesin at the plasma membrane demarcates the division site in plant cells. Curr Biol 16: 308–314.
Zachariadis M, Quader H, Galatis B, Apostolakos P (2001) Endoplasmic reticulum preprophase band in dividing root-tip cells of Pinus brutia. Planta 213: 824–827.

Repression of Plant Gene Expression via Chromosomal Remodelling Using Histone Deacetylases

Lining Tian[1,*], Keqiang Wu[2], Brian Miki[3], and Dan Brown[1]

[1]Agriculture and Agri-Food Canada, London, Ontario, Canada.
[2]Department of Biology, West Virginia University, WV, USA.
[3]Agriculture and Agri-Food Canada, Ottawa, Ontario, Canada.
[*]Email: tianl@agr.gc.ca

Posttranscriptional modification of nucleosomal histones is an important mechanism in the regulation of eukaryotic gene expression. Acetylation and deacetylation is a major process of such regulation. Acetylation of the histones by acetyltransferases leads to a relaxed status of the chromosomes, and thus, enhances accessibility of transcription components to the promoters and activates the gene expression. Deacetylation of histones by histone deacetylases (HDACs) remove acetyl groups from the core histones, which results in chromatin condensation and restricts access of the transcription machinery to gene regulatory elements. This leads to the repression of transcription (Landry et al., 2000; Murfett et al., 2001). Several different types of HDACs have been identified in eukaryotic organism (Dangl et al., 2001). Plants possess a unique type of HDACs, namely, the HD2 family. HD2 family is not related to other classes of HDACs.

In this study, we report the repression of transgene expression in plants by HDACs, including HD2 and RPD3 classes.

Z. Xu et al. (eds.), Biotechnology and Sustainable Agriculture 2006 and Beyond, 125–128.

Materials and methods

Different HDACs genes were isolated from Arabidopsis plant via PCR using appropriate primers. The genes were confirmed via sequence analysis. Various DNA vectors were constructed using standard molecular techniques. For transient gene expression, the vectors were introduced into tobacco plant tissues via bombardment. Transgenic Arabidopsis plants were generated using flowering dip approach. Southern blot and Northern blot were used for transgenic plant analysis. Expression of GUS reporter gene was analyzed via histochemical assay or fluorometric assay.

Results and discussion

Gene expression repression using HDACs was initially evaluated via transient gene expression. A reporter vector was constructed in which a GUS reporter gene was driven by a constitutive promoter. Two copies of yeast transcription GAL4-binding sites (UAS_{GAL4}) were placed in front of the promoter in the reporter vector. A second vector, effector contains a HD2 protein, HD2A. The protein was fused with the DNA-binding domain of the yeast transcription factor GAL4 (GAL4BD). The fused protein was driven by another constitutive promoter. The reporter and effector plasmids were simultaneously introduced into tobacco leaves by co-bombardment. Analysis of GUS expression showed that HD2A proteins strongly repressed GUS gene expression when compared with the controls. Different types of HDAC proteins, including HD2A, 2B, 2C, and RPD 3A were also tested in the same manner and all these HDAC proteins exhibited repression effect.

The repression of gene expression by HDACs was further evaluated in transgenic plants. Reporter transgenic plants containing UAS_{GAL4}-Promoter-GUS and effector transgenic lines carrying GAL4/HD2A were generated separately using different selectable markers. The reporter lines were crossed with the effector lines and F1 seedlings were generated. Analysis showed that the GUS gene expression in F1 progeny plants containing both the effector and the reporter was significantly reduced. Northern analysis indicated that little GUS transcript was detected in the F1 progeny. Decreased GUS transcript in the F1 progeny was correlated with strong GAL4/HD2A expression in these plants. This demonstrates that gene repression system using HDAC protein is applicable in stable gene expression manner.

The seed specific promoter, Napin promoter (NAP2), was used to replace the constitutive promoter to drive the GAL4/HD2A fusion gene. Transgenic Arabidopsis plants containing effector which contained seed specific promoter or reporter lines were generated separately. Analysis of F1 plants showed that the GUS expression was repressed only in the seeds, not in the other organs, of the F1 hybrid progeny. The new expression pattern indicates that specific repression by HDACs can be achieved by directing the expression of the GAL4/HD2 protein in a particular manner as desired.

The research furthered to evaluate if HDACs can mediate gene repression through interacting with a plant transcription factor. Pti4 is a transcription factor that belongs to the ERF family. The genes regulate the expression of the GCC box genes. The effector construct was made which contained the HD2A fused with the Pti4 DNA binding domain (Pti4BD). The reporter vector was constructed in which five copies of GCC box were placed in front of the promoter which drives the GUS gene. The reporter vector and the effector vector were co-introduced in to tobacco leaves by bombardment. Results showed that the expression of GUS reporter gene containing GCC box was significantly reduced. The results demonstrate that HDACs can mediate gene repression through interacting with a plant transcription factor. Expression repression using plant transcription factor can lead to possibilities of mediating the expression of genes residing in the genome without cloning and alteration of their sequence.

Chromosome remodeling is an important mechanism in regulating plant gene expression. We demonstrated that plant gene expression can be repressed by histone deacetylases. Repression or altering of gene expression via chromosome remodelling using HDACs provides a useful and alternative avenue for gene silencing. The system is useful for gene function studies, for manipulation of biochemical processes and for control of plant developmental processes.

References

Dangl M, Brosch G, Haas H, Loidl P, Lusser A (2001) Comparative analysis of HD2 type histone deacetylases in higher plants. Planta 213: 280–285.

Landry J, Sutton A, Tafrov ST, Heller RC, Stebbins J, Pillus L, Sternglanz R (2000) The silencing protein SIR2 and its homologs are NAD-dependent protein deacetylases. Proc Natl Acad Sci USA 97: 5807–5811.

Murfett J, Wang X, Hagen G, Guilfoyle TL (2001) Identification of Arabidopsis histone deacetylase HDA6 mutants that affect transgene expression. Plant Cell 13: 1047–1061.

Functional Genomics

DNA-METHYLTRANSFERASE 1 is a Member of FIS Polycomb Complex and is Involved in Seed Development in *Arabidopsis*

Ming Luo[1], Mingzhu Luo[3], Fred Berger[2], E.S. Dennis[1], Jim W. Peacock[1], and Abed Chaudhury[1]

[1] CSIRO, Plant Industry Canberra ACT.
[2] Chromatin and Reproduction Group, Temasek Lifesciences Laboratory, National University of Singapore, 1 Research Link, 117604, Singapore.
[3] College of Agronomy, South China Agricultural University, China, 510642.
Email: ming.luo@csiro.au

FIS class mutants such as *medea* (*mea*) show two distinct seed phenotypes; the growth of the central cell into endosperm in the absence of fertilization, and the arrest of the embryo at the heart stage of development following fertilization (Chaudhury et al., 1997). The embryo arrest can be rescued by pollination with a hypo-methylated paternal gene (Vielle-Calzada et al., 1999; Luo et al., 2000; Vinkenoog et al., 2000). In this study, we show that reduced methylation of the female genome cannot rescue the *fis* embryo arrest phenotype.

Plants with low levels of DNA-methylation develop seed-like structures in the absence of fertilization. Emasculated Arabidopsis flowers from anti-*MET1* plants initiate autonomous development of fruit and seed-like structure. After 8 days, siliques of emasculated anti-*MET1* plants had elongated relative to siliques of wild-type emasculated plants. Such autonomous siliques contained enlarged ovules of variable size which resembled developing seeds. We noted autonomous endosperm development in only a single case among

Z. Xu et al. (eds.), *Biotechnology and Sustainable Agriculture 2006 and Beyond*, 131–133.

about 200 enlarged ovules. The seed-like structures show some of the phenotypes of *fis* class autonomous seeds. C24 with low levels of DNA-methylation produce large, misshapen and shriveled seeds (Luo et al., 2000) which show over-proliferation of endosperm when pollinated with wild type pollen (Adams et al., 2000), and resemble the post-fertilization phenotype of *fis* mutants (Kiyosue et al., 1999).

Because of the phenotypic similarity between low-methylation plants and *fis* mutants, we determined if DNA METHYLTRANSFERASE1 (MET1) is a member of the FIS polycomb complex. We tested the interaction of MET1 with three members of the FIS polycomb complex, FIS2, MEA, and FIE using a yeast two hybrid system. We showed that MET1 interacts with MEA in a yeast two-hybrid system suggesting that MET1 is directly involved in FIS polycomb repression. Similar interaction has been found in human cell (Viré et al., 2006).

In contrast to the observation that hypomethylation of the female genome cannot rescue the *fis* seeds, a complete rescue of *mea* mutant seed by paternal hypomethylation can occur by suppression of genes down-stream of the *FIS* class genes. This rescue is independent of *MEA* function. *PHE1*, which is up-regulated in *fis* mutants, was down regulated in the siliques of L*er* pollinated with C24 anti-*MET1* relative to wild type C24, indicating *PHE1* activity is reduced in a hypomethylated background. Thus hypomethylation has distinct roles in seed development based on the parent of origin.

References

Adams S, Vinkenoog R, Spielman M, Dickinson HG, Scott RJ (2000) Parent-of-origin effects on seed development in *Arabidopsis thaliana* require DNA methylation. Development 127: 2493–2502.

Chaudhury AM, Ming L, Miller C, Craig S, Dennis ES, Peacock WJ (1997) Fertilization-independent seed development in *Arabidopsis thaliana.* Proc Natl Acad Sci USA 94: 4223–4228.

Kiyosue T, Ohad N, Yadegari R, Hannon M, Dinneny J, Wells D, Katz A, Margossian L, Harada JJ, Goldberg RB, Fischer RL (1999) Control of fertilization-independent endosperm development by the *MEDEA* polycomb gene in Arabidopsis. Proc Natl Acad Sci USA 96: 4186–4191.

Luo M, Bilodeau P, Dennis ES, Peacock WJ, Chaudhury A (2000) Expression and parent-of-origin effects for FIS2, MEA, and FIE in endosperm and embryo of developing Arabidopsis seeds. Proc Natl Acad Sci USA 97: 10637–10642.

Vielle-Calzada JP, Thomas J, Spillane C, Coluccio A, Hoeppner MA, Grossniklaus U (1999) Maintenance of genomic imprinting at the Arabidopsis media locus requires zygotic DDM1 activity. Genes Dev 13: 2971–2982.

Vinkenoog R, Spielman M, Adams S, Fischer RL, Dickinson HG, Scott RJ (2000) Hypomethylation promotes autonomous endosperm development and rescues postfertilization lethality in *fie* mutants. Plant Cell 12: 2271–2282.

Viré E, Brenner C, Deplus R, Blanchon L, Fraga M, Didelot C, Morey L, Eynde AV, Bernard D, Vanderwinden JM, Bollen M, Esteller M, Croce LD, de Launoit Y, Fuks F (2006) The polycomb group protein EZH2 directly controls DNA methylation. Nature 439: 871–874.

Integrating Transcriptional and Metabolic Profiling to Unravel Secondary Metabolite Biosynthesis in Plants

Heiko Rischer[1], Matej Oresic[1], Alain Goossens[2], Dirk Inzé[2], and Kirsi-Marja Oksman-Caldentey[1]

[1]VTT Technical Research Centre of Finland, P. O. Box 1000 (Tietotie 2), FIN-02044 VTT, Finland.
[2]Department of Plant Systems Biology, VIB-Ghent University, Technologiepark 927, B-9052 Gent, Belgium.

Plants exhibit an overwhelming metabolic capacity and the quest to discover all secondary metabolites has barely begun. Only a fraction of the 500,000 plant species estimated to exist has been formally described and only about 10% of those have been chemically investigated to some extent. In the recent years many biological functions of secondary metabolites have been revealed. The historical term 'secondary metabolites' implied that these were inessential compounds as compared to essential i.e., 'primary metabolites' involved in the basic processes in the cell. Now we know that secondary metabolites have for example many stress related functions, act as pollinator attractants or defence compounds etc. Nevertheless for most of them their biological function is not yet understood.

Besides their functions in plants, the secondary metabolites are also important for humans in the form of pharmaceuticals (Rischer and Oksman-Caldentey, 2005). In developing countries 80% of the population rely on traditional medicine (employing mostly plant extracts) as the primary healthcare. In the western world about 25% of all prescribed drugs contain actually plant-derived compounds

Z. Xu et al. (eds.), Biotechnology and Sustainable Agriculture 2006 and Beyond, 135–138.

and 121 such compounds are still in use – among them several anticancer blockbuster drugs such as vincristine and paclitaxel. The monetary value of plant-derived pharmaceuticals exceeded 30 billion dollars in the United States in recent years.

Chemical synthesis is usually preferred by the pharmaceutical companies. Although nearly all molecules can be synthesized this is often not economically feasible. Extracting plants from the wild or the field is possible but associated with problems regarding geographical, political, seasonal and environmental influences. Often the content is quite low. It has been tried to overcome some of these problems by using tissue cultures because this could theoretically offer sustainable and fully controlled conditions. In practice, however, one has to fight against slow growth and low productivity among others. It is clear therefore that genetic and metabolic engineering is needed to reach more productive systems. New methods are needed to speed up the whole pathway elucidation process. For this we have created a strategy and the necessary technology platform which is called SoluCel. It is based on functional genomics in combination with metabolic profiling and it allows genome wide gene discovery in any kind of plant irrespective of whether it has been sequenced or not (Oksman-Caldentey and Inzé, 2004). The genes can then be functionally analysed. As a result we get the tools for metabolic engineering either to enhance the production of known compounds or even to produce novel compounds in safe, reliable and contained systems as we are using plant cell and organ cultures.

Two central elements of the SoluCel technology (www.solucel.com) are the integrated application of metabolic and transcriptional profiling which are described here in detail using a case study. *Catharanthus roseus* cell suspension cultures were chosen as a test system because this plant contains pharmaceutically important terpenoid indole alkaloids (TIAs) the biosynthesis of which has been investigated during the past 50 years and the knowledge is summarized in publicly available databases (e.g., http://www.genome.jp/kegg/pathway/map/map00901.html).

In a first test differential conditions were created by methyl-jasmonate elicitation and both profiling methods were examined separately. Metabolic profiling, i.e., the analysis of selected compounds predefined by the extraction method, was performed by LC-MS. After data processing using the in-house developed software MZmine (mzmine.sourceforge.net) principal components analysis (PCA) clearly demonstrated significant differences between the elicited and the control samples (Orešič et al., 2006). A following factor analysis could identify the metabolites contributing mainly to this separation (Katajamaa and Orešič, 2005). Also the sensitive, genome-wide and quantitative cDNA-AFLP method (Breyne et al., 2003) used for transcript profiling performed very well resulting in the visualization of ca. 600 differential tags. Therefore in a subsequent experiment more parameters were introduced (control, elicited, with/without auxin) and samples taken at different intervals were profiled with both methods. Because of their similar structure (identifier and quantitative value) the data were combined in one set for further analysis. A PCA clearly showed that the experimental conditions were reflected in the data. Visualization of the data is crucial for biological interpretation. For known metabolites and genes the expression kinetics can be directly marked in a pathway representation as a snapshot for a given sampling time. More interesting is nevertheless the integration of all known or unknown compounds and genes. Correlation network analysis allowed the generation of priority lists for genes which influence key metabolites (Rischer et al., 2006). Genes from such lists are good candidates for further functional analysis and combinatorial biochemistry approaches.

As a summary we conclude that the integrated metabolite and transcript profiling allows the visualization of the majority of all known genes in a given pathway as well as the establishment of the kinetics of their expression in a single experiment. Furthermore an ample collection of novel genes is generated and the most interesting genes can be conveniently selected from priority lists based on co-regulation with key metabolites.

References

Breyne P, Dreesen R, Cannoot B, Rombaut D, Vandepoele K, Rombauts S, Vanderhaeghen R, Inze´ D, Zabeau M (2003) Quantitative cDNA-AFLPanalysis for genome-wide expression studies. Mol Genet Genomics 269: 173–179.

Katajamaa M, Orešič M (2005) Processing methods for differential analysis of LC/MS profile data. BMC Bioinformatics 6: 179.

Oksman-Caldentey KM, Inzé D (2004) Plant cell factories in the post-genomic era: New ways to produce designer secondary metabolites. Trends Plant Sci 9: 433–440.

Orešič M, Rischer H, Oksman-Caldentey KM (2006) Metabolics of plant secondary compounds: Profiling of Catharanthus cell cultures. In: Biotechnology in Agriculture and Forestry, Vol. 57, Plant Metabolomics, pp. 277–290.

Rischer H, Oksman-Caldentey KM (2005) Biotechnological utilization of plant genetic resources for the production of phytopharmaceuticals. Plant Genet Resour 3: 83–89.

Rischer H, Oresic M, Seppanen-Laakso T, Katajamaa M, Lammertyn F, Ardiles-Diaz W, Van Montagu MC, Inze D, Oksman-Caldentey KM, Goossens A. (2006) Gene-to-metabolite networks for terpenoid indole alkaloid biosynthesis in *Catharanthus roseus* cells. Proc Natl Acad Sci USA 103: 5614–5619.

Plant Transformation Technology

Application of Two New Selectable Marker Genes, *dsdA* and *dao1* in Maize Transformation

Fang-Ming Lai*, Kangfeng Mei, Luke Mankin, and Todd Jones

BASF Plant Science L.L.C., 26 Davis Drive, Research Triangle Park, NC 27709, USA.
*Email: fang-ming.lai@basf.com

D-amino acids are naturally present in many higher plant species including dicots and monocots (Bruckner and Westhauser, 2003). When exposed to concentrations of D-serine and D-alanine in the millimolar range, plant growth is inhibited and this led Erikson et al. (2004, 2005) to demonstrate the new plant selection marker system using the genes involved in D-amino acid metabolism in plants.

In this study, we have evaluated using D-amino acids and the *dsdA* (D-serine dehydratase) gene from *E. coli* and the *dao1* (D-amino acid oxidase) gene from *Rhodotorula gracilis* as selectable markers in maize transformation. D-serine dehydratase is a substrate-specific enzyme that metabolizes D-serine, while the *dao1* enzyme is substrate non-specific and capable of metabolizing a range of D-amino acids including D-serine and D-alanine. D-serine and D-alanine, in a concentration range between 2 and 15 mM, inhibited germination of isolated non-transgenic immature embryos and the growth of maize calli *in vitro*. Transgenic plants were recovered under the preferred selection conditions with either *dsdA* or *dao1* as the selection marker, at efficiencies comparable to a mutated *ahas* selection control (Table 1). As another control, immature embryos infected with a mutated *ahas* gene construct did not yield any putative transgenic events on the selection media with either 10 mM D-serine or 10 mM D-alanine indicating the two D-amino acids provide effective selection (Table 1).

Z. Xu et al. (eds.), Biotechnology and Sustainable Agriculture 2006 and Beyond, 141–142.

Molecular analysis has confirmed the integration of these two markers in the transgenic plants that do not exhibit any adverse phenotypes in the greenhouse (data not shown). These data indicate the *dsdA* and *dao1* genes are properly expressed in maize, and their proteins function appropriately to metabolize D-amino acids and hence remove their effect on growth inhibition. These genes, *dsdA* and *dao1*, in combination with the appropriate D-amino acids, represent a new set of effective selectable markers for maize transformation.

Table 1. Corn transformation with the *dsdA* and *dao1* gene as the selectable markers. Agrobacterium-mediated transformation experiments were conducted with a spit-cob experimental design for all three constructs in a BASF Plant Science hybrid line. The insertion of T-DNA into plant genome and the insertion copy numbers were determined with a routine Taqman analysis method

Construct Name	Selectable markers	Selection agents	Number of replicates	Number of immature embryos infected	TE%± StdEr
MM232	ahas	750 nM Pursuit	3	71	31.3 ± 4.5
MM232	ahas	10 mM D-alanine	3	69	0
LM205	dao1	10 mM D-alanine	3	171	30.8 ± 11.6
MM232	ahas	10 mM D-serine	3	67	0
LM166	dsdA	10 mM D-serine	3	165	41.6 ± 15.3

(1) The descriptions of the transformation constructs used in this experiment: MM232 construct contains the mutated maize acetohydroxyacid synthase (*ahas*) gene as the selectable marker; LM205 and LM166 contain the *dao1* and *dsdA* as the selectable markers, respectively. (2) The chemical name of Pursuit is imazethapyr. (3) TE – transformation efficiency.

References

Bruckner H, Westhauser T (2003) Chromatographic determination of L- and D-amino acids in plants. Amino Acids 24: 43–55.

Erikson O, Hertzberg M, Torgny N (2004) A conditional marker gene allowing both positive and negative selection in plants. Nat Biotechnol 22: 455–458.

Erikson O, Hertzberg M, Torgny N (2005) The dsdA gene from *Escherichia coli* provides a novel selectable marker for plant transformation. Plant Mol Biol 57: 425–433.

Efficient *Agrobacterium*-Mediated Transformation of Various Barley (*Hordeum vulgare* L.) Genotypes

Goetz Hensel[1,*], Vladimir Valkov[1,2], Conny Marthe[1], and Jochen Kumlehn[1]

[1]Leibniz Institute of Plant Genetics and Crop Plant Research (IPK), Plant Reproductive Biology, 06466 Gatersleben, Germany.
[*]Email: hensel@ipk-gatersleben.de
[2]Present Address: Istituto di Genetica Vegetale, Via Universita, 133, 80055 Portici NA Campania, Italy.

Reliable and efficient barley genetic transformation technology is indispensible for contemporary functional analyses of regulatory and coding nucleotide sequences as well as for the generation of genetically engineered breeding material. Biolistic gene transfer has long been the preferred system for the integration of foreign genes into barley (Wan and Lemaux, 1994). Compared with this method, *Agrobacterium*-mediated gene transfer predominantly results in transgenic plants of higher quality with regard to several aspects. *Agrobacterium*-mediated transformation of barley was first published by Tingay and co-workers in 1997, however, only a few labs have established this method to date. Hence, further improvements as well as an increased reproducibility of the method are desirable.

To this end, an appropriate binary vector carrying the selectable marker gene *HPT* along with the reporter gene *GFP* was constructed (Fang et al., 2002) and introduced into the *A. tumefaciens* strains AGL1 and LBA4404. A comparison of different protocols was performed and the influence of factors investigated which had been anticipated to be crucial. In contrast to earlier published methods for

Z. Xu et al. (eds.), Biotechnology and Sustainable Agriculture 2006 and Beyond, 143–145.

barley transformation, co-culture in liquid medium was tested. As a result, we have been able to process a larger number of immature embryos at a time without detectable influence on transformation efficiency.

Several publications showed the positive effect of copper sulfate on regeneration of plants (Trifonova et al., 2001). In our study, the number of regenerants was also increased following callus induction on medium with increased copper sulfate content. A further improvement with regard to the generation of green plants was achieved by decreasing the ammonium nitrate content of the regeneration medium to one tenth.

Cysteine has been shown to influence transformation efficiencies of soybean and other crop plants. This effect of cysteine has been attributed to its possible implication in the plant's response to wounding and pathogen attack (Olhoft and Somers, 2001). In our study, less browning of the embryos was observed upon addition of cysteine to the liquid co-culture medium. Furthermore, the number of transgenic plants obtained per number of immature embryos was increased. Interestingly, no such effect was observed following co-culture on solid medium.

The resultant transformation protocol includes liquid co-culture, callus induction on medium with increased copper sulfate content, and less ammonium nitrate supply in the regeneration medium. Applying this method, about twenty transgenic barley plants were routinely produced per 100 immature embryos.

The standard method developed was subsequently employed to test five spring type and three winter type barley genotypes for their amenability to transformation. Upon co-culture, embryos of all genotypes showed transient reporter gene expression. Unfortunately, the calluses of three lines did not survive the selection process. However, transgenic plants were generated from the remaining five genotypes.

Taken together the results obtained, the established method results in more efficient transformation of cv. Golden Promise is also useful to transform a number of further barley genotypes.

References

Fang Y-D, Akula C, Altpeter F (2002) *Agrobacterium*-mediated barley (*Hordeum vulgare* L.) transformation using green fluorescent protein as a visual marker and sequence analysis of the T-DNA: Barley genomic DNA junctions. J Plant Physiol 159: 1131–1138.

Olhoft PM, Somers DA (2001) L-Cysteine increases *Agrobacterium*-mediated T-DNA delivery into soybean cotelydonary-node cells. Plant Cell Rep 20: 706–771.

Tingay S, McElroy D, Kalla R, Feig S, Wang M, Thornton S, Brettell R (1997) *Agrobacterium tumefaciens*-mediated barley transformation. Plant J 11: 1369–1376.

Trifonova A, Madsen S, Olesen A (2001) *Agrobacterium*-mediated transgene delivery and integration into barley under a range of in vitro culture conditions. Plant Sci 162: 871–880.

Wan Y, Lemaux PG (1994) Generation of large numbers of independently transformed fertile barley plants. Plant Physiol 104: 37–48.

In Vitro System as the Model for Discovery in Molecular Biology

Use of Alfalfa In Vitro Cultures in Studies on Regulation of Cyclin-Dependent Kinase (CDK) Functions

Dénes Dudits*, Mátyás Cserháti, Pál Miskolczi, Attila Fehér, Ferhan Ayaydin, and Gábor V. Horváth

Institute of Plant Biology, Biological Research Center, Hungarian Academy of Sciences, Szeged, Hungary 6726.
*Email: dudits@brc.hu

The production of new cells by division and their subsequent elongation are key cellular events in the plant life cycle. The orderly progression of cells through the various phases of the cell cycle and their appropriate responses to nutritional, hormonal, developmental and environmental signals are governed by multiple regulatory mechanisms, including reversible protein phosphorylation, interactions of proteins, and specific protein degradation. Since protein phosphorylation is widely recognized to be the major mechanism controlling cell cycle progression, the roles of different kinases, and particularly the family of cyclin-dependent protein kinases (CDKs), are critical for cell division control. After discovery of plant homologues of cdc2/CDC28 kinases by complementation of the yeast temperature sensitive mutants, the functional characterization of the plant CDK genes and protein complexes was largely based on the use of tissue culture systems. As example, the alfalfa cultured cells were synchronized with either aphidicolin or hydroxyurea and the enrichment of cell populations at G_1/S or G_2/M phases provided experimental material for isolation of RNA molecules and kinase complexes. The Northern hybridization studies uncovered the existence of plant specific CDKs that showed preferential transcript

Z. Xu et al. (eds.), Biotechnology and Sustainable Agriculture 2006 and Beyond, 149–152.

accumulation in G2/M cells of synchronized alfalfa cultures (Magyar *et al.*, 1997). These studies revealed two representatives of the non-PSTAIRE kinase genes. A recent analysis of the mitosis-specific promoter of the alfalfa CDK (Medsa;CDKB2;1) gene detected several putative *cis*-elements, such as MSA (M-specific activator), ABRE (ABA-responsive element), E2F-binding site, ERE (ethylene-responsive element) and TCA-box (wound-responsive element). Functional characterization of a short (360 bp) promoter region of this gene in transgenic lines showed that the listed putative *cis*-elements could simultaneously ensure G_2/M phase-specific gene activity and responses to wounding, ethylene and 2,4-D treatments. The combination of *cis*-elements predicted to regulate cell cycle phase specificity as well as hormonal, environmental and developmental responses is a general feature of CDK promoters in plants (Zhiponova *et al.*, 2006).

Also *in vitro* cultures of alfalfa organs or cultured cells provide ideal system for studies on the hormonal responses of the CDK genes. Activation of CDK A genes was observed in the presence of 2,4-D. In accordance with activation of CDK genes, the histone H1 phosphorylation activity of the cognate PSTAIRE kinase (Medsa; CDKA1;2) was also elevated by treatment of leaf explants with 2,4-dichlorophenoxyacetic acid (2,4-D) and kinetin. The stress hormone abscisic acid (ABA) abolished this hormone-stimulated kinase function.

In alfalfa, the leaf mesophyll protoplasts represent an excellent cell type for studies on the reactivation of the cell cycle under *in vitro* conditions (Fehér *et al.*, 2003). The protoplast-derived cells enter into the division cycle in a synchronous manner that can be monitored by BUdR incorporation and counting the number of divided cells. The type of cell division as the symmetry of the newly formed cells can be influenced by hormone content of the culture media. High concentration of 2,4-D generates smaller and highly cytoplasmic asymmetrically dividing cells that can develop to somatic embryos. In the presence of low concentration of 2,4-D, the enlarged cells divide symmetrically and the subsequent cell divisions result in unorganized callus growth. As described in the recent publication of our laboratory, in alfalfa leaf protoplast cultures the nitric oxide (NO) promoted the auxin concentration-dependent division and embryogenic cell formation accompanied by increase of MedsaCDKA1,2 kinase activity (Ötvös *et al.*, 2005).

Plant CDKs are themselves under phosphorylation control, so they serve as substrates for kinases and phosphatases. We used the treatment of cultured alfalfa cells with chemical inhibitor (endothal) to reduce the PP2A phosphatase activity. This modification resulted in an early increase in Medsa;CDKB2;1 (cdcMsF) kinase activity, disturbing the coordination between chromosomal and microtubule events during the G_2/M transition in cultured alfalfa cells (Ayaydin *et al.*, 2000).

At present tissue culture systems are basic tools for transformation technologies. Both *Agrobacterium*-mediated gene transfer and the particle bombardment target cultured cells and after selection of transgenic cell population, they serve as a starting point for plant regeneration. Functional characterization of cell cycle genes can also be carried out by production of transgenic genotypes carrying gene construct for overproduction of proteins or vectors for gene silencing.

References

Ayaydin F, Vissi E, Mészáros T, Miskolczi P, Kovács I, Fehér A, Dombrádi V, Erdődi F, Gergely P, Dudits D (2000) Inhibition of serine/threonine-specific protein phosphatases causes premature activation of cdc2MsF kinase at G_2/M transition and early mitotic microtubule organization in alfalfa. Plant J 23: 85–96.

Fehér A, Pasternak T, Dudits D (2003) Transition of somatic plant cells to an embryogenic state. Plant Cell Tissue Organ Cult 74(3): 201–228.

Ötvös K, Paternak T, Miskolczi P, Domoki M, Djorgotov D, Szűcs A, Bottka S, Dudits D, Fehér A. (2005) Nitric oxide is required for, and promotes auxin-mediated activation of, cell division and embryogenic cell formation but does not influence cell cycle progression in alfalfa cell cultures. Plant J 43: 849–860.

Magyar Z, Mészáros T, Miskolczi P, Deák M, Fehér A, Brown, S, Kondorosi É, Athanasiadis A, Pongor S, Bilgin M, Bakó L, Koncz CS, Dudits D (1997) Cell cycle phase specificity of putative cyclin-dependent kinase variants in synchronized alfalfa cells. The Plant Cell 9: 223–235.

Zhiponova MK, Pettkó-Szandtner A, Stelkovics É, Neer Zs, Bottka S, Krenács T, Dudits D, Fehér A, Szilák L (2006) Mitosis-specific promoter of the alfalfa cyclin-dependent Kinase Gene (Medsa;CDKB2;1) is activated by wounding and ethylene in a non-cell division-dependent manner. Plant Physiol 140: 693–703.

Phytohormone Regulation of Cotton Fiber Development In Vitro

Barbara A. Triplett[1,2,3], Hee Jin Kim[2], Doug Hinchliffe[1], Sing-Hoi Sze[3], Peggy Thaxton[4], David Stelly[3], and Z. Jeffrey Chen[5]

[1]USDA-ARS, Southern Regional Research Center, New Orleans, LA 70124, USA.
[2]Department of Biological Sciences, University of New Orleans, New Orleans, LA 70148, USA.
[3]Texas A&M University, College Station, TX 77843, USA.
[4]Delta Research and Extension Center, Stoneville, MS 38776, USA.
[5]University of Texas, Austin, TX 78712, USA.

Cotton fibers (cottonseed trichomes) are a well-established model system for investigating cellulose biosynthesis. Cotton ovules, excised from ovaries prior to fertilization, will continue to develop and produce trichomes *in vitro* under defined conditions when auxin and gibberellic acid are present. Our team is also using cotton fiber development *in vitro* as a model to investigate plant cell differentiation at the single-cell level and to identify the regulatory components of the transition from cell elongation to secondary wall biogenesis. These two developmental phases determine key properties of fiber quality and yield.

Fiber cells elongate to 1–1.5 cm in culture and then produce a thick, cellulosic secondary cell wall. During the transition from cell expansion to secondary wall thickening, the rate of cellulose biosynthesis rises nearly 100-fold. A similar increase in cellulose biosynthetic capacity occurs in fibers produced *in vitro* by ovule cultures. Although the genes for the cellulose synthase catalytic subunit, *CesA*, were first described from cotton fiber, little is known about

Z. Xu et al. (eds.), Biotechnology and Sustainable Agriculture 2006 and Beyond, 153–156.

how *CesA* expression is regulated. We have identified a group of cotton *CesA* genes that are expressed during cell elongation and another set of *CesA* genes that are expressed during secondary wall thickening. We investigated changes in culture conditions that alter the timing of secondary cell wall *CesA* expression using real-time quantitative RT-PCR. Twenty-four h treatment with exogenous indoleacetic acid and/or abscisic acid at a time prior to initiation of secondary cell wall synthesis stimulated the premature expression of *CesA1* and *CesA2*, genes responsible for secondary wall synthesis in cotton fiber. Simultaneous treatment with auxin and abscisic acid had an additive effect on relative transcript abundance for *CesA1* and *CesA2*. Similar phytohormone treatments had little effect on the expression of genes predominantly expressed during cell elongation or constitutively expressed throughout fiber development (i.e., α-tubulin 4, α-tubulin 5, actin, expansin 1, and ubiquitin conjugating enzyme). Furthermore, addition of exogenous GA_3, an essential phytohormone for fiber elongation, down-regulated expression of secondary wall *CesA* genes.

In silico analysis of three cotton *CesA* promoters shows an abundance of phytohormone-responsive *cis*-acting elements. Interestingly, the phytohormone response of a cotton *CesA* promoter is maintained in the leaf trichomes of transgenic Arabidopsis transformed with *GhCesA4::GUS* constructs. Emerging evidence from a number of laboratories suggests conserved mechanisms for cottonseed trichome and Arabidopsis leaf trichome development. Efforts to characterize the promoters of several additional *CesA* paralogs in the cotton genome are continuing in our laboratory.

By transcript profiling with oligo-microarrays we have begun to identify genes that are differentially regulated in ovule cultures established with auxin and gibberellin compared to cultures where the phytohormones are absent. This approach will identify essential targets for increasing fiber number per seed, an important component of yield.

The cotton microarray utilized in this study contained 12,006 ESTs from 30 cDNA libraries generated from various sources including fiber, ovule, and leaf tissues of *Gossypium arboreum* L. and *G. hirsutum*, and whole seedling and whole flowers tissues of

Gossypium raimondii L. (Udall et al., 2006). Ovules were harvested from day-of-anthesis bolls of *G. hirsutum* L., Texas Marker-1 and placed immediately in Beasley-Ting media either supplemented with or lacking 1-naphthaleneacetic acid and GA_3.

Preliminary results were obtained from direct comparison of phytohormone-treated ovules and untreated ovules at a 6-h time-point. The microarray data was normalized using Lowess normalization and a 't' test was performed on the microarray data using a Benjamini-Hochberg correction of the obtained P values to control the false discovery rate (FDR). A total of 63 statistically significant genes were identified with expression up-regulated two-fold or greater in response to phytohormone treatment. Among the up-regulated genes were ten genes known to encode gene products that are responsive to auxin and/or gibberellic acid, including a GA_3-induced protein involved in cell elongation. Other gene products of interest included several transcription factors and cell wall modification enzymes.

The recent release of 32,789 ESTs representing 8540 unique sequences from −3 to +3 DPA cotton ovules from *G. hirsutum* L. (Yang et al., 2006) and the addition of these sequences to a new cotton gene chip will greatly enhance our ability to identify key genes involved in the initiation of fiber growth. Once completed, our microarray studies will provide novel insights into the process of cotton fiber cell initiation, and will identify key proteins that regulate early development of cotton fiber cells.

Analysis of cotton ovule cultures using contemporary techniques in plant molecular biology offers the potential for uncovering fundamental aspects of plant growth and development that will also be useful for designing applied approaches for cotton improvement.

References

Udall JA, Swanson JM, Haller K, Rapp RA, Sparks ME, Hatfield J, Yu Y, Wu Y, Dowd C, Arpat AB, Sickler BA, Wilkins TA, Guo JY, Chen XY, Scheffler J, Taliercio E, Turley R, McFadden H, Payton P, Klueva N, Allen R, Zhang D, Haigler C, Wilkerson C, Suo J, Schulze SR, Pierce ML, Essenberg M, Kim H, Llewellyn DJ, Dennis ES, Kudrna D, Wing R, Paterson AH, Soderlund C, Wendel JF (2006) A global assembly of cotton ESTs. Genome Res 16: 441–450.

Yang SS, Cheung F, Lee JJ, Ha M, Wei NE, Sze SH, Stelly DM, Thaxton P, Triplett B, Town CD, Chen ZJ (2006) Accumulation of genome-specific transcripts, transcription factors and phytohormonal regulators during early stages of fiber cell development in allotetraploid cotton. Plant J 47: 761–775.

Patterns of Growth and Cell Death in Diploid *Arabidopsis* Cell Cultures with Synchronised Cell Cycle

Ranjith Pathirana*, Phillip West, and Jocelyn Eason

Crop & Food Research Ltd, Batchelar Road., Palmerston North 5301, New Zealand.
*Email: pathiranar@crop.cri.nz

Introduction

Populations of cultured cells with synchronised cell cycle provide large numbers of cells at defined points of the cell cycle for manipulation and investigation of molecular events associated with plant function. *Arabidopsis thaliana* is a preferred model plant for genetic and molecular studies of various metabolic events due to its rapid life cycle and fully sequenced genome. However, one constraint for the routine use of Arabidopsis cell cultures is the limited availability of cultures suitable for cell cycle synchronisation. Here we report the establishment of a rapidly growing diploid Arabidopsis cell culture, synchronisation of the cell cycle, and use of the cultured cells to study cell death.

Materials and methods

A liquid media consisting of Murashige and Skoog (1962) salts supplemented with 2,4-D (4.5 μM), NAA (5.37 μM), IAA (5.71 μM), BAP (2.2 μM) and 3% sucrose was used to establish and maintain cell cultures for cell cycle synchronisation as previously reported

Z. Xu et al. (eds.), Biotechnology and Sustainable Agriculture 2006 and Beyond, 157–160.

(Pathirana and Eason, 2006). *Arabidopsis thaliana* cv. Columbia cell cultures were established from callus derived from leaf tissue induced on the same medium solidified with 0.7% agar. The cultured cells were sieved through 200–400 µm nylon mesh filters and the filtrate containing single cells or small cell clusters (up to ten cells) was transferred to fresh medium at 7 day intervals. This achieved rapid division and good cell dispersion of the cultures. The cells were grown in the dark at 24°C, 110 rpm with a 50 mm orbit. After 10–12 weeks culture establishment late-exponential stage cells were subcultured in 250 mL Erlenmeyer flasks (10 mL into a final volume of 50 mL) for synchronization.

Aphidicolin treatment (4 $\mu g.mL^{-1}$, 24 h) was used to synchronize the cell cycle, and the cell cycle phase for populations of cells was monitored by flow cytometry as previously described (Pathirana and Eason, 2006). Cells were induced into PCD using ethylene (8 mL in 250 mL flask headspace), mannose (100 mM), polyethylene glycol (PEG – 20%), and hydrogen peroxide (2 mM). The synchronised cells were subjected to these treatments after 2 and 12 h of release from the aphidicolin block, corresponding to G2 and G1 phases of the cell cycle respectively. Cell death was assessed at 2 h intervals after treatment by staining with Evan's blue. The stain was extracted from a known weight of cells in methanol and quantified by spectrophotometry at 595 nm.

Results and discussion

A complete block of the cell cycle in the G1-S interphase and a high degree of cell cycle synchronization (ca 78%) was achieved when exponentially growing cells (3 days after subculture) were treated with 4 mg L^{-1} aphidicolin. Our rapidly growing diploid cell culture has advantages over previously reported polyploid cultures in genetic studies, as the cells contain only one set of chromosomes.

Cell death in the synchronized cultures was induced at two different points of the cell cycle; in the G2 phase after 2 h of release of cells from the aphidicolin block and at G1, after 12 h of release from the cell cycle block. We used four different treatments (ethylene, mannose, PEG, and hydrogen peroxide) to cause the death rate of cultured cells to increase. Both ethylene and PEG caused an increase in Evans Blue-stained cells after 2 and 12 h treatments (Figure 1A and B), however mannose and hydrogen peroxide only caused an elevation of cell death at 12 h compared to the control cultures (Figure 1C and D). The patterns of cell death were also different when cells were treated with different chemicals. For instance, ethylene and hydrogen peroxide treatments caused a peak in Evan Blue-stained cells, after which the cultures recovered and returned to baseline cell death rates (the same as the control cultures). PEG and mannose treatments caused a continuing increase in the amount of Evans Blue-stained cells in cultures, and the cell population did not recover (Figure 1). The increase in Evans Blue-stained cells correlated with an increase in mRNA accumulation for AtSAG (an *Arabidopsis thaliana* Senescence Associated Gene). Northern analysis indicated that the expression of this cell death-associated gene preceded cell death as measured by Evan's blue staining.

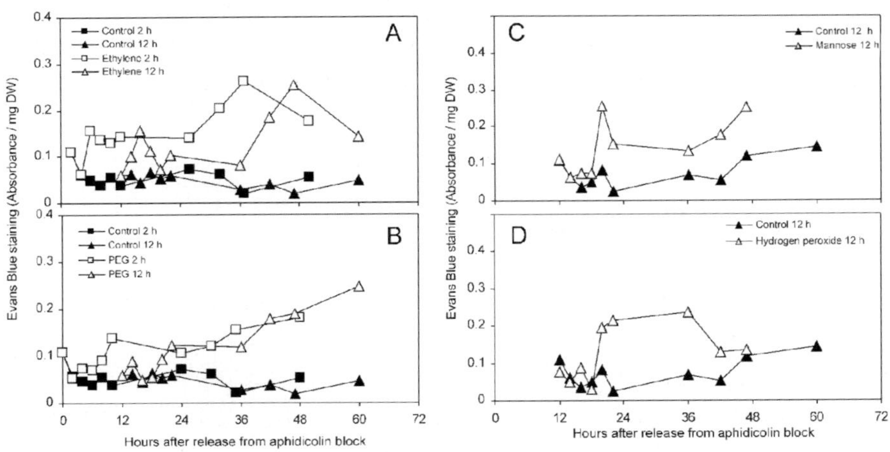

Figure 1. Evans blue staining of cell cultures treated with ethylene (A), PEG (B), mannose (C) and hydrogen peroxide (D) after 2 and 12 h release from aphidicolin cell cycle block. Control cultures were blocked with aphidicolin then released from the block by washing.

Acknowledgements

We wish to thank Dr. Ross Bicknell for the use of his equipment and technical expertise in flow cytometry. We thank Andrew Mullan and Margaret Young for tissue culture media preparation, and Duncan Hedderley for statistical support. This research is part of the Vital Vegetables collaboration supported by the New Zealand Foundation for Research Science and Technology, and Horticulture Australia Ltd.

References

Murashige T, Skoog F (1962) A revised medium for rapid growth and bioassays with tobacco tissue cultures. Physiol Plant 15: 473–497.

Pathirana R, Eason J (2006) Establishment and characterization of a rapidly dividing diploid cell suspension culture of *Arabidopsis thaliana* L. suitable for cell cycle synchronisation. Plant Cell Tissue Organ Cult 85: 125–136.

Molecular Control of Abiotic Resistance

Engineering of Reactive Species Detoxification Pathways for Increasing Stress Tolerance in Plants

Gábor V. Horváth*, Zoltán Turóczy, Mátyás Cserháti, Petra Kis, Katalin Török, László Sass, and Dénes Dudits

Institute of Plant Biology, Biological Research Center HAS, Szeged, Hungary.
*Email: hvg@brc.hu

The productivity of plants is greatly affected by environmental stresses such as drought, high or low temperature, high salinity and UV-B irradiation, therefore there is a continuous need for the genetic improvement of stress tolerance in the agriculture. The abiotic stresses can disturb the homeostasis between assimilation and oxidative reactions, negatively influencing the photosynthetic yield of higher plants and resulting in oxidative damage. Reactive compounds produced under such conditions significantly increase the cytotoxic effect of environmental stresses. Beside the generated reactive oxygen species (ROS), reactive aldehydes (as 4-hydroxy-nonenal and methylglyoxal) can further increase the cellular damages mainly due to their better penetration through biological membranes and their rapid reaction with biomolecules such as proteins and DNA. Improvement of intracellular scavenging capacity of such toxic compounds provenly leads to increased stress tolerance. Plant aldo-keto reductases (AKRs) are important enzymes for such function since they have a wide range of activity on lipid peroxidation and glycolysis generated reactive aldehydes. AKRs can detoxify those lipid peroxidation products (e.g., 4-hydroxynon-2-enal) and glycolysis-derived reactive aldehydes (e.g., methylglyoxal) that contribute significantly to

Z. Xu et al. (eds.), Biotechnology and Sustainable Agriculture 2006 and Beyond, 163–166.

cellular damages caused by environmental stresses. Moreover, the specific members of this NADPH-dependent aldo-keto reductase superfamily are able to catalyze the production of sugar alcohols (like sorbitol or mannitol). The products of these reactions can act as radical scavengers even at low concentration and their accumulation as osmolytes can lead to improvement of osmotic adaptation.

Our results have further supported the important role of aldo-keto reductases in plant stress responses. We have isolated the *MsALR* gene from *Medicago sativa* and showed the accumulation of the transcript at higher levels in response to wide range of stress treatments: osmotic stress, toxic metal (cadmium), abscisic acid or chemicals generating free radicals (Horváth et al., 1999; Oberschall et al., 2000). The recombinant alfalfa protein exhibits specific enzymatic activities: it reduces aldehyde substrates with much higher activity than aldoses using NADPH cofactor but not NADH. In its biochemical properties, the alfalfa enzyme resembles the human aldose reductase in several aspects (substrate preferences, sensitivity to inhibitors) rather than the recombinant barley aldose reductase and similarly to the human enzyme the recombinant alfalfa protein could also metabolize 4-hydroxy-trans-2-nonenal.

Transgenic tobacco plants ectopically expressing the *MsALR* cDNA under the control of the CaMV 35S promoter were more tolerant to dehydration stress and recovered better from damages caused by water deficit than the untransformed wild type plants and were more tolerant to heavy metal, salt or dehydration stress (Hegedűs et al., 2004). The transgenic plants showed significantly reduced damage (measured by chlorophyll fluorescence) when exposed to oxidative stress-inducing agents such as paraquat or hydrogen peroxide. This improved tolerance could be attributed to the decreased level of ROS and reactive aldehydes, because the concentration of such compounds was considerably lower in tobacco plants over-producing the alfalfa enzyme than in non-transformed plants. Due to the increased synthesis of MsALR the plants were more resistant to UV-B irradiation as well (Hideg et al., 2003). Although hydrogen peroxide accumulation decreased in the transgenic lines, the most significant change was observed in the concentration of the thiobarbituric acid reactive species (TBARS). This result also supported the importance of lipid peroxidation and the reactive aldehyde formation in cellular damage under stress condition.

Complex analysis of AKR genes from *Arabidopsis thaliana*, rice and wheat focusing on their use in stress tolerant transgenic plants will help not only to understand the role of different detoxification pathways in plant stress responses but can lead to the production of stress resistant transgenic plants. Our gene discovery program was extended by searching for putative stress-related transcriptional factor binding sites (TFBSs) in rice promoterome. After ranking the putative dyads found in stress-activated genes, we analyzed 24 rice aldo/keto reductase genes to correlate the presence of TFBS dyads with the stress response of the tested genes.

These computer-assisted promoter analyses revealed a correlation between the number of stress motifs found and the level of stress response represented by the expression pattern. Moreover, comparative gene expression profiling of methylglyoxal treated vs. methylglyoxal adapted rice suspension cultures revealed novel transcription factors and enzymes that may also play important role in reactive aldehyde detoxification in plants. Quantitative Real Time PCR was used to determine the expression pattern of selected genes in abscisic acid (ABA), H_2O_2, NaCl and mannitol treated rice cell suspensions. Among the genes studied, OsALR1 showed the highest inducibility and transcript level during the treatments, suggesting its important role in stress tolerance. The *in vitro* enzyme kinetic constants of the expressed GST-OsALR1 fusion protein revealed a high affinity for reducing toxic aldehydes like methylglyoxal (MG) in the presence of NADPH. Results of *in vivo* assays in *E. coli* for protection against methylglyoxal toxicity has also strengthened our hypothesis that one important function of these AKR proteins is the elimination of reactive aldehydes from plant cells expletively besides the glyoxalase system.

Acknowledgements

This work was supported by NKFP Grant No. 4-064-2004 and National Scientific Research Found (OTKA) Grant No. K-46495; Gábor V. Horváth is grateful for the support of the 'János Bólyai' Research Fellowship.

References

Hegedüs A, Erdei S, Janda T, Tóth E, Horváth G, Dudits D (2004) Transgenic tobacco plants overproducing alfalfa aldose/aldehyde reductase show higher tolerance to low temperature and cadmium stress. Plant Sci 166: 1329–1333.

Hideg É, Nagy T, Oberschall A, Dudits D, Vass I (2003) Detoxification function of aldose/aldehyde reductase during drought and ultraviolet-B (280–320 nm) stresses. Plant Cell Environ 26: 513–522.

Horváth et al. (1999). J Plant Biotechnol 1: 61–68.

Oberschall A, Deak M, Torok K, Sass L, Vass I, Kovacs I, Feher A, Dudits D, Horvath GV (2000) A novel aldose/aldehyde reductase protects transgenic plants against lipid peroxidation under chemical and drought stresses. Plant J. 24: 437–446.

Orchid Flower Fragrance Biosynthesis Pathway and Its Key Enzyme Geranyl Diphosphate Synthase

Yu-Yun Hsiao[1], Wen-Chieh Tsai[1], Chang-Sheng Kuoh[1], Tian-Hsiang Huang[2], Hei-Chia Wang[2], Yann-Lii Leu[4], Tian-Shung Wu[5], Wen-Huei Chen[6], and Hong-Hwa Chen[1,3]

[1]Department of Life Sciences, National Cheng Kung University, Tainan 701, Taiwan.
[2]Institue of Informational Management, National Cheng Kung University, Tainan 701, Taiwan.
[3]Institute of Biotechnology, National Cheng Kung University, Tainan 701, Taiwan.
[4]Graduate Institute of Natural Products, Chang Gung University, Taoyuan 333, Taiwan.
[5]Department of Chemistry, National Cheng Kung University, Tainan 701, Taiwan.
[6]Department of Life Sciences, National University of Kaohsiung, Kaohsiung 811, Taiwan.

Floral scent is a key modulating factor in plant-insect interactions and thus plays a central role in successful pollination. Closely related plant species that rely on different insects for pollination produce different odors. The floral scent is of paramount importance to plant reproduction and evolution. *Orchidaceae* is one of the largest monocotyledon families, containing more than 25,000 species. In orchids, large quantities of pollen are formed in masses spread by animals (bees, moths, flies and birds) and the floral scents serve as attractants for species-specific pollinators. These pollinators have played a major role in orchid evolution. The range of odors produced by orchids

Z. Xu et al. (eds.), Biotechnology and Sustainable Agriculture 2006 and Beyon, 167–169.

is enormous, providing an inexhaustible basis for specificity. Such diversity is advantageous in the evolution of an obviously successful family.

We investigated how flower fragrance is generated in certain *Phalaenopsis* orchids by identifying floral expressed-sequence-tags (ESTs) and deducing the floral scent biosynthesis pathway of *Phalaenopsis bellina* with bioinformatics analysis. The major chemical components in the *P. bellina* flower were found to be monoterpenoid, benzenoid, and phenylpropanoid as revealed by gas chromatography-mass spectrometry. The encoded enzymes collection in the biosynthetic pathway from glycerolaldehyde-3-phosphate (G3P) to geraniol and linalool were recognized through data mining of the *P. bellina* floral EST database (dbEST). Transcripts preferentially expressed in the *P. bellina* were distinguished by a comparison of the scent floral dbEST to that of a scentless species *P. equestris*, including those encoding for lipoxygenase, epimerase, diacylglycerol kinase and geranyl diphosphate synthase. Furthermore, EST filtering results showed that transcripts encoding for signal transduction, Myb transcription factor, and methyltransferase, in addition to those for scent biosynthesis, were detected with *in silico* hybridization of *P. bellina* unigene database against those of the scentless species, rice and *Arabidopsis.* Altogether, we pinpointed 66% of the biosynthetic steps from G3P to geraniol, linalool and their derivatives.

We then focused on one of the key enzymes in this pathway, geranyl diphosphate synthase (GDPS) which catalyzes the formation of GDP for further synthesis of geraniol, linalool and their derivatives. The full-length cDNA of *P. bellina* GDPS (*PbGDPS*) was cloned and sequenced with a length of 1102 bp and encoding an open reading frame of 693 bp for a protein of 231 amino acids. The amino acid sequence of PbGDPS showed low similarity (30%) to that of other plants. In addition, similar to the GDPS of *Mentha piperita*, *Antirrhinum majus* and *Clarkia breweri*, the PbGDPS lacked the conserved aspartate-rich motif $DD(X)_{2-4}D$. Based on the genomic sequence, there were no introns resided in the *PbGDPS* gene. The *PbGDPS* was expressed in flower buds, but not detected in vegetative tissues. Expression of the *PbGDPS* was first detected at day 3 pre-anthesis flower buds in northern blot, and its expression peaked at day 5 post-anthesis flowers coincident with the optimal scent emission. This result suggested that the *PbGDPS* was expressed in a pattern correlated with monoterpene emission during blooming of the flower. In contrast, the expression of *PbGDPS* was much reduced in the scentless species of *P. bellina* hybrids. By using *in situ* hybridization, the *PbGDPS* signals were concentrated in the epidermis of perianth where the scent emits. Altogether, our results showed that the *PbGDPS* was the first *GDPS* isolated from monocots, and it was a flower-specific GDPS catalyzing the formation of GDP for further scent production in orchid flowers.

In this work, we have shown how EST research can be usefully applied to the construction of a putative scent metabolism pathway in *P. bellina* and the identification of the genes encoding the enzymes involved in this pathway. For a non-model plant with a very large genome ($1.5 \sim 8.1 \times 10^9$ bp for *Phalaenopsis* spp.), which cannot easily be accessed for whole genome sequencing, EST analysis of its transcriptome profile becomes a very efficient and informative tool. A combination of volatile analysis, EST database mining and EST filtering can be applied to deduce the scent biosynthesis pathway in *P. bellina* and to identify scent-related genes.

Cytokinin Function in Drought Stress Response and Subsequent Recovery

M. Novakova[1,*], P. Dobrev[1], V. Motyka[1], A. Gaudinova[1], J. Malbeck[1], J. Pospisilova[1], D. Haisel[1], H. Storchova[1], J. Dobra[1], M.C. Mok[2], D.W.S. Mok[2], R. Martin[2], and R. Vankova[1]

[1] Institute of Experimental Botany AS CR, Rozvojová 135, 165 02 Praha 6, Czech Republic.
[2] Department of Horticulture and Center for Gene Research and Biotechnology, Oregon State University, Agricultural and Life Science 4017, Corvallis, OR 97331, USA.
[*]Email: vankova@ueb.cas.cz

Introduction

Plants respond to exogenous stimuli, including abiotic stress, by modulation of their behavior, which is partially mediated by plant hormones. Understanding of the role of different hormones could improve the choice of a suitable strategy for enhancement of plant stress tolerance. Cytokinins (CKs) affect multiple processes in plants, including stimulation of cell division, enhancement of sink strength and promotion of photosynthesis. As CKs counteract many processes induced by water stress which are mediated by the 'stress hormone' abscisic acid (e.g., stomata closure or senescence acceleration) their role in drought stress response does not seem straightforward. Exogenous CKs were reported to increase the tolerance to mild stress and to speed up the recovery (e.g., Itai et al., 1978). As reported by Zhang et al. (2000) and Huynh et al. (2005) the increased cytokinin content following over-expression of CK biosynthetic gene (*IPT*) driven by senescence-inducible promoter *SAG12* positively correlated with elevated flooding tolerance. This together with

Z. Xu et al. (eds.), Biotechnology and Sustainable Agriculture 2006 and Beyond, 171–174.

reported role of CKs in the sink-source polarization during mild water stress in tobacco (Cowan et al., 2005) indicates that CKs play an important in plant responses to drought.

In present contribution changes in CK pool during drought stress and subsequent recovery were investigated. The stress response of tobacco plant over-expressing *trans*-zeatin O-glucosyltransferase (*ZOG1*) gene from *Phaseolus lunatus* (Martin et al., 2001) under either constitutive (*35S*) promoter (causing uniform elevation of total CKs in whole plant already before stress initiation) or senescence-inducible (*SAG12*) promoter (inducing CK increase related to the time and location of the stress) was compared with the corresponding wild type plants (WT).

Results and discussion

After mild stress an elevation of physiologically active CKs was observed in the upper leaves of WT plants. The increase of content of physiologically active CKs correlated with the stimulation of stomatal conductance and net photosynthesis rate. The *35S:ZOG1* transgenics, containing elevated levels of total CKs (mainly due to CK O-glucosides), exhibited after mild stress neither elevation of active CKs in the upper leaves nor photosynthetic gradient in favor of these leaves. Analysis of xanthophyll cycle pigments revealed that uniform increase of total CKs and/or stimulation of CK turnover were associated with delayed stress sensing and postponed activation of defense mechanisms.

At moderate drought stress all tested genotypes exhibited gradient of active CKs in favor of the upper leaves. Potential positive effect of CKs on stomata conductance and photosynthesis was probably ameliorated by tremendous accumulation of abscisic acid. As enhancement of sink strength by CKs has already been postulated by Leopold and Kawase (1964), we assume that CK elevation corresponds to the increase of the sink strength of upper leaves. CK gradient was further promoted at severe stress by stimulation of CK degradation by cytokinin oxidase/dehydrogenase (CKX) in lower leaves. This conclusion is supported by the results obtained using *SAG12:ZOG1* plants. After moderate stress *SAG12* driven *ZOG1* expression resulted in the formation of the second sink in lower leaves. Simultaneously CKX activity maximum appeared in the middle leaves (a new source tissue).

During moderate and severe stress substantial increase of total CKs was represented predominantly by physiologically inactive CK N- and O-glucosides including those of cis-zeatin type. This accumulation might be the result of CK deactivation caused by enhanced CK turn-over. Moreover, some of these compounds may play a physiological role *per se*. *cis*-Zeatin is very promising candidate in this respect.

Drought induced elevation of CK content in roots of both wild type and *ZOG1* transgenic plants, mainly by limited CK xylem transport and by decrease of CKX activity. Simultaneously auxin level (free indolylacetic acid, IAA) was strongly increased in lower leaves and roots of all tested genotypes. After rehydration content of CKs and IAA started decreasing. The results indicate that drought may stimulate auxin transport from growth suppressed shoots to growing roots, where it may support, together with elevated CK content, development of primary roots. The change in root architecture improves the ability of plant to reach the underground water.

Conclusion

Elevation of levels of active CKs in the upper leaves and stimulation of CK degradation by CKX in lower leaves indicate that CK role in drought response of tobacco plants is based on the establishment of strong sink-source polarization in favor of the upper leaves.

Uniform elevation of total CKs and stimulation of CK turn-over achieved by constitutive *ZOG1* expression resulted in delayed stress sensing and postponed activation of defense mechanisms (as indicated by DEPC analysis).

Accumulation of both auxin and CKs during prolonged stress in roots seems to be a necessary prerequisite for adaptation of roots to drought.

Acknowledgements

This work was supported by GA CR, project No. 522/04/0549.

References

Cowan AK, Freeman M, Björkman PO, Nicander B, Sitbon F, Tillberg E (2005) Effects of senescence-induced alteration in cytokinin metabolism on source-sink relationships and ontogenic and stress-induced transitions in tobacco. Planta 221: 801–814.

Huynh LN, Van Toai T, Streeter J, Banowetz G (2005) Regulation of flooding tolerance of SAG12: ipt Arabidopsis plants by cytokinin. J Exp Bot 56: 1397–1407.

Itai C, Benzioni A, Munz S (1978) Heat stress: Effects of abscisic acid and kinetin on response and recovery of tobacco leaves. Plant Cell Physiol 19: 453–459.

Leopold AC, Kawase Y (1964) Benzyladenine effects on bean leaf growth and senescence. Am J Bot 51: 294–298.

Martin RC, Mok DWS, Smets R, Van Onckelen HA, Mok MC (2001) Development of transgenic tobacco harboring a zeatin *O*-glucosyltransferase gene from *Phaseolus*. In Vitro Cell Dev Biol Plant 37: 354–360.

Zhang J, Van Toai T, Huynh L, Preiszner J (2000) Development of flooding-tolerant Arabidopsis thaliana by autoregulated cytokinin production. Mol Breed 6: 135–144.

Xenogenomics: Bioprospecting for Genetic Determinants of Cold and Freezing Stress Tolerance in the Cryophilic Antarctic Hair Grass *Deschampsia antarctica* E. Desv.

U.P. John[1,2,3], R.M. Polotnianka[1,2,3], A. Sivakumaran[1,2], L. Mackin[1], M.J. Kuiper[4], J.P. Talbot[1,2], O. Chew[1,2], E. Ribarev[1], D. Azria[1], D. Gunawardana[1], and G.C. Spangenberg[1,2,3]

[1]Primary Industries Research Victoria, Victorian AgriBiosciences Centre, La Trobe University R&D Park, 1 Park Drive, Bundoora, Victoria 3083, Australia.
[2]Australian Centre for Plant Functional Genomics, Australia.
[3]Victorian Centre for Plant Functional Genomics, Australia.
[4]Victorian Partnership for Advanced Computing Ltd, 110 Victoria Street, Carlton South, Victoria 3053, Australia.

The world's plant based agriculture is based on a very small proportion of the total number of vascular plant species. Some non-domesticated species inhabit extreme environments and are uniquely adapted to a range of abiotic stresses including: salinity, drought, temperature extremes, nutrient deprivation, and metal toxicity. However very little of this genetic diversity can be exploited by conventional breeding, nor have many of these species been subjects of genomics programs. We have coined the term 'Xenogenomics' to describe structural and functional genomics specifically targeting non-model and non-crop plants with enhanced tolerance to abiotic stresses, for the discovery of novel genes, gene variants and promoters that can be employed to enhance these characteristics in agronomically important plant species.

Z. Xu et al. (eds.), Biotechnology and Sustainable Agriculture 2006 and Beyond, 175–177.

With respect to low temperature and freezing stress we have looked to one of the most extreme environments inhabited by plants, Antarctica. *Deschampsia antarctica* E. Desv. (Antarctic hair-grass) is one of only two vascular plants to have overcome the geographical and environmental impediments to colonisation of the antarctic continent. It is an over-wintering species that grows in sheltered locations along the western coast of the Antarctic Peninsula. Laboratory studies have demonstrated that *D. antarctica* is truly freeze tolerant: significant cellular damage only occurs in plants exposed to temperatures substantially below those at which they freeze.

D. antarctica expresses potent ice recrystallisation inhibition (RI) activity that inhibits the growth of small ice crystals into potentially damaging large ones. This RI activity is highly induced by cold acclimation, is found in the apoplastic (extracellular) spaces where freezing occurs, and is resistant to boiling. *D. antarctica* harbours multiple ice recrystallisation inhibition protein (IRIP)-encoding genes, whose transcript levels are greatly elaborated in response to cold-acclimation. *IRIP* genes appear to be confined to the sub-family Pooideae of the grass family Poaceae, and exhibit sequence similarity to phytosulfokine receptors. IRIPs are apoplastically-targeted proteins with two potential ice-binding motifs, are predicted to adopt structures with two ice-binding surfaces, and when purified are sufficient to reconstitute activity, present in cold-acclimated *D. antarctica* leaves. We propose that elaboration of IRIP-mediated RI activity has contributed to the cryo-tolerance of *D. antarctica* and thus its unique ability to colonise Antarctica.

Thus the activity of IRIPs in *D. antarctica* can coherently account for this species cryo-tolerance, offering the prospect of being able to enhance this capacity in economically important crop and pasture plants, and illustrating the utility of the Xenogenomics approach for identifying novel sources of genetic diversity.

Acknowledgements

This work was funded by the Australian Research Council, the Grains Research and Development Corporation, the Victorian Department of Innovation, Industry and Regional Development, and the Victorian Department of Primary Industries.

Plant Insect Interaction

Biotechnological Control of Plant Parasitic Nematodes

P.E. Urwin

Centre for Plant Sciences, University of Leeds, LEEDS LS2 9JT, UK.

Nematode biology

Cyst and root-knot nematodes share both similarities and differences in their interactions with a host plant such as *Arabidopsis*. A central, common feature of the parasitism is the establishment and maintenance of a permanent nematode feeding site that sustains the nematode throughout its growth in the plant (Barthels et al., 1997). Nematodes modify plant cells with products released from three gland cells (two sub ventral and one dorsal, pharyngeal gland cells). The products from the two sub-ventral gland cells are particularly important in root invasion before the corresponding dorsal cell is active. Using their stylet, cyst nematodes migrate intracellularly through cortical cells towards the vascular cylinder where an initial feeding cell is selected. Nematode secretions that are probably from the dorsal pharyngeal gland are secreted through the bore of the stylet into the initial feeding cell and induction of the feeding site is triggered. This results in dramatic changes in gene expression and considerable reprogramming of root cell development. Interestingly, although there are overarching morphological similarities and a shared function, the nature of these transformations differs among cyst and root-knot nematodes (Davis et al., 2000). Cyst nematodes modify a single cell by cell wall dissolution and fusion to form a syncytium that progressively enlarges to eventually involve up to

Z. Xu et al. (eds.), Biotechnology and Sustainable Agriculture 2006 and Beyond, 181–190.

200 neighbouring cells (Favery et al., 1998). The progressively enlarging syncytium attains transfer cell status in contact with xylem and phloem elements by about 7 days post infection (dpi). The continually stimulated feeding cell furnishes the growing cyst nematode with all its metabolic requirements throughout development to an egg-laying female, a process which takes 3–6 weeks.

Economic impact of plant parasitic nematodes

Cyst and root knot nematodes cause damage worth billions of dollars to crops worldwide. In the absence of natural resistance in commercially viable cultivars, growers are forced to use crop rotations and environmentally damaging nematicides, many of which are being phased out in the United Kingdom and worldwide. Fumigant nematicides are, in effect, biocides while non-fumigant nematicides target a broad range of animals including (target and non target) nematodes, insects, birds and mammals. New nematicides need to have a nematode specific mode of action, be low cost compounds for soil application, have the ability to remain available in soil water and need to show persistence that is no longer than required for an effective treatment period. These exacting requirements mean that little effort has been made by the pesticide industry to develop novel compounds. This is a concern given that the most widely used compound, aldicarb, will be withdrawn by the EU from use on potato crops in 2007. The future of other carbamates and organophosphate nematicides with a similar anti-acetylcholinesterase activity is uncertain given that they too have high mammalian, avian and insect toxicities.

The potato cyst nematodes (PCN) *Globodera rostochiensis* and *G. pallida* are the most economically important nematode problems of UK arable agriculture. They occur in 65% of UK potato growing land with *G. pallida* present at 92% of these sites (Minnis et al., 2002). They impose an annual cost in excess of £50 million on UK farmers and threaten the future of the crop for many growers. Full and durable resistance was achieved rapidly for *G. rostochiensis* using the H1 resistance gene. Although first identified in the 1950s, the H1 gene remains a considerable commercial success for plant breeding as a trait in Maris Piper, the most planted cultivar in the United Kingdom. The lack of a comparable single, dominant natural resistance gene for *G. pallida* has resulted in an emphasis on multi-trait quantitative resistance that is difficult to breed and is more readily overcome by virulent pathotypes. The absence of potatoes naturally resistant to *G. pallida* means that growers need to use nematicides in order to protect yield. Lack of control options forces growers to use prolonged rotations to reduce populations of the highly persistent *G. pallida*. The hidden cost of this has led to increased use of PCN-free land including loss of permanent grassland to potato cropping. This has contributed to 637×10^3 ha of grassland lost in the United Kingdom to agriculture since 1950. Deep ploughing of previously permanent grassland for potato crops is also a major cause of damage to previously unknown, unscheduled archaeological sites under grass (English Heritage, 2003). To maintain economic viability, specialist potato growers have shortened their rotations and crop potatoes, on average, every 5 years, which is too frequent to prevent *G. pallida* populations increasing and spreading, despite using other control measures, such as nematicides.

Biotechnological control of plant parasitic nematodes using cystatins

Cysteine proteinases are important digestive enzymes in many nematodes, but not humans. Cysteine proteinase activity is inhibited in female cyst nematodes after incubation with a cystatin (Lilley et al., 1996). A gene encoding a rice cystatin, Oc-I, was engineered to have an enhanced inhibitory activity. Expression of the engineered variant (OcIΔD86) in root cultures conferred higher levels of resistance against nematodes than the unaltered molecule (Urwin et al., 1995). Expression of OcIΔD86 in *Arabidopsis* plants was the first transgenic technology shown to work against both root-knot and cyst nematodes (Urwin et al., 1997a). Resistance has also been shown against *Rotylenchulus reniformis* (Urwin et al., 2000) and to be effective in rice (Vain et al., 1998) and banana (Atkinson et al., 2004a). Additive transgenic resistance has been demonstrated using two inhibitors of different proteinase classes in the same transgenic plant (Urwin et al., 1998). The work has culminated in successful field and containment trials of transgenic potatoes expressing a cystatin. The best transgenic lines were shown to have commercially useful resistance (Urwin et al., 2001). Full resistance to *G. pallida* was observed in the United Kingdom field by stacking natural and transgenic resistance (Urwin et al., 2003). Transgenic potato plants with cystatin expression restricted largely to roots have been field trialled with similar resistance levels to those achieved with constitutive expression (Lilley et al., 2004). We have demonstrated the environmental safety of the defence (Cowgill et al., 2002a, Cowgill, 2002b, Cowgill and Atkinson, 2003) and a *prima facie* case has been established for the food safety of cystatin-expressing potato plants (Atkinson et al., 2004b).

Biosafety of anti-nematode protein in biotech-cystatin potato plants

We have established a lack of environmental impact of biotech-cystatin potato plants on above ground non-target associates of the crop. Initial work concentrated on the worst-case scenario of constructs in which the nematode resistance was under control of the CaMV35S constitutive promoter providing expression in both roots and leaves. Species at potential risk as leaf feeders such as *Eupteryx aurata* (a leafhopper) and *Myzus persicae* were not adversely affected in the field for plants that provided control of nematodes (Cowgill et al., 2002a, Cowgill and Atkinson, 2003). The approach also has no adverse effects on the parasitoid *Asaphes vulgaris* that attacks *M. persicae*. In contrast to systemic nematicides with anti-insect effects such as aldicarb, our biotech-cystatin potato plants are fully compatible with biocontrol for insect pests (Cowgill and Atkinson, 2003).

We have demonstrated promoters of root-preferentially expressed plant genes deliver effective levels of cystatin to feeding cyst nematodes (Lilley et al., 2004). There is no requirement for expression of the transgenes in aerial tissue to control root parasites. To assure complete biosafety, we have examined effects on soil organisms of cystatin delivered from constitutively expressed promoters. We found that the microbial community was not affected in its abundance, evenness or metabolic activity as determined by ester-linked phospholipid fatty acid analysis (PLFA) and rates of litter decomposition were not affected (Cowgill et al., 2002b). Subsequent work has established no change in microbial communities based on community level physiological profiles (CLPP) using BIOLOG plates that provide 31 different substrates for microbes. In contrast, analysis based on PLFA and CLPP establishes that potato genotype and the farmer's choice of crop has an impact on soil microbial communities.

Biotech-cystatin plants had no significant effect on the abundance of soil microarthropods or free-living nematodes and offer environmental benefits in avoiding the earthworm mortality that results from use of nematicides. They also have no effect on the soil nematode community whereas aldicarb disturbs the balance between rapid and slow life-cycle soil nematodes 3 months after treatment (Celis et al., 2004). This effect has also been detected for another nematicide fosthiazate (Sturz and Kimpinski, 1999) which is a candidate to replace aldicarb when that is withdrawn from EU use in 2007. The work establishes a *prime facie* case for environmental safety of GMNR-cystatin potato in contrast to nematicides.

Potato is an important and safe crop in which to develop GM technology. It has only two wild relatives in the United Kingdom and can only be forced to hybridise with one of them, *Solanum nigrum* (black nightshade) artificially. It is not naturally invasive of natural or semi-natural habitats. PCN is also an alien species to the United Kingdom and its distribution here is restricted to potato fields. This ensures no ecological advantage to any GM potato plants establishing in field margins. Volunteer (weed potatoes) within a field are already controlled as a matter of routine farm management. Potato does not provide nectar and so is not visited by butterflies or hive bees to any great extent. The use of root specific promoters prevents expression of anti-nematode proteins in pollen.

Toxicological and allergenicity testing of biotech-cystatin potato tubers

Safety assessment of the newly introduced heterologous proteins has already been carried out for the cystatin (OcIΔD86) used in our biotech-cystatin crops. Our principal conclusion from a broad ranging toxicological study was that daily administration by oral gavage of up to 10 mg OcIΔD86/kg body weight for 28 consecutive days had no significant toxic effect on male rats (Atkinson et al., 2004b). The results suggest that the no adverse effect level provides a margin of dietary exposure of >2000 fold using a feeding-site preferential promoter to express the cystatin in potato plants. In addition, work has established that OcIΔD86 is not an allergen which is consistent with both its rapid digestion by simulated gastric fluid (Atkinson et al., 2004b) and its lack of sequence similarity with known allergens.

Peptide mimetics cause disruption of nematode chemosensory response

We have established a second, novel non-lethal approach to nematode control additional to the use of cystatins. Peptides have been selected by biopanning that are functional mimetics of compounds with known sub-lethal effects on nematodes (Winter et al., 2002). In both cases the peptides disrupt chemosensory-mediated behaviour of cyst nematodes at very much lower concentrations than required to affect other processes such as locomotion.

Our proof of principle work has been based on a peptide that inhibits acetylcholinesterase (AChE), the known mode of action of aldicarb. Aldicarb is effective at ultra low levels when it disrupts nematode orientation to chemical attractants and against locomotion when present at $10^6 \times$ this concentration. Low levels of the peptide we developed also do not affect locomotion. They are probably taken up by retrograde transport specifically along certain chemoreceptive neurones and lack the additional ability of aldicarb to enter nematodes by a transcuticular route (Winter et al., 2002). Root exudates of lines of potato plants that express and secrete this peptide inhibit acetylcholinesterase by up to 63.9% with a highly significant correlation between this inhibition and ability to suppress root invasion by cyst nematodes soaked in the solution (R^2= 0. 0.63; $P < 0.05$). In containment trials the best line suppressed establishment of females of *G. pallida* compared to the controls by 47.9% and 61 ± 4% at 21 and 56 dpi respectively (Liu et al., 2005). Soil water extracted by lysimeters from the rhizosphere of these plants contained sufficient peptide to inhibit acetylcholinesterase to the same extent of c0.6 nM aldicarb. This is very much lower than the amount present in soil water after nematicide application. This work demonstrates continual low-level plant release of the peptide provides efficacy but not the environmental damage of nematicides.

Although we have established a high level of environmental safety of the AChE-inhibiting peptide we do not intend to advance this variant of the technology beyond the proof of principle phase. Instead we will use a peptide that binds to nicotinic receptors in the same manner as levamisole. Levamisole is used as a veterinary anthelmintic and an adjuvant in certain drug therapies. We have shown that the peptide mimic of levamisole also perturbs nematode establishment (Liu et al., 2005).

Promoters for targeted expression of anti-nematode proteins

The CaMV35S promoter provides sufficient expression of a cystatin for partial resistance to *G. pallida* in the field (Urwin et al., 2001) but it is progressively down-regulated in the syncytial feeding cells of cyst nematodes (Goddijn et al., 1993; Urwin et al., 1997b).

Promoters that direct expression preferentially in root tissue offer a clear food safety benefit by reducing exposure of non-target organisms feeding on aerial tissues or presence in the yield. We have demonstrated several root-specific promoters (e.g., those of *TUB-1, RPL16* and *ARSK1*) deliver anti-nematode effectors to feeding nematodes with little expression elsewhere in the plant (Lilley et al., 2004). Such promoters are not active at sites of nematode invasion and would not be suitable for driving expression of nematode disorienting peptides. Therefore we have characterised appropriate promoters using *Arabidopsis* sequences in the database that have homology to maize genes expressed only in the distal region of roots. Using transgenic promoter-reporter lines we have established the patterns of activity of a number of these *Arabidopsis* sequences. One of these, the *mdk4-20* promoter had a particularly appropriate pattern of expression for peptides that disrupt nematode orientation to roots.

A fluorescent protein that shifts colour from green (at 500 nM) to red (at 580 nM) over time, (Fluorescent Timer, Clontech) demonstrated that the *mdk4-20* promoter remained more active than CaMV35S in border cells after their detachment from roots. The pattern of *mdk4-20* promoter activity is restricted to the outer cell layers at the root tip and border cells. It is still able to deliver AChE-inhibiting peptide to suppress root invasion of *Arabidopsis* by the cyst nematode *Heterodera schachtii* (>90% in initial containment trial). Its restricted pattern of activity enhances targeted delivery of peptide to sites of nematode invasion plus the rhizosphere.

RNAi as a basis for defining gene function in plant parasitic nematodes

We have used RNAi to study the consequences of targeted knockout for plant parasitic nematode genes. The effect is systemic in nematodes so genes expressed in many of their tissues can be targeted. Prospects for basing novel resistance on the approach depend on the plant generating an effective form of double-stranded RNA in the absence of an endogenous target gene without detriment to itself. These RNA molecules must then become available to the nematode and not be excluded by the feeding tube. If these requirements can be met, a plant delivering a dsRNA that targets a nematode gene and induces a lethal or highly damaging RNAi effect on the parasite could achieve crop resistance. It would be a specific approach without the need for novel protein expression (Bakhetia et al., 2005)

Changes in transcript abundance in response to nematode infection

Microarrays allow the opportunity to identify those genes that are differentially expressed in response to nematode infection. This will provide new leads for biotechnological control strategies. We have used the ATH1 Affymetrix arrays, representative of the complete *Arabidopsis* genome, to compare transcript abundance in sections of root harbouring the syncytium of a female cyst nematode (*H. schachtii*) with similar uninfected root lengths. We found 1167 genes changed in expression by greater than two-fold with 460 showing increased and 707 showing decreased expression in response to nematode infection. Some of these genes have been selected and confirmed as nematode responsive by qPCR plus *in situ* hybridisation and GUS reporter studies for those of high interest. A number of genes involved in development and differentiation were seen to alter in their expression in response to nematode infection (Haeger, 2005). We have shown similar changes in response to infection of *Arabidopsis* with root knot nematode. Comparison of the two sets of data has determined the differences and commonalities in the gene expression following infection with these nematode species.

References

Atkinson HJ, Grimwood S, Johnston K, Green J (2004a) Prototype demonstration of transgenic resistance to the nematode Radopholus similes conferred on banana by a cystatin. Transgenic Res 13: 135–142.

Atkinson HJ, Johnston KA, Robbins M (2004b) Prima facie evidence that a phytocystatin for transgenic plant resistance to nematodes is not a toxic risk in the human diet. J Nutr 134: 431–434.

Bakhetia M, Charlton WL, Urwin PE, McPherson MJ, Atkinson HJ, (2005) RNA interference and plant parasitic nematodes Trends in Plant. Trends Plant Sci 10: 362–367.

Barthels N, Leevan der FM, Klap J, Goddijn OJM, Karimi M, Puzio P, Grundler FMW, Ohl SA, Lindsey K, Robertson L, Robertson WM, Montagu M, Gheysen G, Sijmons PC (1997) Regulatory sequences of Arabidopsis drive reporter gene expression in nematode feeding structures. Plant Cell 9: 2119–2134.

Celis C, Scurrah M, Cowgill SE, Chumbiauca S, Franco J, Main G, Keizenbrink DT, Green J, Atkinson HJ (2004) Environmental biosafety and transgenic potato in a centre of diversity for this crop. Nature 432: 222–225.

Cowgill SE, Atkinson HJ (2003) A sequential approach to risk assessment of transgenic plants expressing proteases inhibitors: effects on non target herbivorous insects. Transgenic Res 12: 439–449.

Cowgill SE, Wright C, Atkinson HJ (2002a) Transgenic potatoes with enhanced levels of nematode resistance do not have altered susceptibility to nontarget aphids. Mol Ecol 11: 821–827.

Cowgill SE, Bardgett RD, Kiezebrink DT, Atkinson HJ (2002b) The effect of transgenic nematode resistance on non-target organisms in the potato rhizosphere. J App Ecol 39: 915–923.

Davis E, Hussey RS, Baum TJ, Bakker J, Schots A (2000) Nematode parasitism genes. Annu Rev Phytopathol 38: 365–396.

English Heritage (2003) 50791, Ripping up History, July.

Favery B, Lecomte P, Gil N, Bechtold N, Bouchez D, Dalmasso A, Abad P (1988) RPE, a plant gene involved in early developmental steps of nematode feeding cells. EMBO J 23: 6799–6811.

Goddijn OJM, Lindsey K, van der Lee FM, Klap JC, Sijmons P (1993) Differential expression in nematode-induced feeding structures of transgenic plants harboring promoter-GUS a fusion constructs. Plant J 4: 863–873.

Haeger A (2005) PhD Thesis, University of Leeds.

Lilley CJ, Urwin PE, McPherson MJ, Atkinson HJ (1996) Characterisation of intestinally active proteinases of cyst-nematodes Parasitology (Cambridge) Parasitology 113: 415–424.

Lilley CJ, Urwin PE, Johnston KA, Atkinson HJ (2004) Preferential expression of a plant cystatin at nematode feeding sites confers resistance to meloidogyne incognita and Globodera pallida. Plant Biotechnol J 2: 3–12.

Liu B, Hibbard JK, Urwin PE, Atkinson HJ (2005) The production of synthetic chemodisruptive peptides in planta disrupts the establishment of cyst nematodes. Plant Biotechnol J 3: 487–496.

Minnis ST, Haydock PPJ, Ibrahim SK, Grove IG, Evans K, Russell MD (2002) Potato cyst nematodes in England and Wales – occurrence and distribution. Ann. Appl Biol 140: 187–195.

Sturz AV, Kimpinski J (1999) Effects of fosthiazate and aldicarb on populations of plant- growth-promoting bacteria, root-lesion nematodes and bacteria-feeding nematodes in the root zone of potatoes. Plant Pathol 48: 26–32.

Urwin PE, Atkinson HJ, Waller DA, McPherson MJ (1995) Engineered oryza-cystatinI expressed in hairy roots confers resistance to Globodera pallida. Plant J 8: 121–131.

Urwin PE, Lilley CJ, McPherson MJ, Atkinson HJ (1997a) Resistance to both cyst and root-knot nematodes conferred by transgenic Arabidopsis expressing a modified cystatin. Plant J 12: 455–461.

Urwin PE, Lilley CJ, McPherson MJ, Atkinson HJ (1997b). Characterization of two cDNAs encoding cysteine proteinases from the soybean cyst nematode Heterodera glycines. MPMI 10: 394–400.

Urwin PE, McPherson MJ, Atkinson HJ (1998) Enhanced transgenic plant resistance to nematodes by dual proteinase inhibitor constructs. Planta 204: 472–479.

Urwin PE, Levesley A, McPherson MJ, Atkinson HJ (2000) Transgenic resistance to the nematode Rotylenchulus reniformis conferred by A. thaliana plants expressing proteinase inhibitors. Molecular Breeding 6: 257–264.

Urwin PE, Troth KM, Zubko EI, Atkinson HJ (2001) Effective transgenic resistance to Globodera pallida in potato field trials. Mol Breed 8: 95–101.

Urwin PE, Green J, Atkinson HJ (2003) Expression of a plant cystatin confers partial resistance to Globodera, full resistance is achieved by pyramiding a cystatin with natural resistance. Mol Breed 12(3): 263–269.

Vain P (1998) Expressing of an engineered cysteine proteinase inhibitor (Oryzsacystain-IND86) for nematode resistance in transgenic rice plants TAG 96: 266–271.

Winter M, McPherson MJ, Atkinson HJ (2002) Neuronal uptake of pesticides disrupts chemosensory cells of nematodes. Parasitology 125: 565–565.

Molecular Basis of Colorado Potato Beetle Adaptation to Potato Defence Mechanisms

N. Slapar[1], M. Pompe-Novak[3], M. Buh[3], M. Ravnikar[3], B. Štrukelj[1,2], M.A. Jongsma[4], K. Gruden[1]

[1]Department of Biochemistry and Molecular Biology, Jožef Stefan Institute, Ljubljana, Slovenia.
[2]Faculty of Pharmacy, University of Ljubljana, Ljubljana, Slovenia.
[3]Department of Plant Physiology and Biotechnology, National Institute of Biology, Ljubljana, Slovenia.
[4]Plant Research International, Wageningen, the Netherlands.

Abstract

Pathogen infection or herbivore attack, the phenomena named plant defence mechanism. The proteins induced after herbivore or insect attack can be classified into several groups including protease inhibitors, different phytoalexins, etc. As a counteract to the natural plant defence response the adaptation of insects has evolved. So far this process was studied only on the level of digestive proteases. When larvae are fed on high levels of proteinase inhibitors, the expression of inhibitor resistant enzymes is induced and plant defence becomes ineffective. To study the mechanism of insect adaptation Colorado potato beetle (Leptinotarsa decemlineata Say) larvae were used as a model. We have tried to assess the molecular basis of this resistance first by biochemical characterisation of different proteases from the adapted guts and later on by homology modelling of the analysed enzymes. To gain a broader knowledge about the process we have used the microarray technology and qPCR. Four hundred genes from subtractive libraries of guts prepared from larvae fed on control leaves versus larvae fed on defence induced leaves were

Z. Xu et al. (eds.), Biotechnology and Sustainable Agriculture 2006 and Beyond, 191–195.

sequenced and printed on microarrays. The expression of those genes was followed in time scale of insect adaptation. Different digestive enzymes are significantly up-regulated. Maybe the most interesting from biotechnological point of view are genes with no significant identity in the database and a gene with potential function in regulation of insect adaptation. Those might have potential for environmentally friendly plant protection agents by disturbing insect adaptation.

Introduction

Colorado potato beetle (Leptinotarsa decemlineata Say, CPB) is a major pest of potato in many areas of the world. It rapidly acquires resistance to pesticides, so there is a need for alternative means of control. One possible approach is to disrupt insect protein digestion by overexpressing protease inhibitors (PIs), which negatively impact insect physiological status (Wolfson and Murdock, 1987). However, in many cases insects can adapt to the challenge of PIs by quantitative and qualitative remodelling their digestive protease complement (Bolter and Jongsma, 1995). Previously, three cysteine proteases (termed intestains) were purified from adapted guts, with different substrate specificities and inhibition profiles. One of the isolated intestains was also able to inactivate one of the cystatin inhibitors by cleaving off its N-terminus (Gruden et al., 2003). cDNA sequences coding for cysteine proteases were isolated and classified under three intestain groups: A, B and C, A and C being regulated in the process of adaptation and all showing some specific structural characteristics in the inhibitor binding sites that could influence enzyme-inhibitor interaction (Gruden et al., 2004). Another approach is to study insect adaptation using systemic approach, like transcription profiling to identify underlying mechanisms (Liu et al., 2004, Moon et al., 2004). The aim of the current study was to analyze additional intestains present in adapted guts by a combination gene expression study and protein modeling. Additionally, combination of subtractive libraries and DNA microarray analysis was used to find other elements contributing to Colorado potato beetle adaptation towards plant defense.

Materials and methods

Defence response (including PIs) in potato plants was induced by methyl jasmonate (MeJA). CPB larvae were reared on MeJA-induced or control potato leaves for short or long periods of time. To simulate field conditions larvae were fed on the whole set of potato plants in growth chamber. Midguts were isolated from control and MeJA-adapted larvae for mRNA isolaupand the other down-regulated genes in MeJA-adapted midguts. Full-length cDNAs of new intestain cDNAs were amplified from MeJA-adapted midgut cDNA library. Expression of intestain genes during the larvae adaptation was analyzed using microarray and qPCR approach. The structure of human cathepsin L was used to model intestain D and E. Putative enzyme-inhibitor interactions were studied by superimposing molecular models of intestains on known enzyme-inhibitor. Four hundred clones were randomly selected from subtractive libraries to prepare cDNA microarrays. Genes, differentially expressed in insect adaptation to potato defence were identified by microarray hybridisations with samples 4 h, 24 h and 3 days after initiation of adaptation. Full length sequences for most interesting of those were cloned.

Results and discussion

cDNA libraries of larvae midgut that were classified under two new intestain groups: D and E. Intestains D and E share app.70% identity with other intestain genes and up to 50% identity to other cysteine proteases from insects and other arthropods and mammalian cathepsins L, K, S and P.

Homology modelling predicts that intestains D and E follow the general fold of papain-like proteases. Unlike other papain-like enzymes, intestains D and E have the insertion of Glu64a (papain numbering) in front of the conserved Gly65 which is a part of the S1 substrate binding site (Turk et. al, 1998). Superimposing of the intestain D and E models to the papainstefin B complex reveals steric clashes of Glu64a with the inhibitor N-terminus. The other possible outcome of such structural change is that the flexible inhibitor N-terminus could adapt its conformation and shift inwards the active side cleft. In this case, the inhibitor N-terminus would come in the proximity of the catalytic residues and would therefore become exposed to the proteolytic cleavage. No major structural characteristics of intestains D and E were observed in the binding sites of tyropin type inhibitors that would influence the enzyme-inhibitor interaction. Intestains D were down-regulated appoximately twofold 4 h after the beginning of larvae feeding on MeJA-induced potato leaves and up-regulated twofold in the long-term adaptation. Expression of intestain E genes remains at approximately the same level in short- or long-term adaptation to potato protease inhibitors induced by MeJA and insect wounding. Feeding of larvae in the field simulated conditions provokes less intensive adaptation than MeJA-induced plant defence.

Probes included on the microarray represented altogether 240 different genes that can be divided into four groups: genes coding for digestive enzymes, cell maintenance and development genes, genes coding regulatory proteins and genes with no significant similarity found or unknown function. More than half were significantly regulated in at least one of the experiments. Several genes coding for digestive enzymes were found to be differentially regulated. The strongest response was however detected with genes with no significant similarity found or with similarity found to the genes in genomic databases with no known function. Also interesting is a gene with potential function in regulation of insect adaptation.

The process of adaptation of insect to the potato defence compounds is very complex. The composition of digestive proteases in CPB gut changes upon feeding on high levels of potato defence PIs in such a way that a set of enzymes resistant to the inhibitors are produced to compensate for the loss of activity of sensitive proteases. In addition, a whole array of genes coding for other digestive enzymes and genes with other functions were found to be regulated in larvae in course of this process. Only when we understand the regulation and structural characteristics of the full process of adaptation we will be able to effectively design alternative pest control management.

References

Bolter CJ, Jongsma MA (1995) Colorado potato beetles (Leptinotarsa decemlineata) adapt to proteinase inhibitors induced in potato leaves by methyl jasmonate. J Insect Physiol 41: 1071–1078.

Gruden K, Kuipers AG, Guncar G, Slapar N, Strukelj B, Jongsma MA (2004) Molecular basis of Colorado potato beetle adaptation to potato plant defence at the level of digestive cysteine proteinases. Insect Biochem Mol Biol 34: 365–375.

Gruden K, Popovic T, Cimerman N, Krizaj I, Strukelj B (2003) Diverse enzymatic specificities of digestive proteases, 'Intestains', enable Colorado potato beetle larvae to counteract the potato defence mechanism. Biol Chem 384: 305–310.

Liu Y, Salzman RA, Pankiw T, Zhu-Salzman K (2004) Transcriptional regulation in southern corn rootworm larvae challenged by soyacystatin N. Insect Biochem Mol Biol 34: 1069–1077.

Moon J, Salzman RA, Ahn JE, Koiwa H, Zhu-Salzman K (2004) Transcriptional regulation in cowpea bruchid guts during adaptation to a plant defence protease inhibitor. Insect Mol Biol 13: 283–291.

Turk D, Guncar G, Podobnik M, Turk B (1998) Revised definition of substrate binding sites of papain-like cysteine proteases. J Biol Chem 379: 137–147.

Wolfson JL, Murdock LL (1987) Entomol Exp Appl 44: 235–240.

Molecular Mechanisms on Plant-Microbe Interactions

An Arabidopsis Nudix Hydrolase Modulates the Innate Immune Response Independent of NPR1 and Salicylate

Yiji Xia*, Xiaochun Ge, Guo-Jing Li, and Huifen Zhu

Donald Danforth Plant Science Center, 975 N Warson Rd, St. Louis, MO 63132, USA.
*Email: yxia@danforthcenter.org

Plants have evolved complicated systems to regulate immune responses. The innate immune response which is triggered following recognition of conserved microbe-associated molecular patterns (MAMPs) provides the first layer of induced defense against an invading pathogen. The innate immune response (also termed basal resistance), together with pre-formed physical and chemical barriers, successfully prevents most infections from becoming established. Pathogens have evolved mechanisms, such as delivering virulence effectors to host cells, to block the basal defense system to promote its infection. In turn, plants have evolved the gene-for-gene recognition mechanism to recognize specific effectors as a signal of invasion, resulting in activation of a second tier of defense against the pathogens.

Activation of induced resistance will likely cause deleterious effects on normal cell functioning. Precise control of the innate immune response is not only critical for defense against potentially pathogenic microorganisms but also essential for cohabiting with non-pathogenic microorganisms. Both positive and negative signaling pathways interplay to coordinate the development of a resistance response with the appropriate amplitude and duration. We have identified an Arabidopsis Nudix hydrolase (AtNUD7) that plays an

Z. Xu et al. (eds.), Biotechnology and Sustainable Agriculture 2006 and Beyond, 199–200.

important role in modulating the basal defense response. AtNUD7 functions as a negative regulator of the defense response. Loss of the AtNUD7 function causes the plants to be primed for a remarkably heightened defense response that can be triggered not only by pathogenic and non-pathogenic microorganisms but also by MAMPs. The hyper-responsive nature of the *Atnud7* mutant enhances its resistance to infection by the bacterial pathogen *Pseudomonas syringae* but reduces the fitness and growth of the plants although the mutant does not exhibit a strong constitutive activation of the defense response. Analysis of double mutants between *Atnud7* and a variety of Arabidopsis disease resistance mutants indicates that the enhanced activation of defense-related genes associated with the *Atnud7* mutation does not require a functional NPR1 or accumulation of Salicylate. AtNUD7 hydrolyzes nucleoside diphosphate derivatives, including ADP-ribose and NADH. Its hydrolysis activity is essential for its biological function and is sensitive to inhibition by Ca^{++}. The study provides novel and important insights into regulation of the immune response mediated by the Nudix hydrolase to prevent excessive cell stimulation.

RNA-Mediated Resistance to Cassava Geminiviruses in Transgenic Cassava

Hervé Vanderschuren[1], Rashid Akbergenov[2], Mikhail M. Pooggin[2], Thomas Hohn[2], Wilhelm Gruissem[1], and Peng Zhang[1,3,*]

[1]Institute of Plant Sciences, ETH Zurich, Universitätstrasse 2, 8092 Zurich, Switzerland.
[2]Institute of Botany, University of Basel, Schönbeinstrasse 6, 4056 Basel, Switzerland.
[3]Institute of Plant Physiology and Ecology, Shanghai Institute for Biological Sciences, Chinese Academy of Sciences, 300 Fenglin Road, 200032 Shanghai, China.
[*]Email: pzhang@ethz.ch

Cassava mosaic disease (CMD) is one of the major problems encountered in cassava fields in Africa where the disease can trigger overall yield losses up to 25%. CMD is caused by whitefly-transmitted geminiviruses. Synergism, recombination and pseudo-recombination between different cassava geminivirus species have led to the recent pandemics of severe CMD in Africa (Legg and Fauquet, 2004). Our research aims at developing different CMD prevention strategies in transgenic cassava and therefore providing local farmers with geminivirus resistant lines to secure cassava production in the subsistence faming system.

Two different virus resistance approaches via long-strand antisense RNAs and small interfering RNAs (siRNA) were applied to engineer broad-spectrum cassava mosaic geminivirus (CMG) resistance in transgenic cassava. The expression of full-length antisense RNAs that are complementary to African cassava mosaic virus (ACMV) mRNAs of *Rep* (*AC1*), *TrAP* (*AC2*) or *REn* (*AC3*) could

Z. Xu et al. (eds.), Biotechnology and Sustainable Agriculture 2006 and Beyond, 201–203.

lead to elevated resistance to ACMV infection with reduced viral DNA accumulation and symptoms both *in vitro* and *in vivo* assays (Figure 1; Zhang et al., 2005). Upon ACMV infection of complete plants by biolistic inoculation, short sense and antisense RNAs specific to AC1, which are not detectable before infection, were identified in transgenic lines expressing AC1 antisense RNA and wild-type plants. This suggests that the small RNAs mediate interference by transcriptional and post-transcriptional gene silencing. In transgenic cassava expressing hairpin double-stranded RNA (dsRNA) cognate to the intergenic or the coding sequences of CMG, improved plant recovery from ACMV infection was observed (Figure 2). A similar pattern of 21–24 nt CMG-derived small RNAs was detected from ACMV infected wild-type plants and non-infected dsRNA transgenic plants (Akbergenov et al., 2006). There are also no qualitative differences among small RNAs from infected wild-type, antisense and dsRNA transgenic plants. Further characterization of these small RNAs revealed that modification of the viral siRNAs includes 5' end phosphorylation, as shown by phosphatase treatments, and methylation at the 3'-nucleotide, as confirmed by ß-elimination. The production of these small RNAs indicates that several distinct silencing pathways are involved in DNA virus-plant interactions. The level of small RNAs was increased by infection in transgenic plants, indicating their potential antiviral defense function. Differences among the small RNAs in quantity were detected from infected wild-type and transgenic plants. The 22–24 nt small RNAs, which are predominant in infected wild-type plants, display relatively strong signals in antisense and dsRNA transgenic cassava plants. No significant changes of 21 nt small RNA was found after infection of wild-type and transgenic plants. Interestingly, the 23 nt small RNAs showed only one polarity in the form of antisense strand when hybridized with a probe designed in the intergenic region. We conclude that siRNAs generated from transgenes, combined with those naturally produced upon viral infection, trigger the transcriptional and post transcriptional gene silencing of CMG, and finally result in reduced viral DNA replication and symptom development in transgenic cassava. Our final goal is to provide local farmers with geminivirus resistant lines to secure cassava production in CMG pandemic regions.

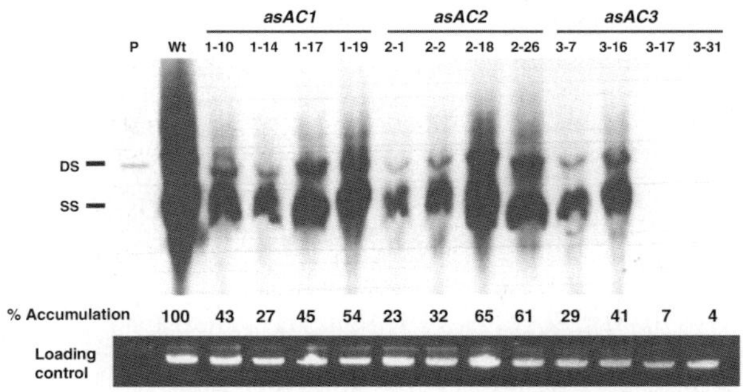

Figure 1. Reduced viral DNA accumulation in cassava leaf disks of transgenic lines in comparison with wild-type (Wt) after biolistic inoculation with ACMV-CM, ss, single-stranded DNA; ds, double-stranded DNA.

Figure 2. Transgenic cassava expressing hairpin RNA homologous to ACMV promoter sequence shows fast recovery from ACMV infection (a) and inhibition of viral DNA accumulation (b). Viral single-stranded (ss), supercoiled (sc) and open-circular (oc) DNA forms are indicated.

References

Akbergenov R, Si-Ammour A, Blevins T, Amin I, Kutter C, Vanderschuren H, Zhang P, Gruissem W, Meins F Jr, Hohn T, Pooggin MM (2006) Molecular characterization of geminivirus-derived small RNAs in different plant species. Nucleic Acids Res 34(2): 462–471.

Legg JP, Fauquet CM (2004) Cassava mosaic geminiviruses in Africa. Plant Mol Biol 56: 585–599.

Zhang P, Vanderschuren H, Fütterer J, Gruissem W (2005) Resistance to cassava mosaic disease in transgenic cassava expressing antisense RNAs targeting virus replication genes. Plant Biotechnol J 3: 385–397.

The Molecular Basis of Cold-Induced Pollen Sterility in Rice

Sandra N. Oliver[1], Xiaochun Zhao[2], Elizabeth S. Dennis, and Rudy Dolferus[2]

CSIRO Plant Industry, G. P. O. Box 1600, Canberra ACT 2601, Australia.
[1]Present Address: Max Planck Institute of Molecular Plant Physiology, 14476 Golm, Germany.
[2]This research is supported by the Rural Industries Research & Development Corporation, P. O. Box 4776, Kingston ACT 2604, Australia.

Although originally a tropical plant, an increasing amount of rice is grown in temperate climate zones of the world. Rice yields in most temperate climate zones are variable and yield losses due to cold-induced pollen sterility (CIPS) are the main contributor to this variation. The problem of CIPS is therefore a major bottleneck for the further improvement of rice yields in those areas. The growing season in temperate climate zones is shorter, confronting rice crops with colder conditions both at the start and end of the season. Cold spells at the reproductive stage are most damaging and an estimated 7 million ha world-wide are prone to this damage (Sthapit and Witcombe, 1998). In Australia, cool temperatures during the early booting stage cause an average yield reduction of 5–10% annually ($A44 million), with cold snaps occurring on average every 3–4 years causing severe yield losses of 20–40% (Jacobs and Pearson, 1994, 1999). Breeding efforts to produce cold-tolerant rice cultivars have focused on improving seedling vigor, shortening the growth season, and improving cold-tolerance at the booting stage (Andaya and Mackill, 2003a, b). Breeding for cold tolerance has traditionally been difficult due to the

Z. Xu et al. (eds.), Biotechnology and Sustainable Agriculture 2006 and Beyond, 205–207.

unpredictability and variability of the weather (timing and severity are beyond control) and due to complex interactions between cold sensitivity and environmental factors (lighting, nutrition). Clearly, breeding for cold tolerant rice would benefit from a better understanding of the underlying molecular mechanism and the development of molecular markers.

Our investigation of the molecular basis of CIPS has focused on the young microspore (YM) stage, the most cold-sensitive stage of pollen development. In the Australian cold-sensitive cultivar *Doongara*, cold treatment at YM leads to accumulation of sucrose and hexoses in anthers, starch depletion in pollen at anthesis and irreversible induction of pollen sterility. Cell wall invertase activity is also decreased, suggesting that cold induces a blockage of sugar transport to the pollen grains. We have identified the genes involved in the apoplastic sucrose transport pathway in rice anthers, cell wall invertase (*OSINV4*) and monosaccharide transporter (*OSMST8*), and we have shown that both genes are repressed by cold at YM. Both genes are expressed in the tapetum at YM and in pollen at later stages of development. In the cold tolerant Chinese upland rice variety *R31* (*Lijiangheigu*) this repression of sucrose transport is not taking place (Oliver et al., 2005).

In order to study how cold represses sugar transport genes we focused on the hormone ABA, which is known to play a role in regulation of gene expression under cold conditions. Both *OSINV4* and *OSMST8* are repressed by ABA in cold sensitive rice but not in the cold-tolerant cultivar *R31*. ABA levels increased very quickly in *Doongara* anthers, while levels of ABA were two-fold lower under normal conditions in *R31* and they remained low following cold treatment. Investigation of the expression of ABA biosynthetic and catabolic genes in *Doongara* and *R31* anthers showed that the zeaxanthin epoxidase gene (*OSZEP1*) and one out of two anther-expressed 9-cis-epoxycarotenoid dioxygenase genes (*OSNCED3*) were up-regulated in *Doongara*, while these genes were not induced in cold-tolerant *R31*. In contrast, compared to *Doongara* the ABA-8-hydroxylase gene *ABA8OH1*was expressed at higher levels under control and cold conditions in *R31*. Our results indicate that the cold-tolerance phenotype of *R31* is associated with lower levels of anther ABA under both control and cold conditions and that this is the consequence of the combined action of lower biosynthesis and faster degradation (Oliver et al., 2007).

Treatment of rice panicles with exogenous ABA induced pollen sterility and a reduction in grain number due to premature spikelet abortion. This was the case for both *Doongara* and the cold-tolerant cultivar *R31*. *R31* appeared to be more sensitive than *Doongara* to ABA treatment. We are currently carrying out mapping of cold tolerance QTLs using a *DoongaraXR31* recombinant inbred line population. Our results indicate that ABA could play a regulatory role in the control of pollen fertility and grain number under abiotic stresses such as cold and drought. ABA interacts with gibberellic acid (GA) and ethylene, which control growth and development and senescence response respectively. And both GA and ethylene in turn are known to interact with cytokinin and auxin respectively, the two main hormones involved in controlling plant development. The mapping of cold QTLs in the variety *R31* may provide an insight into the regulation of ABA biosynthesis and how this contributes to cold tolerance in reproductive organs of rice.

References

Andaya VC, Mackill DJ (2003a) Mapping of QTLs associated with cold tolerance during the vegetative stage in rice. J Exp Bot 54: 2579–2585.

Andaya VC, Mackill DJ (2003b) QTLs conferring cold tolerance at the booting stage of rice using recombinant inbred lines from ajaponicax indica cross. Theor Appl Genet 106: 1084–1090.

Jacobs BC, Pearson CJ (1994) Cold damage and development of rice: a conceptual model. Aust J Exp Agric 34: 917–919.

Jacobs BC, Pearson CJ (1999) Growth, development and yield of rice in response to cold temperature. J Agron Crop Sci 182: 79–88.

Oliver SN, Dennis ES, Dolferus R (2007) ABA regulates apoplastic sugar transport and is a potential signal for cold-induced pollen sterility in rice. Plant Cell Physiol 48: 1319–1330.

Oliver SN, Van Dongen JT, Alfred SC, Mamun EA, Zhao XC, Saini HS, Fernandes SF, Blanchard CL, Sutton BG, Geigenberger P, Dennis ES, Dolferus R (2005) Cold-induced repression of the rice anther-specific cell wall invertase gene OSINV4 is correlated with sucrose accumulation and pollen sterility. Plant Cell Environ 28: 1534–1551.

Sthapit BR, Witcombe JR (1998) Inheritance of tolerance to chilling stress in rice during germination and plumule greening. Crop Sci 38: 660–665.

Regulation of the Rice NADPH Oxidase by Binding of Small GTPase Rac and Ca^{2+} to Its N-terminal Extension

Hann Ling Wong[1,*], Reinhard Pinontoan[1], Kana Hasegawa[1], Takashi Yaeno[2], Koh Iba[2], Ryo Tabata[1], Kokoro Hayashi[1], Chojiro Kojima[1], Tsutomu Kawasaki[1], and Ko Shimamoto[1]

[1]Nara Institute of Science and Technology, Japan.
[2]Department of Biology, Kyushu University, Japan.
[*]Email: h-wong@bs.naist.jp

Reactive oxygen species (ROS) produced by NADPH oxidase have been shown to play many important roles in signaling and development in plants. For example, they have been shown to be involved in processes, such as plant defense response, cell death, abiotic stress, stomatal closure and root hair development. Plant NADPH oxidase genes, termed *rboh* (*respiratory burst oxidase homolog*), encoding homologs of the mammalian NADPH oxidase catalytic subunit $gp91^{phox}$ have been isolated from many plants species. In phagocytes, the NADPH oxidase forms a multi-protein complex consisting of $gp91^{phox}$, $gp22^{phox}$, $gp47^{phox}$, $gp67^{phox}$, $gp40^{phox}$ and the small GTPase Rac2. However, with the exception of *rboh* and *Rac* (also known as *Rop*), plants do not possess the homologs of other subunits of the mammalian NADPH oxidase complex. Furthermore, unlike the mammalian $gp91^{phox}$, plant Rboh proteins possess an extended N-terminus, which contains two Ca^{2+}-binding EF-hand motifs. A role for Ca^{2+} in ROS signaling in plants has also been suggested. However, how Ca^{2+} signaling integrates into Rac/Rop and ROS signaling to modulate plants defense response and development remains unknown. The mechanism by which Rboh is activated also

Z. Xu et al. (eds.), Biotechnology and Sustainable Agriculture 2006 and Beyond, 209–210.

remains unknown. In the absence of other homologs of the mammalian NADPH oxidase subunits, the small GTPase Rac/Rop becomes a prime candidate for being a regulator of plant NADPH oxidase.

Previously, we found that overexpression of the constitutively active (CA) form of the small GTPase OsRac1 in transgenic rice plants enhanced resistance to virulent rice fungus and bacterial blight. Elicitor-induced hydrogen peroxide production was also enhanced in CA-OsRac1 overexpressing cultured cells.

In this study, using yeast two-hybrid and *in vitro* binding assays, we found that the CA-OsRac1, but not the dominant negative form, interacted with the N-terminus of OsrbohB. The N-terminus contained two Ca^{2+}-binding EF-hand motifs, was sufficient for OsRac1-OsrbohB interaction. Fluorescence resonance energy transfer microscopy showed that OsRac1 interacted with the N-terminus of OsrbohB predominantly in the GTP-bound form *in vivo* and the interaction was suppressed by cytosolic Ca^{2+} ions accumulation. Furthermore, *in vitro* NADPH oxidase activity assay also showed that GTP-bound OsRac1 enhanced NADPH oxidase activity and the activity was suppressed by Ca^{2+} ions.

Taken together, the results showed that the direct binding OsRac1 to the N-terminus of OsrbohB regulates NADPH oxidase activity. Furthermore, our studies also suggest how Rac/Rop and Ca^{2+} ions may interact with Rboh to modulate ROS production.

Post-Transcriptional Gene Silencing of the p23 Silencing Suppressor of *Citrus tristeza* Virus Confers Resistance to the Virus in Transgenic Mexican Lime

Carmen Fagoaga[1], Carmelo López[2], Alfonso Hermoso de Mendoza[1], Pedro Moreno[1], Luis Navarro[1], Ricardo Flores[2], and Leandro Peña[1,*]

[1]Instituto Valenciano de Investigaciones Agrarias, Apdo. Oficial, Moncada 46113, Valencia, Spain.
[2]Instituto de Biología Molecular y Celular de Plantas (UPV-CSIC), Universidad Politécnica de Valencia, Avenida de los Naranjos, Valencia 46022, Spain. The two first authors contributed equally to this work.
[*]Email: lpenya@ivia.es

Citrus is the most important fruit tree crop in the world, with a cultivated surface of more than 7 million ha distributed in about 100 countries, mainly in tropical and subtropical areas, and a fruit production of more than 100 million tons per year (FAO, 2005). *Citrus tristeza virus* (CTV) causes the most destructive viral disease and is considered a major threat for the citrus industry worldwide. CTV produces decline and death of scion varieties, except lemons (*Citrus limon* (L.) Burm.), grafted on sour orange (*C. aurantium* (L.). Additionally, severe CTV strains cause stem pitting, stunting, low yield and poor fruit quality of some varieties of sweet orange (*C. sinensis* (L.) Osb.), limes (*C. aurantifolia* (Christ.) Swing.; *C. latifolia* Tan.) and grapefruits (*C. paradisi* Macf.), regardless the rootstock used (Bar-Joseph et al., 1989).

CTV has a plus-strand RNA genome of ~20 kb organized in 12 open reading frames (ORFs) and two 5' and 3' unstranslated regions

Z. Xu et al. (eds.), Biotechnology and Sustainable Agriculture 2006 and Beyond, 211–213.

(UTRs) (Karasev et al., 1995). The 5' half of the genome encompasses two ORFs encoding proteins associated with viral replication that are expressed from the genomic RNA. The ten 3' proximal ORFs are expressed via 3' co-terminal subgenomic RNAs and they encode the minor and major coat proteins (CPs) of 27 and 25 kDa (p27 and p25), respectively, and several other proteins (p33, p6, p65, p61, p18, p13, p20 and p23) (Karasev et al., 1995). Protein p23, which has no homologue in other closteroviruses, binds in vitro to RNA in a non-sequence-specific manner (López et al., 2000), and it is involved in regulating the balance of plus and minus RNA strands during replication (Satyanarayana et al., 2002). Recently, p23, p20 and p25 have been found to act as RNA silencing suppressors in *Nicotiana tabacum* and *N. benthamiana* plants (Lu et al., 2004).

We have previously reported that ectopic expression of the p23 gene from mild or severe CTV strains in Mexican lime (*Citrus aurantifolia* (Christ.) Swing.) induces aberrations resembling viral leaf symptoms, the intensity of which is independent on the pathogenicity of the CTV strain but correlates with p23 accumulation. Transformation with p23 of other CTV-susceptible and -resistant citrus genotypes (sweet orange, sour orange, and trifoliate orange) also leads to CTV-like symptoms that correlate with the levels of p23 transcripts, although p23 is barely detectable in these hosts. In contrast, transgenic expression of p23 in CTV non-host *N. tabacum* and *N. benthamiana* does not induce any phenotypic aberration albeit they consistently accumulate p23. Altogether, these results indicate that p23 is an important CTV pathogenicity determinant that interferes with plant development specifically in citrus species and relatives (Ghorbel et al., 2001; Fagoaga et al., 2005).

In the course of these experiments, three out of 60 lines of Mexican lime carrying the p23 transgene of the severe CTV isolate T36, and two out of 20 lines carrying the p23 transgene of the mild isolate T317, were visually normal and developed similarly to controls transformed with the empty vector or non-transformed plants. Here we report that five independent transgenic lines having normal phenotype displayed characteristics typical of post-transcriptional gene silencing (PTGS): multiple copies of the transgene, low levels of the corresponding mRNA, methylation of the silenced transgene, and accumulation of p23-specific small interfering RNAs (siRNAs). When graft- or aphid-inoculated with CTV, some propagations of these silenced lines were immune: they neither expressed symptoms nor accumulated virions and viral RNA as estimated by DAS-ELISA and Northern blot hybridization, respectively. Other propagations were moderately resistant because they became infected later and showed attenuated symptoms compared to controls. The susceptible propagations, in addition to symptom expression and elevated virus titer, accumulated p23-specific siRNAs at levels significantly higher than immune or non-inoculated propagations, and showed transgene demethylation. This variable response among clonal transformants indicates that factors other than the genetic background of the transgenic plants play a key role in PTGS-mediated resistance.

Reference

Bar-Joseph M, Marcus R, Lee RF (1989) The continuous challenge of citrus tristeza virus control. Annu Rev Phytopathol 27: 291–316.

Fagoaga C, López C, Moreno P, Navarro L, Flores R, Peña, L (2005) Viral-like symptoms induced by the ectopic expression of the p23 gene of citrus tristeza virus are citrus-specific and do not correlate with the pathogenicity of the virus strain. Mol. Plant Microbe Interaction 18: 435–445.

FAO (2005) http//faostat.fao.org.

Ghorbel R, López C, Fagoaga C, Moreno P, Navarro L, Flores R, Peña L (2001) Transgenic citrus plants expressing the citrus tristeza virus p23 protein exhibit viral-like symptoms. Mol Plant Pathol 2: 27–36.

Karasev AV, Boyko VP, Gowda S, Nikolaeva OV, Hilf ME, Koonin EV, Niblett CL, Cline K, Gumpf DJ, Lee RF, Garnsey SM, Lewandowsky DJ, Dawson WO (1995) Complete sequence of the citrus tristeza virus RNA genome. Virology 208: 511–520.

López C, Navas-Castillo J, Gowda S, Moreno P, Flores R (2000) The 23 kDa protein coded by the 3’-terminal gene of citrus tristeza virus is an RNA-binding protein. Virology 269: 462–470.

Lu R, Folimonov A, Shintaku M, Li W-X, Falk BW, Dawson WO, Ding SW (2004). Three distinct suppressors of RNA silencing encoded by a 20-kb viral RNA genome. Proc Nat Acad Sci USA 101: 15742–15747.

Satyanarayana T, Gowda S, Ayllón MA, Albiach-Martí MR, Rabindran S, Dawson WO (2002). The p23 protein of citrus tristeza virus controls asymmetrical RNA accumulation. J Virol 76: 473–483.

Phloem Feeding Regulates the Plant Defense Pathways Responding to Both Aphid Infestation and Pathogen Infection

Yinghua Huang

USDA-ARS Plant Science Research Laboratory and Oklahoma State University, 1301 N. Western Street, Stillwater, OK 74075, USA.
Email: yinghua.huang@ars.usda.gov

Abstract

Molecular basis of plant defense response to diseases has already been explored in great detail, but little is known about the molecular mechanisms controlling host defense against insect attack. Our research aims to improve understanding of plant defense against aphid, a major pest on many cereal crops. In this study, transcriptome of sorghum in response to greenbug feeding was investigated with microarrays carrying thousands of cDNA clones from the SSH libraries of plants treated by greenbug aphid. Analysis of genome-wide transcriptional regulation allowed the discovery of all defense-related genes and these resultant expression profiles helped in elucidating their function. These data showed that phloem-feeding induced defense response pathways associated with both aphid attack and pathogen infection and indicated signaling pathways were partially overlapping among plant responses to pathogens, insects, and wounding.

Z. Xu et al. (eds.), Biotechnology and Sustainable Agriculture 2006 and Beyond, 215–219.

Introduction

Plants have evolved diverse defense mechanisms to combat various enemies, including phytopathogens and herbivorous insects. Activation of host defense responses is controlled by a complex signal transduction network. Plant responses to phloem-feeding insects, such as aphids which make little wounding while feeding, are not as well understood as those to chewing insects or plant pathogens. It is believed that plant response to aphids is complicated, involving several defense strategies (Thompson and Goggin, 2006). Thus, it is important to explore the genetic mechanisms underlying plant defense against aphids and to identify the factors that regulate resistance or susceptibility of the host plants. In our current studies, we have used sorghum/aphid as the model system to examine aphid-induced responses in crop plants through genome-wide transcript analysis. After infestation with a virulent biotype of greenbug, substantial changes in gene expression (up- and down-regulation) in the host plants were observed based on the parallel analyses between various treatments (Park et al., 2006). This report focuses on the recent research progress in understanding of sorghum plant defense responses to greenbug aphid and other stresses at the molecular level.

Materials and methods

Plants have evolved diverse defense mechanisms to combat various enemies, including phytopathogens and herbivorous insects. Activation of host defense responses is controlled by a complex signal transduction network. Plant responses to phloem-feeding insects, such as aphids which make little wounding while feeding, are not as well understood as those to chewing insects or plant pathogens. It is believed that plant response to aphids is complicated, involving several defense strategies (Thompson and Goggin, 2006). Thus, it is important to explore the genetic mechanisms underlying plant defense against aphids and to identify the factors that regulate resistance or susceptibility of the host plants. In our current studies, we have used sorghum/aphid as the model system to examine aphid-induced responses in crop plants through genome-wide transcript analysis. After infestation with a virulent biotype of greenbug, substantial changes in gene expression (up- and down-regulation) in the host plants were observed based on the parallel analyses between various treatments (Park et al., 2006). This report focuses on the recent research progress in understanding of sorghum plant defense responses to greenbug aphid and other stresses at the molecular level.

Sorghum (*Sorghum bicolor*) seedlings were used to study plant defense response to attack by greenbug aphid (*Schizaphis graminum*), a phloem-feeding insect. The induction of genes associated with host defense responses to greenbug was investigated using cDNA microarrays and gene expression profiles of the greenbug challenged plants were obtained. The differentially expressed genes were validated by Northern analysis and quantitative real-time PCR (qRT-PCR), and their function was analyzed by sequence annotation against Genbank databases. The transcript profiles and their expression patterns were also compared to those induced by *Pseudomonas andropogonis*, a causing bacterial leaf stripe, and wounding.

Results and discussion

Similar to plant responses to other attackers, plants respond to phloem-feeding aphids with a complicated arsenal of defensive responses. Analysis of gene expression profiles generated from our studies indicated that the aphid induced a surprisingly complex set of transcriptional products, of which are often known as anti-aphid proteins, insecticidal metabolites, and other stress-related compounds. Gene annotation for over 100 differentially expressed cDNA clones were categorized into ten major function groups, including a large portion of genes with unknown function. The two large groups of candidate genes were for defense-related functions and signal transduction pathways.

There were some specific defense responses to greenbug challenge in sorghum plant. A cDNA coding for cyanogenic β-glucosidase was identified in the microarray experiment. This enzyme catalyzes the process of generation of hydrogen cyanide (HCN) from a cyanogenic glucoside precursor. The toxic HCN is usually associated with plant defense against herbivores. Another sorghum β-glucosidase gene (glucan endo-1,3-β-glucosidase) also responded to greenbug infestation. The gene encoding glucan endo-1,3-β-glucosidase expressed at a relatively higher level in resistant sorghum seedling tissues after exposed to greenbug. Other cDNAs coding for insect-specific defense compounds such as mannose binding lectin precursor and polyphenol oxidase (PPO) were also detected as up-regulated in the expression profile of greenbug-challenged resistant sorghum tissues. Lectins are anti-nutritive proteins that bind reversibly to specific mono- or oligosaccharides, becoming toxic plant proteins which are very effective in combating herbivores. The anti-nutritive effect of PPO has been well documented for defense against herbivores. RT-qPCR analysis showed that PPO transcripts were induced to a significant level (982-fold). This strong up-regulation of PPO in the resistant sorghum plants further supports the defensive role of this protein against greenbug aphid.

In addition, in response to attack by greenbug aphids, sorghum plants expressed a subset of plant defensive compounds which are typically induced by pathogen attack. For example, the expression level of β-1,3-glucanase (BGL) increased 3.22 folds, and the class III chitinase had 3.44-fold increase in sorghum seedling tissues 72 h after treatment with greenbug. Another cDNA coding for a pathogenesis-related protein (PR-5) was also identified as an up-regulated gene in greenbug-infested seedlings. Detection of disease-responsive genes in greenbug-challenged tissues suggests their functions in defense response against greenbug infestation as well.

Comparison of plant transcriptional responses to the greenbug and a bacterial pathogen, *Pseudomonas andropongonis*, provided further evidence for a coordinated regulation of plant defense pathways in the host. Several signal molecules and regulatory proteins were identified in the transcript profiles, including lipoxygenase (LOX) and mitogen-activated protein kinases (MAPK). These data suggest that the regulation of crop plant's response to greenbug aphid involved multiple signaling systems, including salicylic acid (SA), jasmonic acid (JA) and ethylene (ET) signal transduction pathways, and the crop plants responded to the two types of attackers, bacterial pathogen and greenbug aphid, with distinctive and partial overlapping defense products involving antimicrobial or insecticidal activities.

References

Park S-J, Huang Y, Ayoubi P (2006) Identification of expression profiles of sorghum genes in response to greenbug phloem-feeding using cDNA subtraction and microarray analysis. Planta 223: 932–947.

Thompson GA, Goggin FL (2006) Transcriptomics and functional genomics of plant defence induction by phloem-feeding insects. J Exp Bot 57: 755–766.

Metabolic Engineering

Towards Metabolic Engineering of Carotenoid Content in Sweet-orange (*Citrus sinensis* (L.) Osb.)

Marcio G.C. Costa[1,*], Amanda F.S. Mendes[1], Luciana C. Cidade[1], Walter S. Soares-Filho[2], Wagner C. Otoni[3], and Gloria A. Moore[4]

[1]Biological Sciences Department, State University of Santa Cruz, Ilhéus, Bahia 45662-000, Brazil.
[2]Embrapa Cassava & Tropical Fruit, Cruz das Almas, Bahia 44380-000, Brazil.
[3]Plant Biology Department, Federal University of Vicosa, Vicosa, Minas Gerais 36570-000, Brazil.
[4]Horticultural Sciences Department, University of Florida, Gainesville, FL 32611, USA.
[*]Email: mcosta@labbi.uesc.br

Plants have a remarkable array of biosynthetic pathways, of which many synthesize products of secondary metabolism with biotechnological and health-related importance. One of these, known as carotenoid biosynthetic pathway, leads to biosynthesis of health-promoting compounds (i.e., beta-carotene, lycopene, lutein, zeaxanthin) which reduce the onset of chronic disease states, such as cancer, macular degeneration and cardiovascular disease. In addition, all the carotenoid species that contain a beta-ring can be converted to retinol and, thus, are precursors of vitamin A. These properties have led to intensive efforts to alter and elevate such isoprenoids in food crops (Taylor and Ramsay, 2005).

Citrus is the main fruit crop produced and consumed in the world. It contains the largest number of carotenoids found in any fruit (Franciullino et al., 2006), making citrus fruit an important target for

Z. Xu et al. (eds.), Biotechnology and Sustainable Agriculture 2006 and Beyond, 223–225.

increasing carotenoids with health-promoting properties. Carotenoid content and composition vary greatly among citrus species and growing conditions. Cryptoxanthin, violaxanthin, and antheraxanthin and their esters are the primary xanthophylls present in most citrus fruits; the hydrocarbon carotenoids, such as phytoene and lycopene, are usually present in low concentrations (Gross, 1976). To enhance nutritionally important carotenoids in citrus plants, the use of gene transfer technologies is essential since improvement of citrus plants by traditional breeding has been hampered by its reproductive biology and prolonged juvenility (Gmitter et al., 1992).

In this work, we developed an optimized protocol for *Agrobacterium*-mediated transformation and regeneration of sweet-orange (*C. sinensis* (L.) Osb.), aiming to produce transgenic plants with higher content of provita min A and other health-promoting carotenoids. More than 30 experiments were set up, which allowed evaluating the *in vitro* responses according to sweet-orange variety, explant source, seedling age, and BAP, kanamycin, and hygromycin concentrations on selective regeneration medium. The data indicated that there were genotypic differences in *in vitro* morphogenesis among sweet-oranges, with 'Midsweet' and 'Pera' showing the best morphogenic responses. As expected, mature stem segments presented higher level of contamination and lower regeneration frequencies as compared to epicotyl explants. Although BAP stimulated bud formation in a range of concentrations from 0 to 2.0 mg L^{-1} in epicotyl explants, the higher number of shoot per explant was achieved only in 1.0 mg L^{-1} BAP, irrespective of the genotype. As the seedling age, the general trend observed was a lower number of regenerating shoots as the age of the seedlings increased. Genotypic differences among sweet-oranges were also observed as to sensitivity of epicotyl explants to the antibiotic kanamycin, with 'Pera' exhibiting greater tolerance as compared to others sweet-oranges varieties. As a result, hygromycin at concentration of 10 mg L^{-1} can be considered a more efficient selective agent, since such noticeable differences in sensitivity were not observed among sweet-oranges. We have used these optimized conditions to transform epicotyl explants of 'Pera' and 'Pineapple' with the carotenoid biosynthesis genes *pds* and *lcy-b* from *Citrus paradisi* (Costa et al., 2002). Several shoots were initially confirmed to be GUS+ shoots, expressing the gene *uid*A. Once these potential transgenic plants are rooted and more developed, further analyses will be performed aiming to confirm their transgenic nature as well as to characterize the types and levels of carotenoids produced by each transgenic line.

References

Costa MGC, Otoni WC, Moore GA (2002) An evaluation of factors affecting the efficiency of *Agrobacterium*-mediated transformation of *Citrus paradisi* (Macf.) and production of transgenic plants containing carotenoid biosynthetic genes. Plant Cell Rep 21: 365–373.

Franciullino A-E, Dhuique-Mayer C, Luro F, Casanova J, Morillon R, Ollitrault P (2006) Carotenoid diversity in cultivated citrus is highly influenced by genetic factors. J Agric Food Chem 54: 4397–4406.

Gmitter Jr FG, Grosser JW, Moore GA (1992) Citrus. In: Hammerschlag EF, Litz RE (ed) Biotechnology of Perennial Fruit Crops. CAB International, Cambridge, pp. 335–369.

Gross, J (1976) Carotenoid pigments in citrus. Citrus Sci Technol 1: 302–354.

Taylor M, Ramsay G (2005) Carotenoid biosynthesis in plant storage organs: recent advances and prospects for improving plant food quality. Physiol Plant 124: 143–151.

Engineering New Crops for Safe Castor Oil Production

Grace Q. Chen*, Yeh-Jin Ahn, and Louisa Vang

U.S. Department of Agriculture, Agricultural Research Service, Western Regional Research Center, 800 Buchanan Street, Albany, CA 94710, USA.
*Email: QHGC@pw.usda.gov

Introduction

Castor plant (*Ricinus communis* L.) produces a unique seed oil with numerous industrial applications. However, castor seed contains toxin ricin and hyper-allergenic 2S albumins detrimental to castor grower and processor. Our project goal is to develop a safe source of castor oil through genetic engineering. The general approach is to generate a safe castor crop by blocking expression of the ricin and 2S albumins in seed. An alternative approach would be transgenic production of ricinoleate from temperate oilseed plants.

Morphological criteria for assessing seed developmental age

Castor seed consists of a mass of endosperm and an embryo with two thin papery cotyledons lying in the center of the endosperm. The endosperm is not absorbed by the embryo until seed germination. Castor plants produce a racemic, monoecious inflorescence, with male and female flowers blooming asynchronously, so we use pollination time as a common starting point for determining seed developmental age. We examined the external and internal morphological features

Z. Xu et al. (eds.), Biotechnology and Sustainable Agriculture 2006 and Beyond, 227–230.

of the seeds associated with specific developmental ages (Chen et al., 2004). We found that the testa color and the endosperm volume were most distinctive in determining the developmental age of castor seed. Based on changes of these two features, we divided seed development into three phases (I to III), each phase spanning approximately 20 days. During phase I of seed development, the capsule and seed developed their basic tissues and grew rapidly to almost full size at 19 days after pollination (DAP). Seeds had ivory white testa color and the endosperm tissue was at the free-nuclear stage and not yet expanded. The majority of the seed volume was filled with ground tissue of inner integument. In phase II (20–40 DAP), the endosperm underwent cellularization and differentiation, expanding and displacing the inner integument, and ultimately occupied most of the seed volume. The testa started color deposition from the caruncle end to the opposite chalazal end, covering the whole seed with uneven shades of purple and brown color. Seeds in phase III (41–61 DAP) can be distinguished by filled cellular endosperm and sclerified, pigmented testa. After 54 DAP, seed testa was mature showing a shiny mosaic color of chocolate and silver. At about 61 DAP, capsules senesced and desiccated. The method of determining seed developmental stage is critical for drawing accurate comparisons between experiments. The use of morphological markers for determining seed age reduces variability and increases reliability of comparisons among experiments and among cultivars, where maturation times may differ.

Expression of ricin and 2S albumin gene during seed development

Ricin is a potent water-soluble protein toxin and is only found in castor seed. Its biochemical activity is well characterized as a Type II ribosome-inactivating enzyme. The ricin molecule is a dimeric glycoprotein composed of a toxic A-chain and a lectin B-chain linked by disulfide bonds. The A-chain is a ribosome-inactivating enzyme which depurinates a specific adenine residue of 28S ribosomal RNA, thereby inactivating eucaryotic protein synthesis (Endo et al., 1987). Castor 2S albumin proteins were identified as the primary allergenic components based on fractionation studies (Machado et al., 1992, Thorpe et al., 1988, da Silva et al., 1996). Analyzing the sequences of the 2S albumin gene and proteins revealed that a single preprotein produces two heterodimeric 2S albumin proteins (Irwin et al., 1990).

As part of a genetic approach to eliminating the ricin and the 2S albumins from castor, we performed comparative Northern analyses between ricin and 2S albumin genes by using the same sets of seeds at specific developmental ages. For the ricin gene (Chen et al., 2005), the mRNA signal was not detectable at early stages before the cellular endosperm emergence (12 and 19 DAP), but increased significantly at the onset of cellular endosperm development (26 DAP) and remained high up to the end of endosperm maturation (54 DAP). When seeds started desiccating (61 DAP), the expression of ricin gene decreased to a trace level. By comparison, low expression levels were detected for the 2S albumin gene in seeds at 12 and 19 DAP before cellular endosperm development (Chen et al., 2004). The mRNA levels rose sharply during 26–40 DAP when the endosperm underwent cellularization and differentiation. Once the cellular endosperm tissue had reached full volume, the level of 2S albumin mRNA started decreasing (47 DAP). After 47 DAP, the mRNA was degraded and did not give a clear signal (data not shown). Our Northern analyses showed that the timing of the major expression of ricin and 2S albumin genes coincided with that of cellular endosperm differentiation. However, we observed different temporal expression patterns between ricin and 2S albumin genes. The ricin gene was up-regulated up to the last stage of the cellular endosperm development, whereas 2S albumin gene had bell-shaped expression pattern. The different temporal expression patterns between ricin and 2S albumin genes indicate distinctive regulatory mechanisms involved in their mRNA accumulation. The results provide us with critical information to develop promoters and antisense constructs that would optimize timing of transgene expression to suppress the ricin and 2S albumins in a safe castor crop.

References

Chen GQ, He X, McKeon TA (2004) 2S Albumin gene expression in castor plant (*Ricinus communis* L.). J Am Oil Chem Soc 81: 867–872.

Chen GQ, He X, McKeon TA (2005) A simple and sensitive assay for distinguishing the expression of ricin and *ricinus communis* agglutinin genes in developing castor seed (*R. communis* L.). J Am Food Chem 53: 2358–2361.

da Silva JG Jr, Machado OL, Izumi C, Padovan JC, Chait BT, Mirza UA, Greene LJ (1996) Amino acid sequence of a new 2S albumin from *Ricinus communis* which is part of a 29-kDa precursor protein. Archives Biochem. Biophys. 336: 10–18.

Endo Y, Mitsui K, Motizuki M, Tsurugi K (1987) The mechanism of action of ricin and related toxic lectins on eukaryotic ribosomes. The site and the characteristics of the modification in 28 S ribosomal RNA caused by the toxins. J Biol Chem 262: 5908–5912.

Irwin SD, Keen JN, Findlay JB, Lord JM (1990) The *Ricinus communis* 2S albumin precursor: a single preproprotein may be processed into two different heterodimeric storage proteins. Mol Gen Genet 222: 400–408.

Machado OL, Silva Jr JG (1992) An allergenic 2S storage protein from *Ricinus communis* seeds which is a part of the 2S albumin precursor predicted by c-DNA data. Brazil J Med Biol Res 25: 567–582.

Thorpe SC, Kemeny DM, Panzani RC, McGurl B, Lord M (1988) Allergy to castor bean. II. Identification of the major allergens in castor bean seeds. J Allerg Clin Immunol 82: 67–72.

Biochemical and Structural Bases for (Iso)flavonoid Biosynthetic Diversity

Chang-Jun Liu[1,*], Wang Cheng[1], Joseph P. Noel[2], and Richard A. Dixon[3]

[1]Biology Department, Brookhaven National Laboratory, Upton, NY 11973, USA.
[2]The Jack H. Skirball Center for Chemical Biology and Proteomics, Salk Institute for Biological Studies, La Jolla, CA 92037, USA.
[3]Plant Biology Division, Samuel Roberts Noble Foundation, Ardmore, OK 73402, USA.
[*]Email: cliu@bnl.gov

(Iso)flavonoids (flavonoids and isoflavonoids) represent one of the most abundant families of polyphenolics in plant kingdom, where they primarily function for flower pigmentation, UV protection, and antimicrobial phytoalexins. In addition, number of (iso)flavonoids have also been demonstrated the potential chemoprevention activities for human cancers, cardiovascular disorders and osteoporosis, constituting a promising group of nutraceuticals (Liu and Noel, 2006).

The wide range of biological activities associated with (iso)flavornoids arise from the rich diversity of their structural characteristics that encompass a variety of substitutions, including hydroxylation, glycosylation, acylation and methylation. S-adenosyl-L-methionine (SAM) dependent *O*-methylation is a typical modification reaction that introduces the active methyl unit into the (iso)flavonoid skeleton to form either biosynthetic intermediates or the active 'end' products.

Z. Xu et al. (eds.), Biotechnology and Sustainable Agriculture 2006 and Beyond, 231–234.

We are employing a combined protein crystallographic and biochemical-genetics approach to understand the details of the structure and functions of the SAM-dependent *O*-methyltransferases that underlie the diversity of (iso)flavonoid biosynthesis.

Substrate promiscuity and regiospecificity of isoflavonoids *O*-methyltransferases

Sequence analysis of model legume *M. truncatula* EST libraries using three function-known OMTs implicated in isoflavonoid biosynthesis in *legume* species (He et al., 1998; Wu et al., 1997; Akashi et al., 2003) resulted in the identification of a number of putative OMT homologs (designated as MtIOMT1-8). Phylogenetic analysis indicated that these putative MtIOMTs cluster into two distinct clades separate from known flavonoid OMTs. Protein sequence identity between OMTs from different clades is less than 50%. One clade, designated as the I7OMT clade, includes MsI7OMT homologs from alfalfa and Glycyrrhiza echinata and MtIOMT1-4. The second clade, designated as the HI4'OMT clade, includes PsHM3OMT1 and 2, HI4'OMT homologs from *G. echinata* and *Lotus japonicus*, and MtIOMT5-8. OMTs within this clade are 73–76% identical to one another at the amino acid level.

Functional analyses reveal that many of the enzymes display moderately promiscuous activities. Besides the established substrate daidzein, MsI7OMT and MtIOMT1 recognize a range of di- or trihydroxyisoflavones responsible for the specific 7-*O*-methylation. While MtIOMT2 and MtIOMT3 also act on isoflavone substrates, they exhibit dual regio-specificity for both the 4′ and 7-hydroxyl moieties of the isoflavonoid nucleus. The HI4′OMT clade homologs (MtIOMT5-8) primarily accept asymmetric (iso)flavanones and pterocarpanoids. MtIOMT5-8 are distinct in their stereoselectivity. Studies also reveal that the MtIOMT5 exhibits almost equivalent steady state kinetics for either the 4′-*O*-methylation of 2-hydroxyisoflavanone or the 3-*O*-methylation of (+)-6a-hydroxypterocarpan.

Dynamic conformational change of isoflavone 7-*O*-methyltransferase confers the substrate promiscuity

The crystal structures of the complex of MsI7OMT with SAM and 6,7,4′-trihydroxyisoflavone; 7,3′,4′-trihydroxyisoflavone; or 5,7,4′-trihydroxyisoflavone (genistein) are solved at near atomic resolution. Superimposition of MsI7OMT-substrate complexes reveals a flexible conformational change from α-helix to loop in the middle region of the ground state helix 8. This conformational change results in the globe rearrangement of the C-terminal Rossmann core related to the N-terminal dimer interface, and the subsequent repositioning of the bound isoflavone substrate in active site. Moreover, the crystals generated from the same co-crystallization setting are formed into distinct space groups and give rise to different global conformations in the holo and the ternary form structures. This observation indicates an accessible dynamic conformational motion during substrate binding, methylating and product releasing.

Structural bases for dual regiospecificity and stereoselectivity of MtHI4'OMT1 (MtIOMT5)

The crystal structures of MtHI4'OMT in complex with the IFS-catalyzed product, 2,7,4′-trihydroxyisoflavanone, and (+)-6a-hydroxymaackiain have also been solved. X-ray crystallographic analysis revealed that 2,7,4′-trihydroxyisoflavanone obtained from *in vitro* IFS reactions, and then complexed to MtHI4′OMT crystals possesses the *2S,3R* configuration. The *2S,3R* configuration of 2,7,4′-trihydroxyisoflavanone bound in the catalytic cavity resides in a conformation in which the plane of the B ring sit nearly perpendicular to the planes formed by the A and C rings. This apparently stable conformation of the 2-hydroxyisoflavanone product fits in the phenolic substrate binding cavity resulting in the proper positioning of the 4′-hydroxyl group of the B ring for transmethylation catalyzed by the histidyl general base, H268, and the nearby reactive methyl group of SAM.

In the HI4′OMT structures solved with either (+) 6a-hydroxymaackiain or (+)-pisatin bound, the plane formed by rings A and B of the pterocarpans orient nearly orthogonal to the C, D and E rings, generating a sharply bent conformation that occupies the same Y-shaped binding cleft previously noted for 2S,3R-2,7,4′-trihydroxyisoflavanone. The close structural similarity of the 6a-hydroxylated 6aR,11aR-pterocarpan, (+)-6a-hydroxymaackiain, with 2S,3R-2,7,4′-trihydroxyisoflavanone accounts for the observed dual activities of HI4'OMT toward two seemingly distinct isoflavonoid-derived substrates. Therefore, our biochemical and structural studies explain the paradoxical recognition and stereospecific turnover of chemically distinct small molecules by orthologous gene products.

References

Akashi T, Sawada Y, Shimada N, Sakurai N, Aoki T, Ayabe S (2003) cDNA cloning and biochemical characterization of S-adenosyl-L-methionine: 2,7,4'-trihydroxyisoflavanone 4'-o-methyltransferase, a critical enzyme of the legume isoflavonoid phytoalexin pathway. Plant Cell Physiol 44: 103–112.

He XZ, Reddy JT, Dixon RA (1998) Stress responses in alfalfa (*Medicago sativa* L.) XXII. cDNA cloning and characterization of an elicitor-inducible isoflavone 7-*O*-methyltransferase. Plant Molecular Biol 36: 43–54.

Liu CJ, Noel JP (2006) Flavonoids: Recent advances in molecular biology, biochemistry, pharmaceutical applications and metabolic engineering. In: Jaiwal PK (ed) Plant Genetic Engineering: Metabolic Engineering and Molecular Farming-1. Studium Press, Houston, pp. 225–259.

Wu Q, Presig CL, VanEtten HD (1997) Isolation of the cDNAs encoding (+)6a-hydroxymaackiain 3-*O*-methyltransferase, the terminal step for the synthesis of the phytoalexin pisatin in *Pisum satium*. Plant Molecular Biol 35: 551–560.

Utilization of Plant Metabolic Engineering for Production of Pharmaceuticals in Tea

A. Borthakur, R.K. Dutta, S. Borchetia, and Sudripta Das*

Department of Biotechnology, Plant Improvement Division. Tea Research Association, Tocklai, Jorhat – 785008.
*Email: sudripta2002@yahoo.co.in, sudriptadas@gmail.com

Abstract

Plant biotechnology combined with genetic engineering have contributed immensely to the development of and increase in the medicinal value of plants. The proven medicinal properties of tea include antioxidative, anti-tumourous, antiallergic, antibacterial, antiviral and antiseptic properties. The present study is on the application of molecular studies to identify and overcome rate-limiting steps in accumulation of secondary metabolites involved in the catechin and terpenoid pathways for production of choice chemicals in tea. The construction of cDNA library, identification, cloning and overexpression of genes related to the catechin and terpenoid pathways in tea are discussed. An increase in the secondary metabolite content in cell suspension cultures of tea was observed in the presence of signal molecules which culminated in the effort to understand the mechanism of regulation of the pathways. For the first time genes from the carotenoid pathway (four) and monoterpenes (two) were cloned from tea. The analysis of modified tea plants containing multiple copies of the various genes cloned would help to better understand the measurements of flux coefficients, metabolite channeling and the interactions between enzymes and other protein partners. The identification of the transcriptors and regulators related to

Z. Xu et al. (eds.), Biotechnology and Sustainable Agriculture 2006 and Beyond, 235–258.

the catechin and terpenoid pathways would help in better manipulation of these pathways for improvement of quality.

Introduction

Tea is the second most popular beverage in the world, next to plain water. Theoretically fresh young shoots of tea plants can be processed into any type of tea. However, in practice young shoots from a specific cultivar may only be suited for processing into one of the six types of tea depending on physical, chemical and biological characteristics of the young shoots. For instance, tea cultivars suitable for black tea (generally belonging to *C. sinensis* cv. *assamica*) are not considered as a good choice for green tea and oolong tea because of high polyphenol contents. The latter two types of tea are usually produced from *C. sinensis* cv. *sinensis*. For long time, it has been well established that the flavor of tea is principally determined by chemical components it contains, such as volatile compounds contributing to the property of aroma and nonvolatile compounds to the taste (Bondarovich et al., 1967; Millin et al., 1969; Nakagawa, 1975; Kubota and Hara, 1976; Nakagawa et al., 1981; Hara et al., 1995). These compounds are either direct metabolites accumulated in young shoots during growth or products of various reactions using these metabolites as substrate during processing. Thus the flavor of made tea is largely affected by abundance of chemical constituents and their relative composition in young shoots.

Chlorophyll and carotenoids are the major pigments in fresh tea shoots. Concentrations of carotenoids in tea shoots range from 0.36 to 0.73 mg g^{-1} dry weight (Ravichandran, 2002). A total of 14 species of carotenoids is identified in fresh shoots (Suzuki and Shioi, 2003). β-carotene, lutein, violaxanthine and neoxanthine are the predominant components (Sanderson et al., 1971; Hazarika and Mahanta, 1983; Ravichandran, 2002; Suzuki and Shioi, 2003). Carotenoids serve as substrate for many volatile compounds found in made tea (Sanderson and Graham, 1973). During processing, carotenoids degrade via oxidation during withering and fermentation or paralytic reaction during firing (Sanderson and Graham, 1973). In green tea a change of structure, conversion of violaxanthin to auroxanthin occurs (Suzuki and Shioi, 2003). Volatile compounds of ionine series are formed as primary oxidation products of carotenoids (Tirimanna and Wickremasinghe, 1965; Sanderson et al., 1971; Sanderson and Graham, 1973; Reynold et al., 1974; Etoh et al., 1980). These compounds smell sweet floral aroma and are believed to have a major effect on the aroma of tea (Sanderson and Graham, 1973).

Carotenoids are integral components of higher plant photosystems, and their composition in photosynthetic plant tissues (lutein, b-carotene, violaxanthin and neoxanthin, in order of abundance) has been remarkably conserved throughout evolution. These are multifunctional compounds serving as structural components of light-harvesting complexes (LHCs), accessory pigments for light harvesting, substrates for abscisic acid synthesis, components of photoprotection and scavengers of singlet oxygen. The biosynthesis of carotenoids in plants has been reviewed extensively in recent years and is only briefly described here. The committed step to carotenoid synthesis is the formation of the first C40 compound phytoene by the head-to-head condensation of two molecules of GGDP (geranylgeranyl diphosphate) by phytoene synthase. Phytoene is subjected to a series of four sequential desaturation reactions, by two separate enzymes, to yield lycopene, which has eleven conjugated double bonds. Lycopene is then cyclized to b-carotene by two b-cyclizations or to a-carotene by one b- and one e-cyclization. The two types of rings (e and b) are produced by distinct enzymatic mechanisms now known to be carried out by two structurally related enzymes, the lycopene b-cyclase and the lycopene e-cyclase. The formation of erings and the production of b,e-carotenoids (a-carotene derivatives) are two of the key differences distinguishing carotenoid biosynthesis in plants from that in cyanobacteria, fungi and bacteria. a-Carotene serves as the precursor for hydroxylation (twice) to lutein, the most abundant carotenoid in green plant tissues. While the carotenoid pathway has been known for several decades, the fact that most enzymes of the pathway are membrane-associated proteins that rapidly lose activity during isolation has hindered attempts at enzyme purification and study. The breakthrough in our understanding of the pathway and enzymes in recent years can be attributed largely to molecular and genetic approaches, which have together allowed the majority of plant carotenoid biosynthetic enzymes to be cloned.

These accomplishments have provided a wealth of information about the encoded proteins and the tools necessary to begin addressing many fundamental biochemical and molecular questions of the pathway. To this end, several studies have demonstrated that messenger RNA (mRNA) levels for early steps of the pathway are strongly upregulated in tissues such as flowers and fruits and to a much lesser extent during leaf development, which holds true in tea. The availability of complementary DNAs (cDNAs) has also greatly aided biochemical studies which can now utilize large quantities of highly purified protein overexpressed in heterologous systems, rather than relying on partially purified, low specific activity natural sources.

Over 600 different carotenoids have been identified from natural sources, although the biological diversity and synthetic capability have not been exploited to develop systems for the production of carotenoids. Currently only P-carotene and astaxanthin are produced from biological sources and sold widely to the food and feed industries. The key limitation to the use of biological systems to produce carotenoids is the low productivity of these systems compared to chemical syntheses. One way to increase the productivity of carotenoid synthesis is to use recombinant DNA technology. However, before recombinant DNA technology and genetic engineering can be used to increase the productivity, genes encoding the carotenoid biosynthetic enzymes need to be isolated and characterized. Until very recently, genes for the carotenoid biosynthetic enzymes had not been isolated. This situation has changed markedly over the past 3 years. Currently a number of the genes coding for carotenoid biosynthetic enzymes have been isolated from bacteria, fungi, cyanobacteria, green algae, and higher plants. With the large number of genes now available for further study, systems can be tested to increase the production of carotenoids from biological sources. Results from these studies will fuel the development of processes for the production of carotenoids from a number of different host organisms and the full range of carotenoid compounds from biological sources will be accessible.

Tea plant breeding has been focusing over the years mainly on increasing crop productivity by pursuing higher yields and better adaptation to biotic and abiotic stresses. However there are still vast populations, mainly in the developing world, whose diet lacks essential nutrients, such as vitamins and minerals. For example, it was estimated that over 124 million children worldwide are vitamin A deficient, and that improved vitamin A nutrition alone could prevent 1.3–2.5 million deaths among late infancy and preschool-age children that occur each year in the developing countries. Modern breeding tools of gene transformation open up the possibility for metabolic engineering targeted at improving nutritional quality of plants. To manipulate complex metabolisms it is necessary to unravel the biochemical pathways and understand the regulation underlying metabolic fluxes. This paper examines the progress made in understanding the biosynthesis and regulation of catechins and carotenoids and describes attempts to manipulate these pathways. So far no report on integrated experimental or applied studies on the terpenoid biosynthetic pathway in tea is encountered. This study may pave the way to application of these biotechnological tools on production of choice nutraceuticals in tea.

Materials and methods

Plant materials

Cell cultures were established from S3/A.3 (garden selection) clone and selected cell line HS26 was used in this study. Friable calli were obtained from immature zygotic embryo cultures and were transferred to 250 ml Erlenmeyer flasks containing 100 ml N/5 media supplemented with 2,4-D and kinetin, and cultured on a rotary shaker (110 rpm) at 24°C either under continuous light or in the dark. The pH of the medium was adjusted to 5.7 before autoclaving. The cell lines were treated with methyl jasmonate and ibuprofen after 28 days of culture, to overexpress and to selectively block production of catechins in culture, for construction of cDNA library and for cloning of the genes and regulators involved in the pathway.

Cell suspension lines were also established from TV1 (TRA released variety) derived filament and young leaf calli. Such calli lines, although slow in growth, were treated with norflurazon, for down regulation and inhibition of the pathway for synthesis of lutein, downstream of the carotenoid pathway. Normal lines were harvested after 21 days of culture while the treated lines were harvested 48 h after treatment, for mRNA isolation and library construction.

Construction of a cDNA library

Poly(A)+ RNA extracted from the various cell lines, treated and untreated, were used as a substrate for the synthesis of a unidirectional cDNA expression library in Uni-ZAP XR (Stratagene, San Diego, Calif.). The library was excised to produce pBluescript II SK(–) containing cDNA inserts ranging over 0.8–1.5 kb and cloned into restriction sites EcoRI and XhoI (Sambrook et al., 1989).

RNA gel blot analysis

Total RNA (16 μg per sample) was separated on a 1% agarose gel containing 17% formamide and transferred onto a Hybond-XL nylon membrane (Amersham Pharmacia Biotech). DNA probes were generated by PCR and labeled using a Rediprime II Kit (Amersham Pharmacia Biotech). Standard procedures were followed for prehybridization, hybridization, and washing.

Bacterial strains and plasmids

E. coli strain TOP10 F' (Stratagene, San Diego, CA) was used for screening tea cDNAs encoding products that affect the carotenoid content of bacteria engineered for carotenoid biosynthesis. Plasmid pACCRT-EIB (courtesy of Prof. E. Wurtzel, CUNY) conferred lycopene accumulation in E. coli; and carotenoid identities were confirmed by high performance liquid chromatography (HPLC) analysis (data not shown). Although we used two gene cassettes for screening, the carotenoid pathway end product was unimportant and functioned only as a reporter of pathway flux.

Identification of cDNAs

The tea filament derived cDNA library was used to transfect E. coli TOP10 F' cells containing pACCRT-EIB. The transformants, selected by the addition of appropriate antibiotics, were plated at a density of about 100 colony-forming units/plate (100-mm Petri dishes) and incubated overnight at 37°C, followed by 5 days incubation at room temperature to screen for colonies having relatively increased or decreased pigmentation. Colony images were collected by scanning on a Hewlett Packard Scan Jet 3700 scanner with HP Scan software and were processed with Adobe Photoshop 7.

PCR primers for amplifying tea CHS (GenBank D 26595) were primer number CasM 25F, 5'-ATGGTSACHGTSGAGGAHRTBHGG-3' and primer number CasM25R, 5'-AGCACCACAGTCTCAACRGTGA-3'; and for amplification of tea DFR (GenBank AB018686), were primer number CasM21F, 5'-TCTCCTACACTTGTGGTCGG-3' and primer number CasM21R, 5'-CTCTTTACCGGGCTCATGTT-3'. The PCR conditions for CHS were one cycle at 94°C (3 min); 35 cycles at 94°C (30 s), 54°C (30 s), 72°C (1 min); and one cycle at 72°C (10 min). The PCR conditions for DFR were one cycle at 94°C (3 min); 40 cycles at 94°C (30 s), 60°C (30 s), 72°C (30 s); and one cycle at 72°C (10 min).

A 521-bp Lycopene ε-cyclase fragment was amplified using primers designed from sequences in the Genebank database (Accession numbers AF486650, AF450280). The primers used were CasM30F 5'-AGTGT ATAGTAAACCAACTAAAGATGG-3' and CasM30R 5'-GCTTGCTA CTGTTGCATCTG-3'. The DNA was denatured at 94°C for 45 s, annealed at 55°C for 1 min and extended at 68°C for 2.5 min. This cycle was repeated 36 times. For *Psy*, a 464 bp fragment was amplified with primers CasM37F 3'-TGCCATGTGGTTCACTGTTT-5' and CasM37R 3'-GAGCGCGTGAGAGAAAAATC-5'. For *Pds*, a 543 bp fragment, amplified with primers CasM2F 3'-ACCGACGAGG TGTTTATTGC-5' and CasM2R 3'-CAGCATTGACCGGTTAGGAT-5', was used as the probe. The PCR cycle was the same used for lycopene ε -cyclase, but the annealing temperature was changed to 60°C and the extension time was changed to 2 min. For probe generation, amplification was done using *Citrus* mRNA for *Lcy-e* and *Arabidopsis* mRNA for *Psy* and *Pds*.

DNA sequencing and analysis

DNA sequences were determined by primer walking for both strands, using the automated DNA Sequencing Facility at the University of Calcutta, Cell and Chromosome Research. Sequence analysis was performed using Vector NTI Suite, ver 5.5 (InforMax, North Bethesda, MD), and BLAST 2.1 (Altschul et al., 1997).

Results and discussion

Production of catechins

The catechin biosynthetic pathway has been one of the most intensively studied metabolic systems in tea. As with any good model, each new piece of information appears to raise a number of unanticipated and intriguing questions. At the same time, new tools are providing the opportunity to consider flavonoid biosynthesis, of which catechins are a part, not as an assemblage of independent components, but as part of a large, complex, and tightly orchestrated metabolic network. The production of catechins and caffeine in cell culture systems of tea have already been reported by various authors including this group (Borthakur et al., 2006).

Effect of elicitors and inhibitors

Production of total catechins by cell cultures under treatment with meJA and ibuprofen were expectedly different. The total catechin content doubled in presence of meJA after 24 h treatment, suggesting that under stress, the catechin biosynthetic genes are overexpressed in tea cultures, leading to production of more catechins (Figure 1a). But when cell cultures were harvested after 72 h treatment or more, most of the catechins are released into the media.

When the media was replaced with fresh one, the cell mass could be revived and cultured again, suggesting that the individual cells were not chronically stressed and it was an adaptive mechanism by which cultured cells released the overproduced metabolites. Also, with removal of catechins from the stretched vacuoles within cells, there was feedback or auto shutdown mechanism for the pathway genes.

Isolation and characterization of catechin genes

DFR is a key enzyme involved in the synthesis of CTs, and the isolation of a full-length DFR cDNA would provide an important tool for analyzing induction of CT synthesis at the molecular level. A cDNA library constructed from meJA treated cell lines was screened with a PCR-generated DFR probe, and nine positive clones were excised and sequenced. Five identical clones had high sequence similarity to DFR genes from other reports. One cDNA clone, DFR22, was completely sequenced on both strands, shown to encode a full-length DFR, and was re-named *CasDFR*. The nucleotide sequence of *CasDFR* is 1413-bp long and predicted to encode a protein of 346 amino acid residues with a molecular mass of 38.8 kDa (details not shown).

In the present experiment, in addition to the numerous clones screened, eight hybridizing to CHS cDNA probes were isolated. According to the classification based on the amplification of a 470-bp fragment from a conserved region, the clones were deduced to represent four different sequences. Two novel genes, *CasS18* and *CasS26*, having a continuous reading frame (except for an intron) showed similarity to the CHS gene. The exon-intron boundaries as well as the start and stop codons of the reading frames were deduced based on the general similarity of CHS enzymes at the amino acid sequence level and by comparison to the CHS cDNA sequence. *CasCHS* is a truncated clone missing approximately the first 40 codons. It has a 28-bp exon, the obtained sequence being of 1130-bp, similar to the clone reported by Park et al. (2001).

In the present experiment, one regulatory gene (clone *Castt2*) over-expressed, along with the catechin synthesis genes, which was identified by subtractive hybridization. This gene had a sequence (partial sequence deduced upto 841 bp, work incomplete) similar to the *tt2* transcription factor of *Arabidopsis* with 89% sequence homology (data not shown). This suggests that the genes involved in catechin biosynthesis are coordinately regulated at the transcription level. Furthermore, the transcription factor controlling expression of the catechin biosynthetic genes seems to be impaired or repressed in inhibited lines (ibuprofen treated lines). Our conclusion is based on the similar genomic organization of the structural genes in the two cell lines which were used for subtractive hybridization. The sequence of *Castt2* would be deposited at NCBI for registration.

Genetic and molecular study of flavonoid biosynthesis in *Arabidopsis* has revealed that the CHS, CHI, F3H, and F3H enzymes are encoded by the *TRANSPARENT TESTA 4* (*TT4*), *TT5*, *TT6*, and *TT7* genes, respectively (Shirley et al., 1992, 1995; Wisman et al., 1998; Schoenbohm et al., 2000). Then, an NADPH-dependent dihydroflavonol reductase (DFR), encoded by the *TT3* gene in *Arabidopsis* (Shirley et al., 1992), leads to the production of flavan-3,4- diols (leucoanthocyanidins), which are the last common intermediates in anthocyanin and proanthocyanidin biosynthesis. And it was demonstrated previously that functional TTG1 (TRANSPARENT TESTA GLABROUS1), TT8, and TT2 proteins are required for the normal expression of at least two flavonoid LBGs, *DFR* and *BAN*, during seed formation (Nesi et al., 2000).

Quantitative variations in transcription factor mRNA, achieved by suppression and over-expression experiments, cause substantial changes in plants. Therefore, accurate regulation of transcription factor genes by their cis- and trans-acting elements is potentially very important and several examples of this are known.

The importance of DFR in aspen defense was confirmed by the co-ordinate induction of other defense proteins, PPO and trypsin inhibitor (Haruta et al., 2001). The overexpression of DFR and CHS along with transcription factor *Castt2* points out their importance in defense mechanism of tea. This underscores the importance of studying plant defense in a variety of experimental systems. It will be interesting to compare defense strategies in other tea varieties, and to explore the molecular basis for the variation in herbivore resistance seen in natural populations in the field.

These recent findings, together with evidence for channeling between PAL and C4H in the general phenylpropanoid pathway (Czichi and Kindl, 1977; Hrazdina and Wagner, 1985; Rasmussen and Dixon, 1999), indicate that the organization of these systems may become an important consideration in understanding how tea metabolism is regulated. Although the flavonoid pathway is likely to be an example of a dynamic, rather than a stable, enzyme complex, it is also possible that this organization could complicate metabolic engineering by limiting the access of intermediates to introduced enzymes (Dixon and Steele, 1999), while at the same time providing opportunities for the efficient redirection of flux into existing or introduced branch pathways. A great deal of work is clearly still needed in this area, including efforts to define protein interaction domains, to determine whether tea cells contain single branched complexes or a variety of complexes dedicated to the production of specific products, and to characterize changes in the organization of enzyme systems in response to organizational, developmental and environmental cues.

Production of carotenoids

Rational metabolic engineering of the specific carotenoid content requires not only isolation and ectopic expression of genes for structural enzymes within the pathway, but also an understanding of many factors affecting carotenogenesis.

Carotenoid formation is a highly regulated process. Concentration and composition of leaf xanthophylls in chloroplasts are affected by light intensity and the accumulation of specific carotenoids in chromoplasts of fruits and flowers is developmentally regulated. Cell lines established from filament and young leaf callus obtained from TRA released clone TV1 were utilized for estimation of carotenoids, specifically lutein and β-carotene. It has been established that expression of the genes *Ggps*, *Psy*, and *Pds* is upregulated during fruit development just prior to color appearance (reviewed in Cunningham and Gantt 1998; Harker and Hirschberg 1998), whereas the expression of the genes for the two lycopene cyclases, *Lcy-b* and

Lcy-e, is abolished at this stage (Pecker et al., 1996; Ronen et al., 1999). That is the rationale for choosing filament calli lines for cloning *Psy* and *Pds* and for young leaf derived calli for cloning and identification of *Lcy-e* in our work.

Constitutive expressions of phytoene synthases in tomato (*Lycopersicon esculentum*) and tobacco (*Nicotiana tabacum*) resulted in dwarfism, chlorosis, and differential coloring of the plants (Fray et al., 1995; Busch et al., 2002). In the tomato plants, the levels of GA were decreased, and in some instances, a small decrease of the levels of abscisic acid (ABA) was also found (Fray et al., 1995). By using tissue-specific promoters for expression of phytoene synthases, many of the problems associated with expression in vegetative parts of the plant that causes dwarfism and chlorosis can be circumvented. For example, expression of a bacterial phytoene synthase in tomato in a fruit-specific manner resulted in elevated levels of phytoene, lycopene, lutein, and β-carotene (Fraser et al., 2002). In canola (*Brassica napus*) seed-specific expression of a bacterial phytoene synthase resulted in a 50-fold increase of the total carotenoid content. The main increase was due to elevated levels of ά- and β-carotene, whereas the level of lutein remained unaltered, and it was noticed that the seed germinated 1–2 d later than the control plant (Shewmaker et al., 1999).

a

Figure 1. Production of secondary metabolites in tea cell cultures (**a**) and production of carotenes in heterologous systems after gene transfer (**b**).

Primers were designed according to the Arabidopsis EST sequence (GenBank AF009954), a 464 bp PCR product from Arabidopsis leaf mRNA was amplified and was used to probe the tea filament derived cell line cDNA library. Nine positive clones were identified. Each clone hybridized to tea genomic DNA digested with HindIII, EcoRI or BamHI with the same restriction fragment pattern (data not shown). The longest, 2265 bp, clone was subcloned into pBluescript II SK (±), designated as pCasPsy48, and used for further sequencing and functional analysis. This tea *Psy* cDNA contains untranslated 5'- and 3'- regions of 185 and 250 bp, respectively, and an open reading frame predicted to encode a 570 amino acid, 63.1 kDa protein. There are no stop codons upstream of the first ATG, and based on sequence comparisons with other *Psy* sequences, it is unlikely that there is another upstream methionine codon. Similarly, the gene and protein of *Arabidopsis Pds* was manually deduced from GenBank AY057669. The deduced amino acid sequences of *Arabidopsi*s and tea *Pds* proteins were compared with other selected carotene desaturase amino acid sequences (ongoing). Based on amino acid and nucleotide primary structure homology, the tea *Pds* shows higher homology to the sequence of the bacterial type (CRTI-) rather than the plant type gene (nucleotide comparison not shown). *Pds* is also distantly related to amino oxidases and to the carotenoid isomerase CRTISO. A phenogram of alignment shows that maize and rice PDS proteins and maize and rice ZDS proteins are highly homologous to each other, while the ZDS for the monocotyledonous plant daffodil (*Narcissus*) is no more closely related to the grass carotenoid desaturases than are ZDS proteins from the selected dicotyledonous plants *Arabidopsis* and *Citrus*. Interestingly, CRTISO is no more related to PDS or ZDS than CRTI: CRTISO is an anciently diverged member of the gene family. The similar enzymatic function of ZDS and PDS was reflected in the sequence similarity of the mature proteins, which showed an amino acid identity of about 32% and a similarity of 50%. A putative transit sequence for plastid targeting was predicted to be minimally the N-terminal residues 1–30 by comparing the tea *Pds* sequence with that of the cyanobacterium *Synechocystis* (GenBank D90914) and by concordance to the predicted transit peptides of the monocotyledonous daffodil, and bell pepper (Albrecht et al., 1995) and *Arabidopsis* (Scolnik and Bartley, 1995), which are 61, 49 and 59 residues, respectively. The computer algorithm ChoroP 1.1 (Emanuelsson et al., 1999) is being used at present to predict the transit peptide cleavage site.

Expression of *Pds* in heterologous system

E. coli cells were transformed with pACCRT-EBP, which carries genes coding for GGPPS (GGPP synthase), Psy from *Erwinia uredovora*, named crtE and crtB, respectively; and the tea *CasPds* gene. Together these three genes conferred accumulation of z-carotene in the *E. coli* host cells under chloramphenicol selection. Next, the *Lcy-e* cDNA was inserted in to produce an in-frame fusion with lacZ. To confirm the enzyme activity or loss of activity coded by the c*rtl-e* cDNA, cells carrying pACCRT-EBP were transformed with pTPds-107 and the accumulation of specific carotenoids was determined (Figure 1b). This system was designed not only to demonstrate the function of the novel tea *Pds* cDNA, but also to examine the products downstream such as lutein and β-carotene, with tea *Pds* and *Lcy-e* acting together (work incomplete).

In vitro inhibition of PDS

The effect of norflurazon on tea *Pds* could be demonstrated *in vitro* after expression in *E. coli*. There was a concentration-dependent inhibition of *ú*-carotene conversion to lycopene. In addition, this reaction was also inhibited by J852 (courtesy of Dr. Michael Kuntz, CNRS), which is a very sensitive inhibitor of PDS. Norflurazon is a very potent inhibitor of PDS with an *I*50 value of 0.12 *i*M.

Recently, much progress has been made towards elucidating these factors, which include: (1) rate-control of flow through isoprenoid precursors pools (Matthews and Wurtzel, 2000; Gallagher et al., 2003), (2) ancillary co-factors and ancillary enzyme activities controlling carotenoid accumulation (Norris et al., 1998; Li et al., 2001), (3) the topology and assembly of the membrane-bound carotenoid complex (Cunningham and Gantt, 1998), (4) the substrate stereospecificities of enzymes (Breitenbach et al., 1999), and (5) requirements for the accumulation and sequestration of products (Cervantes-Cervantes et al., 1990; Li et al., 2001). These factors may be specific to species and plastid types.

Tissue-specific, developmental, and rhythmic gene expression

To determine if the monoterpene synthase gene/s are expressed with spatial and temporal patterns correlated with floral scent emission, we analyzed monoterpene synthase gene expression in different floral tissues. Total RNA was isolated from leaves and different floral parts of three-day-old flowers (sepals, pistils, stamens, and different regions of the corolla: the upper petal lobes, the lower petal lobes, and the tube) and used in RNA gel blot analysis. A single monoterpene synthase probe was used initially for all RNA gel blot analyses. The highest level of monoterpene synthase gene expression was found in the upper and lower lobes of petals, the parts of the flower that were shown previously to be primarily responsible for scent production and emission. A very low level of transcripts was detected in the tube and stamens. No detectable signals were found in pistils, sepals, and leaf tissues. Because the signals detected in this RNA gel blot experiment could represent a sum of transcripts, the contribution of each gene to the total expression level was analyzed further by reverse transcriptase (RT)-PCR using gene-specific primers. When total RNA isolated from upper and lower petal lobes was used in RT-PCR, different levels of expression were found. Gene expression was stable during flower development and during a light/dark cycle.

To examine the steady state levels of linalool synthase mRNA in upper and lower petal lobes during flower development, total RNA was extracted from the petal lobes of mature flower buds at 1 day before anthesis and from open flowers at 1–12 days after anthesis. A plot showing the changes in the relative amounts of transcripts for linalool synthases during the life span of the flower, normalized to equal amounts of 18S rRNA, was obtained. Linalool synthase mRNA was detected first in mature flower buds, and its level increased until it peaked on day 4. Over the next 3 days (days 5–7), mRNA levels declined sharply by 40% and decreased slowly thereafter. Next, we determined whether the oscillations in linalool/geraniol emissions during a light/dark cycle were the function of rhythmic regulation of the corresponding gene expression. Accumulation of mRNA was examined in upper and lower petal lobes at nine time points during a 27-h interval. The mRNA levels were quantified and normalized for any loading differences, and the relative values were noted. We found only slight variations in linalool synthase mRNA levels during the daily light/dark cycle. Levels of transcripts for *LIS* increased during the light period, and their maximum accumulation was detected at 3 PM; they then declined at least twofold during the dark period (12 AM), showing a weak diurnal oscillation pattern. The rhythmic pattern of mRNA accumulation was retained under continuous dark, suggesting that the cyclic expression of *LIS* is under circadian control. Although changes in transcript levels may not directly determine protein levels or enzyme activities due to possible posttranscriptional, post-translational, or enzyme-regulatory mechanisms, the positive correlation between transcript levels and volatile emission suggests that changes in transcript level are an important determinant of aroma production.

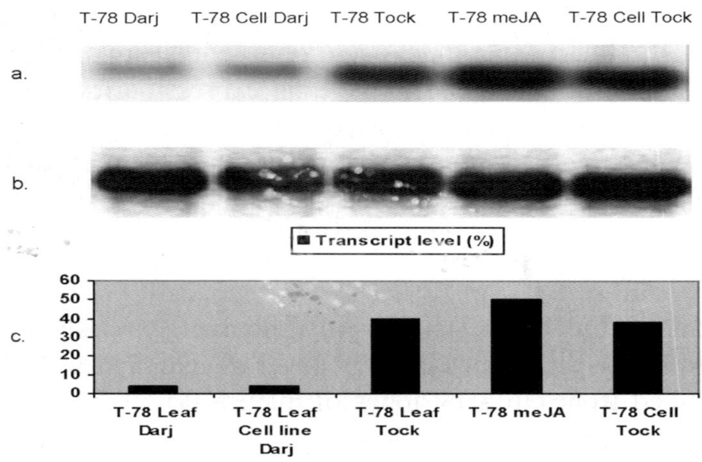

Figure 2. Northern blot analysis of dihydroflavonol reductase (*CasDFR*) mRNA from fresh leaves of T-78 and T-378 collected from Darjeeling and from Tocklai. **a**. Hybridisation with dihydroflavonol reductase transcript (*CasDFR*). **b**. The same blot hybridized with ribosomal DNA (25S). **c**. Quantified expression levels using densitometry (standardized with ribosomal DNA).

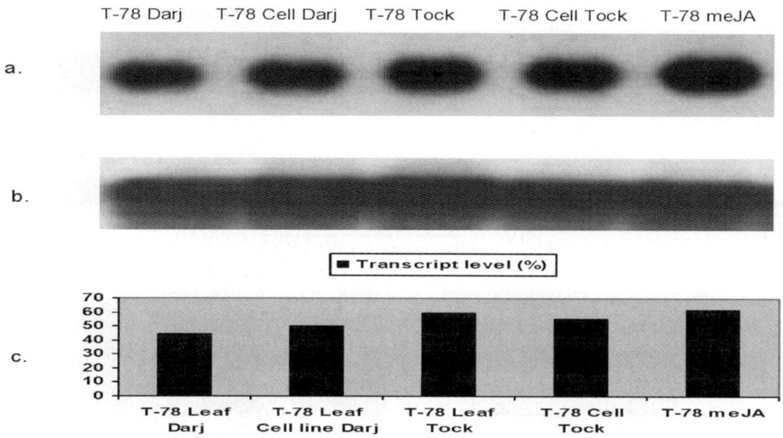

Figure 3. Northern blot analysis of phytoene synthase (*CasPsy*) mRNA from filaments of flowers of T-78 and T-378 collected from Darjeeling and from Tocklai. **a**. Hybridisation with phytoene synthase transcript (*CasPsy*). **b**. The same blot hybridized with ribosomal DNA (25S). **c**. Quantified expression levels using densitometry (standardized with ribosomal DNA).

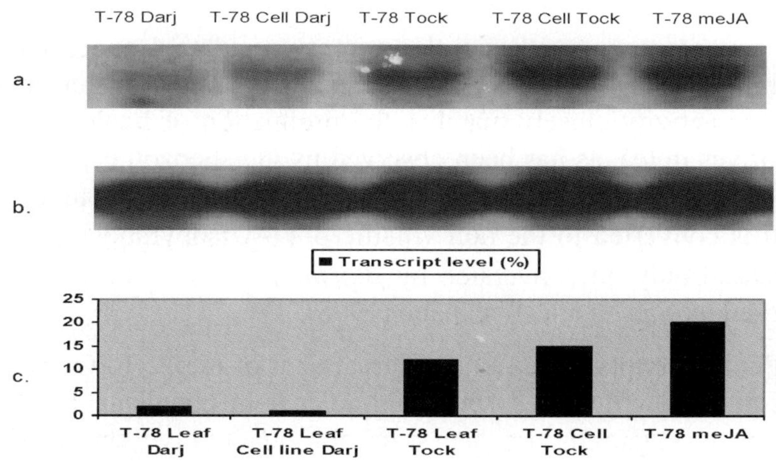

Figure 4. Northern blot analysis of polyphenol oxidase (*CasPPO*) mRNA from fresh leaves of T-78 and T-378 collected from Darjeeling and from Tocklai. **a**. Hybridisation with polyphenol oxidase transcript (*CasPPO*). **b**. The same blot hybridized with ribosomal DNA (25S). **c**. Quantified expression levels using densitometry (standardized with ribosomal DNA).

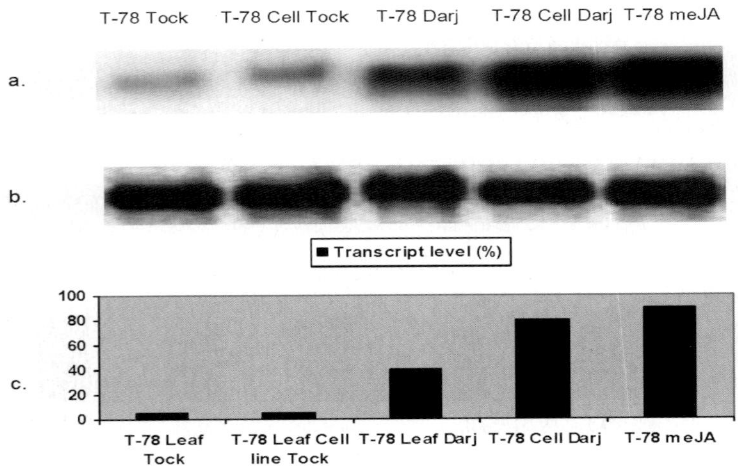

Figure 5. Northern blot analysis of linalool synthase (*CasLIS*) mRNA from flower petals of T-78 and T-378 collected from Darjeeling and from Tocklai. **a**. Hybridisation with linalool synthase transcript (*CasLIS*). **b**. The same blot hybridized with ribosomal DNA (25S). **c**. Quantified expression levels using densitometry (standardized with ribosomal DNA).

In monoterpene synthesis, the primary regulation appears to reside in the first step of the pathway (results not shown). But linalool synthase activity closely matches the production of linalool and its derivatives in tea, as has been observed by Gershenzon et al. in mint (1989). However, with expression of the genes, the linalool produced is converted to the non volatile β–D-glucopyranoside, which is released only on conversion by *β-primeverosidase* in tea, as was observed by Sakata and Watanabe (1998).

PPO transcripts increased with treatment of meJA. PPO is an enzyme putatively involved with plant defense mechanisms. However, further investigations are necessary for quantitative estimation of PPO supported by biochemical data. The results of *Psy* gene expression suggests that the carotenoid pathway genes are not stress regulated in tea and act independently, regardless of the signal transduction regulation mechanism (Figures 2 through 5).

In this study, when the individual cell lines were treated with meJA (details in Das et al., 2005), the increase in transcript level was apparent, indicating that the perceived signal of stress, which induces reduction in transcript level, due to induced drought or low temperature, is overcome by meJA. Many jasmonate responsive genes (JRG's) are known to be involved in stress responses, but only few of them have been found to be related to other physiological events such as flower and fruit development. In this study, since meJA induced the JRG's, it can be inferred that genes involved in physiological events such as flower development (as some of the treated cell lines are from floral parts) are found to respond to meJA, in addition to those related to stress responses such as the DFR from the catechin biosynthetic pathway and PPO. One exception was the effect of stress or meJA on the transcript level of *Psy*, involved in the carotenoid synthesis in tea. Hardly any variation in transcript was observed (although the method of densitometry is too crude), which would be verified by RT-PCR. JRG's are also implicated in phytohormone signaling, disease and wound response, amino acid metabolism and senescence, besides signal transduction (direct involvement) and secondary metabolism (as observed by our group, Das et al., 2005). In *Arabidopsis*, 41 JRG's have been detected by macroarray, including 6 genes of unknown function. Probably, a similar scenario exists in tea for jasmonate induced signal transduction.

Salicylic acid, which is also assumed to be a signal molecule that induces the synthesis of pathogenesis related proteins, show antagonistic effect on jasmonates, as reported in tobacco. But in tea, we have not been able to observe such antagonism in cell cultures (Das et al., 2005). Probably, salicylate acts synergistically in tea, particularly in case of production of catechins in cell cultures or may not play any significant role (Das et al., 2005; Borthakur et al., 2005).

Since the elucidation of the carotenogenic pathway in plants, there has been a steady increase in understanding the complexities of regulation of the pathway, especially in non-photosynthetic tissues such as tomato fruit. However, far more information is needed in complex systems like tea, before these control mechanisms can be fully understood. The biochemical analysis of transgenic lines in tea containing additional carotenoid genes should be a major part of these endeavors, especially measurements of flux coefficients, metabolite channeling and the interactions between carotenogenic enzymes and other protein partners.

Within plant cells, some genes are expressed constitutively, whereas others respond to specific stimuli. Both patterns depend on the interaction of transcription factors with cis-acting elements and/or with other transcription factors required for gene expression, and they are important in the regulation of cell activities. Therefore, alteration in the expression of transcription factor genes normally results in dramatic changes to a plant and structural changes to these genes may represent a significant evolutionary force (Li et al., 1996). As a practical consequence, engineering of transcription factor genes provides a valuable means for manipulation of plants, but success in such endeavors depends on how well the genes are understood.

One strategy would be to overcome the environmental stress effect by over-expressing genes which induce tolerance/resistance to such stresses. The other option which seems to be more viable, is to go for identification of the jasmonate pathway genes and a strategy for coordinate expression along with the specific metabolic pathway genes.

The ability to now consider flavonoid enzymes, for the very first time, in three dimensions and to examine the interdependence of the pathways of secondary metabolism using genomic, proteomic, and metabolic profiling methods are likely to move us much more rapidly toward this end. The development of genomics datasets for organisms that include *Arabidopsis*, the moss *Physcomitrella patens* (http://www. moss.leeds.ac.uk/), and the model legume *Medicago truncatula* (http:// www.noble. org/medicago/), are also offering opportunities to examine this metabolically and nutraceutically model system from entirely new perspectives. It is quite clear that, even for much-studied ‘old’ pathways like flavonoid and carotenoid biosynthesis, these are exciting times.

Acknowledgements

The authors acknowledge the generous funding of the work by CSIR, Govt. of India. SD acknowledges the encouragement and support of the Director, TRA and the facilities provided by the Coordinator, CAS, University of Calcutta, where the work was initiated as part of PhD thesis of SD.

References

Albrecht M, Klein A, Hugueney P, Sandmann G, Kuntz M (1995) Molecular cloning and functional expression in E. coli of a novel plant enzymemediating z-carotene desaturation. FEBS Lett 372, 199–202.

Altschul SF, Madden TL, Schaffer AA, Zhang J, Zhang Z, Miller W, Lipman DJ (1997) Gapped BLAST and PSI-BLAST: a new generation of protein database search programs. Nucleic Acids Res 25: 3389–3402.

Bondarovich HA, Giammarino AS, Renner JA, Shephard FW, Shingler AJ, Gianturco MA (1967) Some aspects of the chemistry of tea. A comparison to the knowledge of the volatile constituents. J Agric Food Chem 15: 36–47.

Borthakur A, Sabhapandit S, Tamuly P, Hazarika M, Das S (2006) Production of catechins in unorganized cell cultures of tea. Phytochemistry. (accepted).

Breitenbach J, Kuntz M, Takaichi S, Sandmann G (1999) Catalytic properties of an expressed and purified higher plant type zetacarotene desaturase from Capsicum annuum. Euro J Biochem 265: 376–383.

Busch M, Seuter A, Hain R (2002) Functional analysis of the early steps of carotenoid biosynthesis in tobacco. Plant Physiol 128: 439–453.

Cervantes-Cervantes M, Hadjeb N, Newman LA, Price CA (1990) ChrA is a carotenoid-binding protein in chromoplasts of Capsicum annuum. Plant Physiol 92: 1241–1243.

Cunningham FX Jr, Gantt E (1998) Genes and enzymes of carotenoid biosynthesis in plants. Annu Rev Plant Physiol Plant Mol Biol 49: 557–583.

Czichi U, Kindl H (1977) Phenylalanine ammonia-lyase and cinnamic acid hydroxylase as assembled consecutive enzymes on microsomal membranes of cucumber cotyledons: cooperation and subcellular distribution. Planta 134: 133–143.

Das S, Borthakur A, Borthakur D, Borchetia S (2005) Cell culture systems in tea for production of nutraceuticals. In: Proceedings of International Symposium on Innovation in Tea Science and Sustainable Development in Tea Industry. Hangzhou, China.

Dixon RA, Steele CL (1999) Flavonoids and isoflavonoids: a gold mine for metabolic engineering. Trends Plant Sci 4: 394–400.

Emanuelsson O, Nielsen H, von Heijne G (1999) ChloroP, a neural network-based method for predicting chloroplast transit peptides and their cleavage sites. Protein Sci 8: 978–984.

Etoh H, Ina K, Iguchi M (1980) 3-S-(+)-3,7-dimethyl-1,5-octadiene-3,7-diol and ionone derivatives from tea. Agric. Biol. Chem. 44: 2999–3000.

Fraser PD, Romer S, Shipton CA, Mills PB, Kiano JW, Misawa N, Drake RG, Schuch W, Bramley PM (2002) Evaluation of transgenic tomato plants expressing an additional phytoene synthase in a fruit-specific manner. Proc Natl Acad Sci USA 99: 1092–1097.

Fray MJ, Evans EJ, Lydiate DJ, Arthur AE (1996) Physiological assessment of apetalous flowers and erectophile pods in oilseed rape (Brassica napus). J Agri Sci Cambridge 127: 193–200.

Gallagher CE, Cervantes-Cervantes M, Wurtzel ET (2003) Surrogate biochemistry: use of Escherichia coli to identify plant cDNAs that impact metabolic engineering of carotenoid accumulation. Appl Microbiol Biotechnol 60: 713–719.

Gershenzon J, Maffei M, Croteau R (1989) Biochemical and histochemical localization of monoterpene biosynthesis in the glandular trichomes of spearmint (Mentha spicata). Plant Physiol 89: 1351–1357.

Hara Y, Luo S, Wickremasinghe RL, Yamanishi T (1995) Flavor of tea. Food Rev Int 11: 477–525.

Harker M, Hirschberg J (1998) Molecular biology of carotenoid biosynthesis in photosynthetic organisms. Methods Enzymol 297: 244–263.

Haruta M, Pedersen JA, Constabel CP (2001) Polyphenol oxidase and herbivore defense in trembling aspen (Populus tremuloides): cDNA cloning, expression, and potential substrates. Physiol Plant 112: 552–558.

Hazarika M, Mahanta PK (1983) Some studies on carotenoids and their degradation in black tea manufacture. J Sci Food Agric 34: 1390–1396.

Hrazdina G (1992) Compartmentation in aromatic metabolism. In: Stafford HA, Ibrahim RK (eds) Phenolic Metabolism in Plants (New York: Plenum Press), pp. 1–23.

Kubota E, Hara T (1976) Evaluating methods of green tea grade by chemical and physical techniques. Study Tea 50: 63–67.

Li L, Paolillo DJ, Parthasarathy MV, Dimuzio EM, Garvin DF (2001) A novel gene mutation that confers abnormal patterns of beta-carotene accumulation in cauliflower. Brassica oleracea var. botrytis). Plant J 26: 59–67.

Li ZH, Matthews PD, Burr B, Wurtzel ET (1996) Cloning and characterization of a maize cDNA encoding phytoene desaturase, an enzyme of the carotenoid biosynthetic pathway. Plant Mol Biol 30: 269–279.

Lu SH (1987) Tea Evaluation and Inspection, 2nd. Agricultural Press, Beijing.

Matthews PD, Wurtzel ET (2000) Metabolic engineering of carotenoid accumulation in Escherichia coli by modulation of the isoprenoid precursor pool with expression of deoxyxylulose phosphate synthase. Appl Microbiol Biotechnol 53: 396–400.

Millin DJ, Crispin DJ, Swaine D (1969) Nonvolatile components of black tea and their contribution to the character of the beverage. J Agric Food Chem 17: 717–722.

Nakagawa M (1975) Contribution of green constituents to the intensity of taste elements of brew. Study Tea 48: 77–83.

Nakagawa M, Anan T, Ishima N (1981) The relation of green tea taste with its chemical make-up. Bull Natl Res Inst Tea 17: 69–123.

Nesi N, Debeaujon I, Jond C, Pelletier G, Caboche M, Lepiniec L (2000) The TT8 gene encodes a basic helix-loop-helix domain protein required for expression of DFR and BAN genes in Arabidopsis siliques. Plant Cell 12: 1863–1878.

Norris SR, Shen X, DellaPenna D (1998) Complementation of the Arabidopsis pds1 mutation with the gene encoding p-hydroxyphenylpyruvate dioxygenase. Plant Physiol 117: 1317–1323.

Park H, Kreunen SS, Curtiss AJ, Dellapenna D, Pogson B (2001) Identification of the carotenoid isomerase provides insight into carotenoid biosynthesis, prolamellar body formation, and photomorphogenesis. Plant Cell 14: 321–332.

Pecker I, Gabbay R, Cunningham FX, Hirschberg J (1996) Cloning and characterization of the cDNA for lycopene beta-cyclase from tomato reveals decrease in its expression during fruit ripening. Plant Mol Biol 30: 807–819.

Rasmussen S, Dixon RA (1999) Transgene-mediated and elicitor-induced perturbation of metabolic channeling at the entry point into the phenylpropanoid pathway. Plant Cell 11: 1537–1551.

Ravichandran R (2002) Carotenoid composition, distribution and degradation to flavour volatiles during black tea manufacture and the effect of carotenoid supplementation on tea quality and aroma. Food Chem. 78: 23–28.

Reynold W, Naff-Muller RS, Keller U (1974) An investigation of tea aroma. Part 1. New volatile black tea constituents. Helvetica Chimica Acta 57: 1301–1308.

Ronen G, Cohen M, Zamir D, Hirschberg J (1999) Regulation of carotenoid biosynthesis during tomato fruit development: Expression of the gene for lycopene epsilon-cyclase is down-regulated during ripening and is elevated in the mutant *delta*. Plant J 17: 341–351.

Sakata K, Watanabe N (1998) Importance of Glycosides as alcoholic aroma precursors in Plants – Molecular basis of alcoholic aroma formation in tea and flower. J Appl Glycoscience 45: 123–129.

Sambrook J, Fritsch EF, Maniatis T (1989) Molecular Cloning: A Laboratory Manual. Cold Spring Harbor Laboratory Press, Cold Spring Harbor, NY.

Sanderson GW, Co H, Gonzalez JG (1971) Biochemistry of tea fermentation: the role of carotenes in black tea aroma formation. J Food Sci 36: 231–236.

Sanderson GW, Graham HN (1973) On the formation of black tea aroma. J Agric Food Chem 21: 576–585.

Schoenbohm C, Martens S, Eder C, Forkmann G, Weisshaar B (2000) Identification of the Arabidopsis thaliana flavonoid 3'-hydroxylase gene and functional expression of the encoded P450 enzyme. Biol Chem 381: 749–753.

Scolnik PA, Bartley GE (1995) Nucleotide sequence of zetacarotene desaturase (Accession No. U38550) from Arabidopsis (PGR95±111). Plant Physiol 108: 1499.

Shewmaker CK, Sheehy JA, Daley M, Colburn S, Ke DY (1999) Seedspecific overexpression of phytoene synthase: increase in carotenoids and other metabolic effects. Plant J 20: 401–412.

Shirley BW, Hanley S, Goodman HM (1992) Effects of ionizing radiation on a plant genome: Analysis of two Arabidopsis transparent testa mutations. Plant Cell 4: 333–347.

Shirley BW, Kubasek WL, Storz G, Bruggemann E, Koornneef M, Auaubel FM, Goodman HM (1995) Analysis of Arabidopsis mutants deficient in flavonoid biosynthesis. Plant J 8: 659–671.

Suzuki Y, Shioi Y (2003) Identification of chlorophylls and carotenoids in major teas by high-performance liquid chromatography with photodiode array detection. J Agric Food Chem 51: 5307–5314.

Tirimanna ASL, Wickremasinghe RL (1965) Studies on the quality and flavour of tea. II. The carotenoids. Tea Quart 36: 115–121.

Wisman E, Hartmann U, Sagasser M, Baumann E, Palme K, Hahlbrock K, Saedler H, Weisshaar B (1998) Knock-out mutants from an En-1 mutagenized Arabidopsis thaliana population generate phenylpropanoid biosynthesis phenotypes. Proc Natl Acad Sci USA 95: 12432–12437.

Improving Nutritional Quality of Crops

Success Towards Alleviating Peanut Allergy: The Major Allergen Ara h 2 Is Silenced via RNA Interference (RNAi)

K.N. Konan[1], O.M. Viquez[2], F.C. Chen[3], and H.W. Dodo[1]

[1]Alabama A&M University, Department of Food and Animal Sciences, Food Biotechnology Laboratory, Normal AL 35762, USA.
[2]Vanderbilt University, Department of Pathology, C3322-MCN, Nashville, TN 37232, USA.
[3]Institute of Agricultural and Environmental Research, Tennessee State University, Nashville, TN 37209-1561, USA.

Introduction

There is evidence that the prevalence of peanut allergy is increasing worldwide. Peanut sensitivity on a standard skin prick test has increased by 55%, while allergic reactions increased by 95% in the past 10 years. Strict avoidance of peanut and peanut-ingredient is the only way to prevent an allergic reaction. However, avoidance of peanuts is very difficult because peanuts are commonly used as an adulterant in food preparation. To date, there is no cure for peanut allergy and no therapy is available to reduce the severity of peanut allergy. Current treatments only address the symptoms of an allergic reaction once it has taken place. Ara h 2 is a 17.5 kDa protein, and identified as the most offending peanut allergen (Palmer et al., 2005). Elimination of Ara h 2 from peanut would significantly reduce peanut allergenicity. In a previous study, we reported the production of several transgenic peanut plants expressing a truncated Ara h 2 transgene (Konan et al., 2004). However, upon maturity, the plants were sterile and no seed was produced, preventing the assessment of

Z. Xu et al. (eds.), Biotechnology and Sustainable Agriculture 2006 and Beyond, 261–264.

silencing Ara h 2 in transgenic peanuts. The objective of this study is to produce fertile transgenic peanut plants using *Agrobacterium tumefaciens* harboring an RNAi transformation vector, specifically designed to silence Ara h 2, and to evaluate the allergenic potency of transgenic peanut seeds.

Materials and methods

A DNA fragment of 265 bp was generated from the coding region of the genomic DNA of Ara h 2 (Viquez et al., 2001) and cloned in the plasmid pHANNIBAL (Wesley et al., 2001) as *Xho* I/*Kpn* sense arm, and *Xba* I/*Hind* III antisense arm to create an inverted repeat transgene. The Ara h 2-specific transformation cassette was spliced out of pHANNIBAL by a *Not* I restriction digest and sub-cloned into the unique *Not* I site of the binary plasmid pART27 (Gleave, 1992) containing the *NPTII* selection marker gene. The resulting plasmid pDK28 was mobilized into *A. tumefaciens* EHA 105 for peanut transformation according to the protocol described by Egnin et al. (1998). Kanamycin-resistant shoots produced from the peanut variety Georgia green were rooted and transferred to an Enconair growth chamber model A60 for plant maturation, flowering and seed setting. Molecular (PCR, Southern) and immunological (SDS-PAGE, Western and ELISA) analyses were performed to detect the presence of the transgene in peanut plants and to evaluate Ara h 2 content and allergenic potency of the transgenic seeds.

Results and discussion

A total of 59 Kanamycin-resistant plants (Ara h 2-T0) were successfully produced within 5–7 months from 555 infected hypocotyls explants. Flowering occurred within 1.5 months upon transfer of the plants to soil and incubation in the Enconair growth chamber. Flowers were produced on both the controls obtained by 1) direct peanut seed planting in soil (wild type, WT) and 2) tissue culture (TCc) without *Agrobacterium* infection. The phenotype, plant growth and reproduction of the kanamycin-resitant plants were indistinguishable from that of WT control. Peg elongated from the plants, and peanut pods were produced within 3–4 months. PCR analyses, performed on leaves of kanamycin-resistant plants by targeting both the NptII selection marker gene and the CaMV 35S promoter, coupled with Southern hybridization revealed the presence of the Ara h 2 RNAi transgene in 26 out off 59 (44%) kanamycin-resistant plants. Thirty nine (39) seeds were collected from ten randomly selected Southern positive peanut plants for further analysis. SDS-PAGE coupled with Western bots using Ara h 2 specific monoclonal (mAb) revealed the absence of the Ara h 2 doublet band in the protein profile of seven of the transgenic samples. Indirect ELISA performed using the same Ara h 2 mAb showed a reduction of Ara h 2 content in all seven samples with a significant decrease ($p < 0.05$) in three transgenic samples in which Ara h 2 was reduced to 2.6, 3.7 and 7.5% of the WT control. The IgE binding capacity of CPE of the three above transgenic peanut seeds was found to be significantly lower ($p < 0.05$) than that of the control WT in sera of five patients with documented history of peanut allergy. Together these data suggest a significant decrease in the allergenic potency of the transgenic peanut seeds, and thus RNAi technology has the potential to eliminate peanut allergy from the source.

References

Egnin M, Mora A, Prakash CS (1998) Factor enhancing *Agrobacterium tumefaciens*-mediated gene transfer in peanut (*Arachis hypogaea* L). In Vitro Cell Dev Biol Plant 34: 310–318.

Gleave AP (1992) A versatile binary vector system with a T-DNA organizational structure conducive to efficient integration of cloned DNA into plants genome. Plant Mol Biol 20: 1203–1207.

Konan NK, Viquez OM, Dodo H (2004) Towards the development of a hypoallergenic peanut through genetic transformation. Appl Biotechnol Food Sci Policy 1: 159–168.

Palmer GW, Dibbern DA, Burks AW, Bannon GA, Bock SA, Porterfield HS, McDermont RA, Dreskin SC (2005) Comaparive potency of Ara h 1 and Ara h 2 in immunochemical and functional assays of allergenicity. Clinical Immunol 115: 302–312.

Viquez OM, Summer CG, Dodo HW (2001) Isolation and molecular characterization of the first genomic clone of a major peanut allergen, Ara h 2 J Allerg Clin Immunol 107: 713–717.

Wesley SV, Helliwell CA, Smith N,. Wang MB, Rouse DT, Liu Q, Gooding PS, Sing SP, Abbott D, Stoutjesdijk PA, Robinson SP, Gleave AP, Green AG, Waterhouse PM (2001) Construct design for efficient, effective high-throughput gene silencing in plants. Plant J 27: 581–590.

Transgenic Poplar for Phytoremediation

Won-Yong Song[1], Young-Im Choi[2], Donghwan Shim[1], Do-Young Kim[1], Eun-Woon Noh[2], Enrico Martinoia[1], and Youngsook Lee[1]

[1]POSTECH-UZH Cooperative Laboratory, Division of Molecular Life Sciences, Pohang University of Science and Technology, Pohang, 790-784, Korea.
[2]Korea Forest Research Institute, Suwon, 441-350, Korea.

Abstract

Heavy metals such as cadmium (Cd) and lead (Pb) are released into the environment by mining, industry, and agriculture. Phytoremediation is a low-cost and environmentally-friendly technology to clean up polluted sites. For effective phytoremediation, plants that can tolerate high level of pollutants are necessary. To engineer plants with enhanced metal tolerance and accumulation properties, we searched for genes that can improve heavy metal resistance through yeast mutant screening and Arabidopsis cDNA library screening, and found *ScYcf1, ScYHL035C, ScPdr13*, and *AtPcr1*. ScYcf1 and ScYHL035C are multidrug resistance-associated protein (MRP) subfamily of ABC transporter. ScPdr13 is a heat shock protein 70(HSP70). AtPcr1 is a plant protein which enhances cadmium resistance by reducing Cd(II) in the cell. We expressed the genes in poplar, a species that grows fast to high biomass, important characters of plants for application of phytoremediation. To prevent pollen dissemination from the transgenic plants, we introduced the genes into non-flowering mutant poplar. We compared growth of the transgenic and wild type poplar

Z. Xu et al. (eds.), Biotechnology and Sustainable Agriculture 2006 and Beyond, 265–271.

plants in Cd- and Pb-containing media, and found that the transgenic poplars were highly improved in the heavy metal resistance. These results suggest that they are useful for phytoremediation of Cd- and Pb-contaminated soils.

Introduction

Our environment is contaminated by diverse pollutants including industrial by-products, waste matters, fertilizers and pesticides. The environment safe and pleasant is an utmost priority for the present and future generations who desire a high quality of living. However, no ideal method to remedy polluted environment is available. The physical or chemical cleanup of the widespread contaminated area is costly and can often further damage the environment. Phytoremediation has recently been proposed as a low-cost, environmentally-friendly way to remove pollutants from contaminated soils (Raskin, 1996). The use of plants costs less than the use of any other form of organisms, since plants use energy from sunlight. Moreover, plants are ascetically pleasing, and have good influence on the physical, chemical and biological aspects of the environment.

Botanists have long been familiar with the fact that some plants can survive in an environment toxic to other plants, and some even accumulate materials very toxic to other plants. The physiology and biochemistry of this unique tolerance and accumulation of toxic matters by some plants began to catch many scientists' attention when the possibility of using these plants to clean up contaminated environment was suggested in the early 1990s. Since then many scientific papers have been published on this subject, and some venture companies actually began to use plants to clean up waste dump, lead-contaminated areas, etc. Useful information on phytoremediation is available from the websites of the venture companies at the addresses of http://www.edenspace.com/, http://www.phytokinetics.com/, and http://www.verdanttech.com/. US environmental protection agency has a web site (http://es.epa.gov/ncerqa/rfa/phytore00.html)

on phytoremediation to help citizens of the United States to understand the concept of phytoremediation and use it for maximum benefit.

At present, plants that naturally hyper-accumulate or tolerate toxic materials (Chaney et al., 1997; Pence et al., 2000) are used for the purpose of remediation of the polluted environment. They include poplar, Indian mustard, and willow trees. However, phytoremediation using these natural plants has limitations. The plants may suffer toxicity symptoms, and it takes long time for the plants to grow to certain size to effectively extract the pollutants. In this respect, the potential for genetic engineering of plants to improve their cleaning capacity is enormous. Plant physiology can be altered relatively easily, since we can generate transgenic plants that can express exogenous genes or homologous genes at the different level and thereby acquire new phenotypes. Many transgenic plants with new characteristics due to expression of genes from microorganisms, yeasts, or mammals have been reported in scientific journals (Bizily et al., 2000; Meagher, 2000; Mitra and Sharma, 2001; Lee et al., 2003; Song et al., 2003). Once an individual plant with suitable phenotypes is produced, it is easy to multiply the numbers of plants by cloning methods, such as cuttings and somatic embryogenesis. Moreover, many trees grow indefinitely as long as the conditions are suitable. We have been developing poplar plants for phytoremediation using genes that showed high potentials for heavy metal tolerance in *Arabidopsis*. Here we report our progress in developing poplar plants for phytoremediation.

Results and discussion

We identified many genes that enhance plants' resistance to heavy metals (Lee et al., 2003, 2005; Song et al., 2003, 2004; Kim et al., 2006). We introduced the genes into poplar (Figure 1), with the aim of producing plants that can effectively extract toxic materials from the environment, thus remedy the environment.

Figure 1. Transformation of poplar plants. The plants were transformed in Petri plates, induced to develop roots in test tubes, acclimated to the green house condition, and then finally tested for their tolerance to heavy metals in a controlled environment.

Poplar was chosen since this species grows very fast into high biomass. For example, a poplar plant can grow as much as 2 m a year under optimal conditions. To minimize the concerns on the impact of transgenic plants to the environment, we used a line of poplar that does not flower. This plant does not produce flowers or seeds, but can be propagated easily by cutting. Our preliminary results from experiments comparing wild type and transgenic poplar plants for their tolerance to heavy metals (Figure 2) show that several genes enhance the plant's tolerance to lead (Figure 3) and cadmium (data not shown).

***ScYCF1* and *ScYHL035C* genes**: We obtained evidence for involvement of an ABC transporter *YCF1* in Pb(II) resistance in baker's yeast (*Saccharomyces cerevisiae*), and showed experimental data for the enhanced Pb(II) and Cd(II) resistance and accumulation in Arabidopsis plants transformed with the gene (Song et al., 2003). The YCF1 transgenic plants were improved in resistance to Cd and Pb due to increased vacuolar compartmentation of glutathione conjugated-heavy metals and accumulated up to two-fold increased amount of Pb and Cd (Song et al., 2003). *YHL035C,* a close homolog

Figure 2. Comparison of transgenic and wild type poplar plants for their tolerance to heavy metals. Similar size seedlings of wild type and transgenic poplar were chosen, placed in a same basket, and subjected to heavy metal treatment which consisted of watering through the bottom of the pots in a tub containing heavy metal solution.

of *YCF1*, when expressed in poplar, enhanced the plant's capacity to tolerate lead (Figure 3).

***AtPcr1* gene**: We isolated *AtPcr1* gene by using a functional screening system in yeasts. Arabidopsis cDNA library was introduced into a yeast strain that has a mutation in a Cd-resistance gene, plated on Cd(II)-containing agar plate, and the surviving colonies were collected. The plasmids in the surviving colonies were rescued, and one of the plasmids contained a gene never studied before. We named it *AtPcr1,* and confirmed its Cd resistance function by expression in yeast and overexpression of the cDNA in sense and antisense orientations in Arabidopsis. On 50 μM Cd(II)-containing half strength MS agar plates, the plants expressing more *AtPcr1* gene grew better than plants expressing less of the gene (Song et al., 2004). When expressed in poplar, this gene improved the plant's tolerance to Cd(II) (data not shown).

***ScPDR13* gene**: We reported that *PDR13* gene enhances yeast's tolerance to high Cu(II) (Kim et al., 2001). When introduced into poplar plants, the gene improved the plant's tolerance to Pb(II) (Figure 3).

Figure 3. Poplar plants that are transformed with an ABC transporter (middle) and *ScPDR13* (right) show better growth than their isogenic wild type (left) in soil soaked in 500 μM $Pb(NO_3)_2$.

Future development

For effective phytoremediation, many steps of transport in plants should be improved. Plants should extract pollutants from the soil, take them up at the root hair, load them into vascular bundle, transfer them to shoot part, unload, and store them at the vacuole. Therefore, it is necessary to identify genes that code for proteins that improve plants at these multiple sites. We will continue to search for new genes that work at the different positions in the process, and express them in combinations to improve several traits together, to obtain ideal plants for phytoremediation. In addition, we plan to introduce the genes into other plants, since many different plants are necessary to meet the requirements of particular sites of contamination with different soil and weather conditions.

Acknowledgment

This research is supported by the global research laboratory program of the ministry of science and technology of Korea.

References

Bizily SP, Rugh CL, Meagher RB (2000) Phytodetoxification of hazardous organo-mercurials by genetically engineered plants. Nat Biotechnol 18: 213–217.

Chaney RL, Malik M, Li YM, Brown SL, Brewer EP, Angle JS, Baker AJ (1997) Phytoremediation of soil metals. Curr Opin Biotechnol 8: 279–284.

Kim DY, Bovet L, Kushnir S, Noh EW, Martinoia E, Lee Y (2006) AtATM3 is involved in heavy metal resistance in Arabidopsis. Plant Physiol 140: 922–932.

Kim DY, Song WY, Yang YY, Lee Y (2001) The role of PDR13 in tolerance to high copper stress in budding yeast. FEBS Lett 508: 99–102.

Lee J, Bae H, Jeong J, Lee JY, Yang YY, Hwang I, Martinoia E, Lee Y (2003) Functional expression of a bacterial heavy metal transporter in Arabidopsis enhances resistance to and decreases uptake of heavy metals. Plant Physiol 133: 589–596.

Lee M, Lee K, Lee J, Noh EW, Lee Y (2005) AtPDR12 contributes to lead resistance in Arabidopsis. Plant Physiol 138: 827–836.

Meagher RB (2000) Phytoremediation of toxic elemental and organic pollutants. Curr Opin Plant Biol 3: 153–162.

Mitra B, Sharma R (2001) The cysteine-rich amino-terminal domain of ZntA, a Pb(II)/Zn(II)/Cd(II)-translocating ATPase from Escherichia coli, is not essential for its function. Biochemistry 40: 7694–7699.

Pence NS, Larsen PB, Ebbs SD, Letham DL, Lasat MM, Garvin DF, Eide D, Kochian LV (2000) The molecular physiology of heavy metal transport in the Zn/Cd hyperaccumulator *Thlaspi caerulescens*. Proc Natl Acad Sci USA 97: 4956–4960.

Raskin I (1996) Plant genetic engineering may help with environmental cleanup. Proc Natl Acad Sci USA 93: 3164–3166.

Song WY, Martinoia E, Lee J, Kim D, Kim DY, Vogt E, Shim D, Choi KS, Hwang I, Lee Y (2004) A novel family of cys-rich membrane proteins mediates cadmium resistance in Arabidopsis. Plant Physiol 135: 1027–1039.

Song WY, Sohn EJ, Martinoia E, Lee YJ, Yang YY, Jasinski M, Forestier C, Hwang I, Lee Y (2003) Engineering tolerance and accumulation of lead and cadmium in transgenic plants. Nat Biotechnol 21: 914–919.

The Africa Biofortified Sorghum Project – Applying Biotechnology to Develop Nutritionally Improved Sorghum for Africa

Zuo-yu Zhao

Pioneer Hi-Bred International, Inc. Johnston, IA 50131, USA.
Email: zuo-yu.zhao@pioneer.com

The Africa Biofortified Sorghum Project Consortium: Africa Harvest Biotech Foundation International; Pioneer Hi-Bred International, Inc.; Council for Scientific and Industrial Research; International Crop Research Institute for Semi-Arid Tropics; African Agricultural Technology Foundation; Forum for Agricultural Research in Africa; University of Pretoria; Agricultural Research Council; University of California, Berkeley. (www.supersorghum.org)

Introduction

Sorghum, the fifth most planted cereal crop in the world, is one of the most important staple crops in Africa and represents the only viable food grain for many of the world's most food insecure people. Sorghum grain has a nutritional profile similar to corn and other cereals (Shewry and Halford, 2003), i.e., it shares the typical nutritional deficiencies of cereal grains, a low content of several essential amino acids, a low vitamin A and E content and a low bio-availability of iron and zinc. Therefore, a diet, based mostly on sorghum, is not adequate to meet the nutritional growth or maintenance requirements for children and adults and needs to be supplemented

Z. Xu et al. (eds.), Biotechnology and Sustainable Agriculture 2006 and Beyond, 273–277.

with essential amino acids and micronutrients. Further, most sorghum food is cooked or heated during preparation. In contrast to other cereal grains, heat treatment results in a severely reduced digestibility of sorghum grain of up to 50%.

A consortium consisting of Africa Harvest Biotech Foundation International led (the grant recipient) by Dr. Florence Wambugu received funding from The Bill and Melinda Gates Foundation in 2005 to develop Nutritionally Improved Sorghum for Africa – The Africa Biofortified Sorghum (ABS). Dr. Paul Anderson from Pioneer is the principal investigator.

The Bill and Melinda Gates Foundation issued the 'Grand Challenges in Global Health Initiative' to focus on 'forgotten' infectious diseases, development of drugs and vaccines and the relationship between nutrition and health in developing countries and awarded $450 million to 43 projects selected from >1500 grant proposals in June 2005 (www.grandchallengesgh.org). 'Nutritionally Enhanced Sorghum for the Arid and Semi-Arid Tropical Areas of Africa' is one of the funded projects in Grand Challenge 9: 'Create a full range of optimal, bioavailable nutrients in a single, staple plant species'.

Methods

The development of the nutritionally enhanced sorghum lines will rely on transgenes and technologies that have shown high efficacy in transgenic maize and that resulted in a significantly improved nutritional quality of maize grain. As a proof of concept, a first generation transgenic sorghum line (ABS #1) that possesses grain with a 50% increase in lysine has already been developed (Zhao et al., 2003).

The goal of the ABS project is to develop transgenic sorghum varieties that will overcome most of the described nutritional deficiencies by substantially improving grain digestibility, by delivering vitamins, the essential amino acids lysine, threonine and tryptophan, and by improving the bioavailability of iron and zinc. This nutritionally enhanced sorghum is named as ABS #2.

The trait targets for ABS #2 sorghum are:

- Improved digestibility: no digestibility decrease as a result of cooking
- Balanced amino acids increases the amount of four of the most limiting essential amino acids: lysine (100%), tryptophan (100%); threonine (20%) and sulfur amino acids (20%).
- Elevated Vit A: 20 mg/g beta-carotene
- Increased Fe and Zn bioavailability: 50% increase Fe availability, 75% increase Zn availability

The approaches used to reach these targets are:

- Improving digestibility through suppression of the prolamin gamma-kafirin (Oria et al., 1995) and expression of Thioredoxin (TRX) (Wong et al., 2002; Vensel et al., 2005)
- Balancing amino acids through expression of an engineered barley high lysine gene (BHL-9) (Roesler and Rao, 2000, 2001) or suppression of prolamin and lysine alpha-ketoglutarate reductase genes
- Elevating Vit A through expression of phytoene synthase and carotene desaturase genes (Hoa et al., 2003).
- Increasing Fe and Zn bioavailability through decreasing the seed phytate content by suppression of Myo-Inositol Kinase

The transformation technology used to generate ABS #2 sorghum:

- *Agrobacterium*-mediated sorghum transformation with immature embryos (Zhao et al., 2000) in the public sorghum line P898012
- Use of a *Agrobacterium* co-transformation vector allowing the segregation of selection marker (bar gene) from the trait genes in the final product (Miller et al., 2002; Zhao et al., 2003)

Results

The Africa Bio-Fortified Sorghum (ABS) project is fully operational and has met the project milestones during the first year of operation. Key accomplishments include the following:

- Permits were obtained and ABS#1 seeds were shipped to Africa for seed increase and backcrossing. African greenhouses were upgraded for GM containment.
- The first two visiting African scientists to work at Pioneer have been extremely productive and moved the early stages of research and product development ahead of schedule.
- Gene introduction work has been carried out with over 42,000 sorghum tissue explants. The first of a large wave of transformed, regenerating sorghum plants are in the greenhouse for seed production.
- First and second generation seeds have been obtained at UC Berkeley from experiments aimed at modification of seed digestibility.
- Experiments have been designed for baseline sorghum gene flow studies.
- The first phase of a 'Biology of Sorghum' document was completed for use in regulatory and public acceptance processes.
- A literature review of sorghum products and processing methods was compiled.
- Ten sorghum varieties were selected for propagation to enable baseline grain and nutritional analyses to be undertaken.
- ABS functional committees and an advisory committee were formed.
- Three full-project planning workshops have been conducted in Pretoria, Nairobi and Johnston.

References

Hoa TT, Al-Babili S, Schaub P, Potrykus I, Beyer P (2003) Golden Indica and Japonica rice lines amenable to deregulation. Plant Physiol 133: 161–169.

Miller M, Tagliani L, Wang N, Berka B, Bidney D, Zhao ZY (2002). High efficiency transgene segregation in co-transformed corn plants using an *Agrobacterium tumefaciens* 2 T-DNA binary system. Transgenic Res 11: 381–396.

Oria MP, Hamaker BR, Chull JM (1995) In vitro protein digestibility of developing and mature sorghum grain in relation to alpha-, beta-, and gamma-kafirin disulphide crosslinking. J Cereal Sci 22: 85–93.

Roesler KR, Rao AG (2000) A single disulfide bond restores thermodynamic and proteolytic stability to an extensively mutated protein. Protein Sci. 9: 1642–1650.

Roesler KR, Rao AG (2001) Rapid gastric fluid digestion and biochemical characterization of engineered proteins enriched in essential amino acids. J Agric Food Chem 49: 3443–3451.

Shewry PR, Halford NG (2003) The prolamin storage proteins of sorghum and millets. http://www.afripro.org.uk/papers/Paper03Shewry.pdf cited 20/05/03.

Vensel WH, Tanaka CK, Nai N, Wong JH, Buchanan BB, Hurkman WJ (2005) Developmental changes in the metabolic protein profiles of wheat endosperm. Proteomics 5: 1594–1611.

Wong JH, Kim YB, Ren PH, Cai N, Cho MJ, Hedden P, Lemaux PG, Buchanan BB (2002) Transgenic barley grain overexpressing thioredoxin shows evidence of communication between the endosperm and the embryo and aleurone. Proc Natl Acad Sci USA. 99: 16325–16330.

Zhao ZY, Cai T, Tagliani L, Miller M, Wang N, Pang H, Rudert M, Schroeder S, Hondred D, Seltzer J, Pierce D (2000) *Agrobacterium*-mediated sorghum transformation. Plant Mol Biol 44: 789–798.

Zhao ZY, Glassman K, Sewalt V, Wang N, Miller M, Chang S, Thompson T, Catron S, Wu E, Bidney D, Kedebe Y, Jung R (2003) Nutritionally improved transgenic sorghum. Plant Biotechnol 2002 and Beyond, Proceeding of the 10th IAPTC&B Congress, Vasil IK (ed). Kluwer Academic Publishers, Orlando, Florida, USA, pp. 413–416.

Genetic Engineering of Strawberry for Taste Improvement and Enhanced Disease Resistance by Introduction of *thauII* Gene

K.A. Schestibratov and S.V. Dolgov

Artificial Climate Station 'Biotron', Branch of Shemyakin and Ovchinnikov Institute of Bioorganic Chemistry RAS. Science Avenue 6, 142290 Pushchino, Moscow region, Russia. Email: schestibratov@rambler.ru, dolgov@fibkh.serpukhov.su; URL: www.fibkh.serpukhov.su

Abstract

Strawberry is an important small fruit crop throughout the world, and its production is growing steadily. Gray mold or *Botrytis* fruit rot, caused by *Botrytis cinerea*, is one of the destructive diseases of strawberry responsible for substantial yield loss before and after harvest. The taste attributes of strawberries (*Fragaria* x *ananassa* Duch.) are regarded as another significant quality factors by both consumers and the food industry. Nineteen independent transgenic lines of cv. Firework and 15 lines of cv. Selekta were obtained via *Agrobacterium*-mediated transformation. Accumulation of thaumatin II was detected by Western blotting analysis. Most of tested F-40-I and S-40-I transgenic lines expressing thaumatin II protein showed significantly smaller area of the necrotic lesions observed on the leaf disks. Organoleptic analysis of transgenic fruits demonstrated sweetness improvement in 8 F-40-I and 5 S-40-I lines accumulating thaumatin II protein.

Z. Xu et al. (eds.), Biotechnology and Sustainable Agriculture 2006 and Beyond, 279–282.

Effect of thaumatin II protein accumulation on sensory characteristic of strawberry fruits

Thaumatin II is a sweet-taste protein isolated from the fruit of the African plant katemfe (Thaumatococcus danielli). In addition to its sweet taste, this protein enhances certain flavours while masking others, binding specifically with taste receptors. The antifungal activity and sweet taste of thaumatin II makes it attractive candidate for use in genetic engineering to produce disease-resistant crop plants with a modified fruit taste. We have produced 19 transgenic lines cv. Firework and 15 lines of cv. Selekta harboring gene of *thaumatin II* (Schestibratov and Dolgov, 2002; Schestibratov and Dolgov, 2004).

The performance of 19 strawberry transgenic lines cv. Firework and 15 lines cv. Selekta was evaluated in the field during 2000–2003 and 2003–2004 respectively. Ten out of 17 F-40-I lines with a detectable thaumatin II expression yielded as well as wild-type cv Firework and have same phenotype. Seven lines exhibited some somaclonal variation. Morphological characteristics of 12 S-40-I lines out of 14 thaumatin II accumulating lines were similar to wild-type cv. Selekta. Two lines had somewhat smaller leaves and plant heights.

Organoleptic analysis of transgenic fruits demonstrated sweetness improvement in 8 F-40-I lines accumulating thaumatin II protein (data not shown). Sensory assessments on homogenized samples derived from each of S-40-I transgenic lines were performed at June 2004. The taste characteristic data of S-40-I lines are presented in Figure 1. The fruits of S-40-I-1, 2, 3, 5 and 12 lines were characterized by higher value of sweet taste and flavour. The obtained sensory evaluation results indicated distinct differences in the sensory quality between fruits of several transformed strawberries lines and Selekta and Firework cultivars.

Effect of recombinant thaumatin II expression on plant disease resistance

Plant resistance to pathogens involves the accumulation in the plant cell of proteins active in defense mechanisms such as PR proteins.

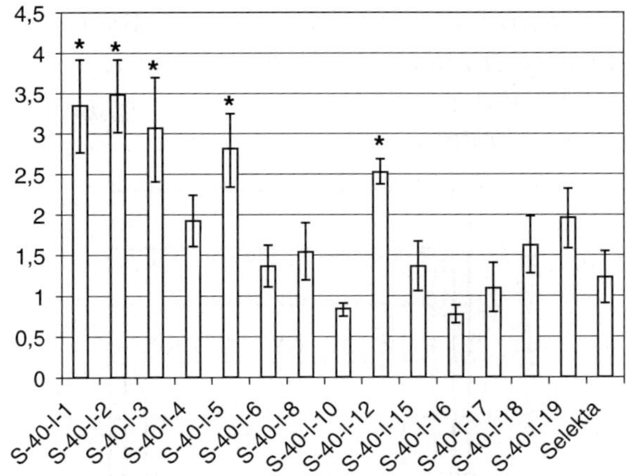

Figure 1. Sweetness evaluation of ripe fruits from S-40-I transgenic plants and wild-type cv. Selekta. *Vertical bars* represent SE of means and *asterisks* indicate a significant difference from the control at the 0.05 level (PLSD).

Typically, these antifungal proteins are expressed constitutively at low levels in cells and accumulate in response to fungal attack or in response to other inducers of acquired resistance. However, these mechanisms are often too weak or appear too late to be effective for protecting the plant. Here we explore an alternative strategy for disease control in crop plants through the use of native and modified PR-5 protein genes. This strategy involves the accumulation of defense proteins that have inhibitory activity against the hyphae and/or germinating spores of the fungal pathogen. Among these are PR proteins including thaumatin II, oatperm1, barperm1, D-34.

Using this strategy we produced transgenic strawberry with enhanced resistance against gray mold. Disease resistance evaluation demonstrated strong correlation between thaumatin II expression and enhanced resistance against *Botrytis cinerea.* All tested transgenic lines expressing thaumatin II protein showed a significant delay of disease development compared to control lines.

Transgenic lines analyzed by Western blotting were further evaluated for resistance to *Botrytis cinerea.* All tested transgenic lines showed only slight disease symptoms at 6 days after inoculation, compared with the moderate symptoms seen in the control plants. Leaf disks without artificial infection exhibited no symptoms (data not shown). The necrotic lesion in transgenic lines expressing thaumatin II protein was significantly lower than that of the control plants respectively (Figure 2). However, the leaves of transformed strawberry plants did not show complete resistance to *B. cinerea* (Schestibratov and Dolgov, 2005).

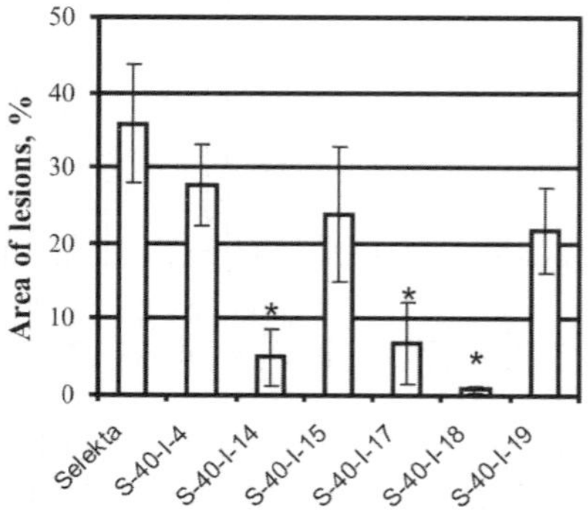

Figure 2. Evaluation of disease resistance of S-40-I transgenic lines with thaumatin II expression and control plants: Necrotic lesion area, estimated 6 days after infection with *Botrytis cinerea. Vertical bars* represent SE of means and *asterisks* indicate a significant difference from the control at the 0.05 level (PLSD).

References

Schestibratov KA, Dolgov SV (2002) Method of producing transgenic plants with increased pathogen resistance. / Patent N 2002128414, 24 October (Russian).

Schestibratov KA, Dolgov SV (2004) Method for producing transgenic plant with the aid of agrobacterium tumefaciens./PCT Patent application N WO 2004/038023, 6 May.

Schestibratov KA, Dolgov SV (2005) Transgenic strawberry plants expressing a thaumatin II gene demonstrate enhanced resistance to *Botrytis cinerea*. Sci Hort 106(2): 177–189.

Genetic Engineering of Crops for Improved Nutritional Quality

Samuel S.M. Sun[1], Qiaoquan Liu[1,2], and Rebecca M.L. Chan[1]

[1]Department of Biology, the Chinese University of Hong Kong, Hong Kong, China.
[2]State Key Laboratory of Plant Functional Genomics of the Ministry of Education, Yangzhou University, Yangzhou, China.

The world population was predicted to increase from 6.1 billion in 2001 to 7.2 billion in 2015 and 8.3 billion in 2030. Facing pressure from this continuous population growth and shrinking agricultural lands, the challenges are not only to meet the future food needs of humans but also to deal with the need of nutritionally balanced crops/foods. Major shortages of basic nutrients in the parts of world restricted to a diet based exclusively on staple foods include vitamin A, iron, iodine, zinc, and quality protein, and deficiencies in these nutrients cause micronutrient and protein-energy malnutrition throughout much of the developing world. Thus enhancing the nutritional quality of crops is an important undertaking for future food security and the nutritional well-being of world population.

Solving nutritional deficiencies can be achieved by supplementation or food fortification. However alternative approach through developing staple food crops with balanced/enhanced nutrients (biofortification) would offer sustainability. Conventional plant breeding method thus has been applied, for a long history, in an effort to enhance a specific nutrient which is deficient in the crop. A well known example is the development of the high-lysine corn through identification of natural variation/mutation. However this effort was hampered by the associated undesirable traits with the mutation

Z. Xu et al. (eds.), Biotechnology and Sustainable Agriculture 2006 and Beyond, 283–287.

(Bright and Shewry, 1983). Thus far, new crop varieties developed through conventional plant breeding method to contain significantly increased specific nutrient(s) for practical use are not common. For mineral nutrients, with progress in elucidating their absorption, transportation, distribution, deposition and homeostatic regulation in plants at physiological and molecular level, and the available methodology including molecular marker-assisted breeding (MAS), quantitative trait loci (QTL) analysis, and genetic intervention, plant breeding and even transgenic approaches can now be attempted to enhance the content of minerals in plants (Ghandilyan et al., 2006). However, increasing the content of a specific mineral may not always be effective to increase its bioavailable content in food, due to the presence of a variety of anti-nutritional compounds in plants, such as phytic acid and polyphenols, which inhibit mineral bioavailability to humans. Efforts thus are necessary also to reduce the content of anti-nutrients in plants. Advances in understanding the biosynthesis and regulation of vitamins in plants, likewise, have allowed the manipulation of their contents in plants (Lucca et al., 2006; DellaPenna and Last, 2006). A good example is the engineering of rice to produce pro-vitamin A (β–carotene) in its endosperm (*GoldenRice*) with an initial concentration of 1.6 μg/g endosperm (Ye et al., 2000). More recently, through further engineering, the β–carotene content was greatly increased (over 20 folds) to 36 μg/g endosperm in the second generation *Golden Rice* (Paine et al., 2005).

Plant proteins are the primary source of dietary protein consumed by humans and livestock. Unfortunately plant proteins, including those of major staple crops, are known for their inherent deficiency in certain essential amino acids (EAAs), making their proteins nutritionally unbalanced. In general, cereals are deficient in lysine while methionine is the first limiting essential amino acid in legumes. Attempts to enhance the EAAs in crops through plant breeding in the past 50 years have been less than satisfaction and the development of high lysine corn is a better known effort and example. With advancements in plant molecular biology and biotechnology, various strategies have been experimented in an effort to correct the deficiency of EAAs in food crops. Three strategies in this regard are recognized, namely 1) to enhance the protein-bound EAAs; 2) to increase the free EAA pool; and 3) to combine strategies 2 and 3 (Sun and Liu, 2004). For Strategy 1, four approaches have been used to increase the protein-bound EAAs, namely i) *Modification of protein sequence*: through engineering an abundant protein of the food crop to contain higher content of specific EAA(s); ii) *Expression of heterologous protein*: by expressing a foreign protein which is particularly rich in a specific EAA but poor in the target food crop; iii) *Manipulation of homologous protein expression*: by boosting the expression of a native protein which is particularly rich in a specific EAA but low overall in the food crop; and iv) *synthetic protein*: through the expression of a synthetic protein composed of high content of specific EAA(s). Many proof-of-concept experiments on these approaches had been reported, with various degrees of success in enhancing the content of specific EAAs; among them, the feasibility of approach ii, i.e., expression of heterologous protein, was clearly demonstrated including the production of proto-type EAA-improved crops. A noted example is the Met-rich Brazil nut 2S protein research and development. For Strategy 2, i.e., enhancing the free EAA pool, major efforts are in the application of two approaches to increase the lysine content in target crops: i) to enhance the biosynthesis of lysine and ii) to reduce the catabolism of lysine. Various degrees of enhancement including significant increases in free lysine content in diverse species of transgenic plants were demonstrated. For Strategy 3, i.e., to combine Strategies 1 and 2, the attempt is to enhance the source of a specific EAA, in free amino acid form, while at the same time to increase the sink for this EAA, in the form of a protein rich in this same EAA, so as to incorporate/trap the rich free EAA into protein-bound form. This combination strategy offers advantages such as reducing the adverse effects that a free amino acid, at high concentration, may bring to agronomy traits. A few reports on this strategy with different degrees of success had appeared.

We have been interested in enhancing the nutritional quality of proteins, especially the essential amino acids in crop plants, for a long period of time. Our early work on introducing and expressing the heterologous Met-rich protein gene from Brazil nut (BNMRP with 18 mol% Met) into tobacco test plant for EAA enhancement led to a 30% increase in the Met content of the total seed protein in transgenic tobacco, demonstrating that it is feasible through this approach to significantly increase the EAA of a target crop (Altenbach et al., 1989; Sun and Liu, 2004). This work represents the first enhancement of the EAA Met in a plant by genetic intervention. The same approach and the Met-rich protein gene were subsequently applied by researchers to plants including crops like canola and soybean and similar enhancement was obtained. Our more recent interest and effort has been focusing on the enhancement of lysine content in rice, which is the staple food of 50% of the Chinese and half of the world population. Rice, consisting of 8% protein by weight, provides 23% dietary protein to the people in Asia. While an inexpensive and relatively good source of protein, rice, like other cereals, is deficient in lysine, about 40% lower than the WHO recommended content, rendering it the first limiting EAA in rice. To apply the 'expression of heterologous protein' approach to enhance the lysine content in rice, we first identified from the edible winged bean a seed protein which contains 10.8 mol% lysine. The gene encoding this lysine-rich protein (LRP) was then cloned and introduced into *japonica* rice varieties under the regulation of rice seed-specific promoter of *Gt1* (glutelin). High level expression and accumulation of the LRP in rice endosperm, up to 12% of total soluble seed protein, was detected, resulting in some 20% increase in protein-bound lysine. To further enhance the content of protein bound-lysine in rice seeds, we fused the LRP with the rice glutelin storage protein, by insertion of the LRP protein into the acidic and basic subunit of the glutelin-1 protein. Expression of the LRP/Gt1 fusion protein was further promoted, with the fusion at the basic subunit yielded the highest expression, leading to some 40% increase in the lysine content of the rice seeds. These high lysine rice lines are currently under field trials for studies on agronomic performance and nutrition enhancement. This lysine enhancement technology and the high lysine rice lines are now being applied to generate a nutrient-rich rice in the Grand Challenges in Global Health project supported by the Bill and Melinda Gates Foundation. The goal is to engineer rice with enhanced content/bioavailability of pro-vitamins A, vitamin E, high quality protein, and Fe and Zn. We also direct our effort to enhance the free lysine content in rice seeds. Approaches i) to promote the synthesis of lysine by introduction and expression of mutant genes encoding aspartokinase (AK) and/or dihydrodipicolinate synthase (DHPS) insensitive to lysine inhibition (provided by G. Galili) and ii) to reduce the catabolism of lysine by down regulating the expression of the genes encoding lysine-ketoglutaric acid reductase (LKR) through anti-sense and RNAi methods. Preliminary results from analysis of the transgenic rice lines showed that combining Approaches i and ii for free lysine enhancement could lead to significant enhancement of the free lysine content in rice.

References

Altenbach SB, Pearson KW, Meeker G, Staraci LS, Sun SSM (1989) Enhancement of the methionine content of seed proteins by the expression of a chimeric gene encoding a methionine-rich protein in transgenic plants. Plant Mol Biol 13: 513–522.

Bright SWJ, Shewry PR (1983) Improvement of protein quality in cereals. CRC Crit Rev Plant Sci 1: 49–93.

DellaPenna D, Last RL (2006) Progress in the dissection and manipulation of plant vitamin E biosynthesis. Physiol Plant 126: 356–368.

Ghandilyan A, Vreugdenhil D, Aarts MGM (2006) Progress in thegenetic understanding of plant iron and zinc nutrition. Physiol Plant 126: 407–417.

Lucca P, Poletti S, Sautter C (2006) Genetic engineering approaches to enrich rice iron and vitamin A. Physiol Plant 126: 291–303.

Paine JA, Shipton CA, Chaggar S, Howells RM, Kennedy MJ, Vernon G, Wright SY, Hinchliffe E, Adams JL, Silverstone AL, Drake R (2005) A new version of Golden Rice with increased pro-vitamin A content. Nat Biotechnol 23: 482–487.

Sun SSM, Liu QQ (2004) Transgenic approaches to improve the nutritional quality of plant proteins. In Vitro Cell Dev Biol Plant 40: 155–162.

Ye X, Al-Babili S, Kloti A, Zhang J, Lucca P, Beyer P, Potrykus I (2000) Engineering the provitamin A (β–carotene) biosynthesis pathway into (carotene-free) rice endosperm. Science 287: 303–305.

Transcriptome Profiling of Wheat Genotypes Differing in Bread Baking Quality under Different Environments

Yongfang Wan, Rowan Mitchell, Tim Wheeler[1], Mike Gooding[1], Rebecca Poole[2], Keith Edwards[2], Clare Mills[3], Peter Skeggs[4], Salvador Gezan, Sue Welham, Christina Shenton[5], Tong Zhu[5], and Peter Shewry

Rothamsted Research, UK.
[1] Reading University, UK.
[2] University of Bristol, UK.
[3] IFR, UK.
[4] RHM Technology Ltd, UK.
[5] Syngenta Inc., USA.

Abstract

Preliminary comparisons showed that the public Affymetrix array platform was more sensitive in detecting changes in gene expression in developing wheat grain than a cDNA array platform. Affymetrix arrays were therefore used to compare gene expression patterns in developing grain of seven doubled haploid lines (DHLs) from a cross between wheat cvs Spark × Rialto, showing that over 440 genes were differentially expressed (significant at $P < 0.05$). The seven DH lines also exhibited differences in processing properties and a number of the differentially expressed transcripts were associated with these.

Z. Xu et al. (eds.), Biotechnology and Sustainable Agriculture 2006 and Beyond, 289–292.

Introduction

Bread baking quality is determined by the amount, composition, structure, and interactions of the major grain components, proteins, starch and cell wall polysaccharides, and these are affected by both genetic and environmental factors. Environmental impacts are particularly important in the cool maritime climate of the United Kingdom and an important goal for plant breeders is to develop varieties with consistently good quality in variable environments. We are therefore using a range of approaches including gene expression analysis to determine the impact of environmental factors on gene expression in developing wheat grain of a series of DH lines selected to show differences in bread-making performance.

Methods and materials

Seven Doubled haploid (DH) lines derived from Spark × Rialto Cross and Hereward were grown in pots in two polytunnels. Differing environments were applied 14 days after anthesis (DAA) (based on Gooding et al., 2003). Tunnel A was cool (approx 23°C day temperature) and watered to 100% field water capacity ('cool wet') while Tunnel B was warmer 20°C –28°C and 44% field water capacity ('hot dry'). Whole grain tissue was harvested at 14DAA and 22DAA. Total RNA samples were profiled with two array platforms: cDNA array (9 K genes) (Wilson et al., 2004) and public wheat Affymetrix GeneChip (61 K genes).

Results and discussion

The two array platforms were initially compared using RNA samples isolated from 14DAA and 21DAA samples of developing grain from the wheat cultivar Hereward (as described in Gooding et al., 2003). A total of 2817 genes showed significant (greater than two-fold) changes in expression between the two stages when the Affymetrix platform was used compared with only 116 when the cDNA array was used. In addition, the magnitude of the differences was greater (134 genes greater than ten-fold) with the Affymetrix chips with no genes showing changes greater than ten-fold change with the cDNA arrays. The lower sensitivity of the cDNA platform could relate to cross-hybridization.

Seven DH lines from the cross Spark × Rialto, selected on the basis of differences in bread-making performance, were grown in 2005 under 'cool wet' and 'hot dry' conditions. Grain from the lines grown in the cool wet environment consistently had lower protein and gave smaller loaf volumes. They also had higher alpha-amylase activity and lower Hagberg Falling Number indicating higher pre-maturity amylase activity. Gene expression profiling using Affy-metrix arrays showed that 446 genes were differentially expressed between the seven genotypes (Figure 1). A large number of genes also differed in expression between the two developmental stages which were selected for analysis and the differences in gene expression between the lines were also generally greater in the hot dry environment. Furthermore, some of the differentially expressed genes were correlated with end use characteristics such as loaf volume. These results

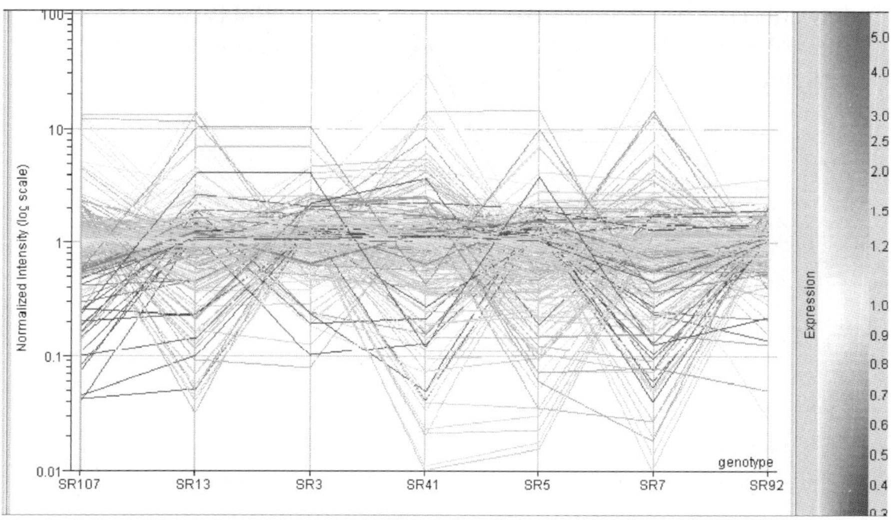

Figure 1. Average expression of 446 transcripts determined on Affymetrix array that are significantly differentially expressed between seven DH lines from the cross Spark × Rialto.

are currently being confirmed using material from a second experiment carried out under the same conditions. Further studies will include the characterisation of the differentially expressed transcripts and determination of their relationships to established QTLs for grain quality.

Acknowledgement

This work was supported by grant EGA-17694 from the Biotechnology and Biological Sciences Research Council, UK (BBSRC). Rothamsted Research and Institute of Food Research receive grant-aided support from the BBSRC.

References

Gooding MJ, Ellist RH, Shewry PR, Schofield JD (2003) Effects of restricted water availability and increased temperature on the grain filling, drying and quality of winter wheat. J Cereal Sci 37: 295–309.

Wilson ID, Barker GLA, Beswick RW, Shepherd SK, Lu CG, Coghill JA, Edwards D, Owen P, Lyons R, parker JS, Lenton JR, Holdsworth MJ, Shewry PR, Edwards KJ (2004) A transcriptomics resource for wheat functional genomics. Plant Biotechnol J 2: 495–506.

Signal Transduction

Cotton *14-3-3L* Gene Is Preferentially Expressed in Fiber

Xue-Bao Li*, Hai-Yan Shi, Xiu-Lan Wang, Wen-Liang Xu, Deng-Di Li, and Hong Wang

College of Life Sciences, Central China Normal University, Wuhan 430079, China.
*Email: xbli@mail.ccnu.edu.cn

14-3-3 protein was originally identified during a systematic classification of bovine brain proteins. Thereafter, a number of 14-3-3 proteins have been found in every eukaryotic organism such as animal, fungi, and plant species. In combination with signal proteins, the 14-3-3s prove to contribute to the regulation of many crucial cellular processes including metabolism, signal transduction, cell-cycle control, apoptosis, protein trafficking, transcription, stress responses, malignant transformation, as well as membrane-transportation of mitochondrion and chloroplast precursor proteins.

14-3-3 proteins are encoded by a multigene family in both animals and plants. *Arabidopsis* contains 13 functional genes encoding typical 14-3-3 proteins. The 14-3-3 genes have been shown to be expressed in cell or tissue specific manners during *Arabidopsis* development. Three 14-3-3 genes have been isolated in barley. In potato, up- or down-regulation of the 14-3-3 gene expression altered the composition of lipids, amino acids, and minerals. Furthermore, over-expression of 14-3-3 gene enhances antioxidant activity by 45%, and consequently, delays leaf senescence in potato. Although

Z. Xu et al. (eds.), Biotechnology and Sustainable Agriculture 2006 and Beyond, 295–298.

14-3-3 genes in a few plant species, such as *Arabidopsis*, have been well characterized, little is known on cotton (*Gossypium hirsutum*) 14-3-3 gene, especially its role in regulation of fiber cell elongation. In this study, a cDNA encoding a putative 14-3-3 protein was isolated from fiber cDNA library of cotton. Here, we reported the molecular characterization of the *Gh14-3-3L* gene and its expression profile in cotton.

Materials and methods

Isolation of *Gh14-3-3L* cDNAs

More than 2,000 cDNA clones were randomly selected from the cotton fiber cDNA library for sequencing. One clone containing complete *Gh14-3-3L* sequence was identified from the cDNAs.

Cotton RNA isolation and Northern-blot analysis

Total RNA was isolated from young fibers, ovules, anthers, petals, leaves, cotyledons, hypocotyls, and roots of cotton using modified CTAB acerbic phenol and hot phenol methods. 3'-UTR fragment of *Gh14-3-3L* cDNA was prepared as a probe by PCR amplification. RNA samples from different cotton tissues were separated on 1.2% (w/v) agarose-formaldehyde gels and transferred onto Hybond-N^+ nylon membranes by capillary blotting. Gene-specific probe was labeled with α-^{32}P-dCTP using the random primer method. RNA northern-blot hybridization was performed as described previously. The membrane was exposed to X-film at −70°C for 1–3 d.

DNA and protein sequence analysis

The open reading frame of *Gh14-3-3L* gene and its deduced protein sequence were analyze by DNASTAR software (DNAStar Co.), and protein sequence alignment and motif analysis were performed with Clustal W (http://www.ebi.ac.uk/clustalw/).

Results and discussion

Isolation of *Gh14-3-3L* cDNA and its homology analysis

To isolate genes involved in cotton fiber development, over 2000 cDNA clones from the fiber cDNA library, were randomly sequenced. One cDNA (designated *Gh14-3-3L*) encoding a putative 14-3-3 protein with 253 amino acids was selected. The Gh14-3-3L shares the highest homology (90% identity) with tomato Le14-3-3, and then high homology (89% identity) with tobacco Nt14-3-3. It also shares 85% identity with Arabidopsis At14-3-3, but relatively lower homology (79–84% identities) with other known plant 14-3-3s. This conserved structure may be related with the maintenance of 14-3-3s functions for regulation of the metabolic pathways in cells.

There are amino acid substitutions occurring at some positions. In total, there are 17 positions generated amino acid substitutions in Gh14-3-3L compared with the other 14-3-3 proteins, of which certain substitutions may influence the protein structure and function. For example, positive charged amino acid was replaced by strongly polar residue at sixth position (Lys_{6}/Gln). The substitutions among Ser_{30}/Ala/Thr/Lys/Asn occurred in CC1 domain may directly influence the function of this domain. Similar substitutions (Asp_{85}/Thr/Glu) are caused in CC2 domain.

Gh14-3-3 contains the typical 14-3-3 domains

Analysis of the protein structure revealed that Gh14-3-3L contained the typical 14-3-3 domains and motifs (i.e., one dimeric domain, one phosphoralated-serine rich motif, four CC domains, and one putative EF hand motif), which existed in all the eukaryotic 14-3-3s. The N-terminus of Gh14-3-3L, including the first CC domain (Val_{9} – Met_{34}), was highly conserved. The second CC domain, containing heptad leucine/isoleucine repeats (Asn_{79} – Ile_{105}) and the dimeric domain (Ile_{83} – Val_{90}—Ile_{97} – Val_{104}), had been shown to be an important domain for 14-3-3 function. On the other hand, the roles of both the first CC domain and the third CC domain (Ala_{134} –Thr_{157}) in 14-3-3s remain unknown so far. The fourth CC domain, including

25 amino acid residues (Glu_{190} – Ser_{214}), overlapped 17 amino acid residues with the putative EF hand motif (Gln_{98} – Leu_{225}). In the C-terminus, the remarkable divergence (including substitutions, insertions and deletions of amino acid residues) was different in eukaryotic organisms, resulting in length diversities of the14-3-3s. As a result, 14-3-3 proteins show low homology in their C-terminus.

Gh14-3-3L gene is preferentially expressed during fiber elongation

Northern blotting analysis showed that *Gh14-3-3L* gene was predominantly expressed during fiber elongation. *Gh14-3-3L* gene expression was detected at early stage of fiber elongation, and the gene transcripts was accumulated at the highest level in 10 DPA (days post anthers) fibers. Thereafter, *Gh14-3-3L* activity decreased gradually to undetectable level. This indicates that *Gh14-3-3L* expression is fiber-preferential and developmental-regulated, suggesting that it may participate in regulating fiber elongation. *Gh14-3-3* mRNAs was also accumulated at higher levels in ovules and petals, but no or very little transcripts were detected in leaves, cotyledons, roots, hypocotyls and anthers of cotton.

Inositol Polyphosphate 6-/3-kinase (*AtIpk2β*), an Early Auxin-Responsive Gene, Positively Regulates Axillary Shoot Branching in *Arabidopsis thaliana*

Zai-Bao Zhang[1], Guang Yang[1], Zhen Chen[1], Yan Li[1], Fernando Arana[2], and Hui-Jun Xia[1,*]

[1]Key Laboratory of MOE for Plant Developmental Biology, College of Life Sciences, Wuhan University, Wuhan, Hubei 430072, China.
[2]Institute of Biochemistry and Biology, University of Potsdam, Karl-Liebknecht-Strasse 24–25, 14469 Potsdam, Germany.
[*]Email: hjxia@whu.edu.cn

Abstract

Arabidopsis inositol polyphosphate 6-/3-kinase gene (*AtIpk2β*) has previously been demonstrated to participate in inositol phosphate metabolism. However, overall little is known about its physiological functions in higher plants. Here, we report that *AtIpk2β* positively regulates axillary shoot branching in Arabidopsis. *AtIpk2β* overexpression plants displayed more axillary shoot branches and produced twice as many total branches than those of the wild-type. This phenotype was confirmed in *ipk2β* mutant transgenic lines. The expression pattern of *AtIpk2β* was similar to *in vivo* IAA distribution. Moreover, *AtIpk2β* can be induced by exogenous IAA. The overexpression plants exhibited IAA-related phenotypes and altered IAA responses. Our experiments showed that the transcription of some auxin-related genes was altered in *AtIpk2β* overexpression plants. Taken together, our data indicate that *AtIpk2β* plays a role in axillary shoot branching via auxin signaling in Arabidopsis.

Z. Xu et al. (eds.), Biotechnology and Sustainable Agriculture 2006 and Beyond, 299–302.

Introduction

Inositol 1,4,5-trisphosphate 3-kinase (IP3K) plays an important role in signal transduction by phosphorylating inositol 1,4,5-trisphosphate (IP_3) to inositoI 1,3,4,5-tetrakisphosphate (IP_4). Both IP_3 and IP_4 are second message which regulate calcium homeostasis in animal cells (Xia and Yang, 2005). Compared with animal field, there are only few reports on plant IP3K. Arabidopsis *IP3K* has been identified and there are two isoforms (*AtIpk2α* and *AtIpk2β*) (Xia et al., 2003). *AtIpk2α* is involved in pollen tube development and root growth. *AtIpk2β* plays a role in phytate synthesis. In this study, we demonstrated a novel role for *AtIpk2β* in axillary shoot branching via auxin signaling.

Results

Overexpressing *AtIpk2β* in wild-type and T-DNA insertion mutants generates more axillary shoots

The *AtIpk2β* overexpression construct was transformed into *Arabidopsis thaliana C24* using *Agrobacterium*-mediated transformation. Compared with wild-type, we observed that branching of axillary shoots was more pronounced in overexpression lines. At maturity, *AtIpk2β* overexpression plants had increased axillary branches. Moreover, when *ipk2β-1* mutant was transformed with the overexpression construct, more branches were produced.

AtIpk2β expression pattern resembles the *in vivo* distribution of IAA

AtIpk2β::GUS fusions revealed that GUS activity was detected in rapidly growing tissues, the distal end of emerging leaf primordial, young stipules, vasculature and the tips of the cotyledons. The staining was also present in leaf axils and the basal part of axillary buds. Notably, *AtIpk2β*-promoter driven GUS activity followed the pattern of *in vivo* IAA distribution: a high level of IAA was found in the first node leaf, stipules of four-day-old seedlings, and the tips of the cotyledons.

AtIpk2β overexpression lines display constitutive and altered IAA responses

The petioles of cotyledons were more elongated in light-grown *AtIpk2β* overexpression plants. In addition, when grown in the dark for 6 days, the hypocotyls of *AtIpk2β* overexpression seedlings were significantly longer than those of wild-type. Moreover, *AtIpk2β* overexpression lines display altered auxin responses. Root elongation and lateral root formation in *AtIpk2β* overexpression plants are affected.

AtIpk2β regulates the expression of auxin-related genes

RT-PCR experiments show that expression of auxin-biosynthesis gene *CYP83B1* is decreased and expression of auxin-transport gene *PIN4* is increased in *AtIpk2β* overexpression plants. Furthermore, we also examined the expression of other auxin-related genes (*MAX4* and *SPS*). Both of them were decreased in overexpression plants.

Discussion

In this study, we presented evidence for an important role of *AtIpk2β* in axillary shoot branching. Our data indicate that *AtIpk2β* affect both axillary meristem initiation and outgrowth. *AtIpk2β* is an early responsive gene in branch signaling pathway via auxin signaling. Moreover, *AtIpk2α*, is not involved in regulating axillary shoot branching. More detailed analysis will lead to further understanding of the molecular mechanism of signaling pathway integration.

Acknowledgements

This work was supported by the National Natural Science Foundation of China, the Program for New Century Excellent Talents in University, the Deutscher Akademischer Austausch Dienst (DAAD), and the Program for Changjiang Scholars and Innovative Research Team in University. We thank the SALK Institute Genomic Analysis

Laboratory for providing T-DNA insertion mutants. We are also grateful to Prof. Hao Yu for providing plasmid of pGreenLC201, Prof. Jianru Zuo for plant care, Aleksandra Skirycz for technical assistance, Dr. Jian Xu for his critical reading of this manuscript. We greatly appreciate Prof. Dr. Bernd Mueller-Roeber for his comments on this manuscript.

References

Xia HJ, Brearley C, Elge S, Kaplan B, Fromm H, Mueller-Roeber B (2003) Arabidopsis inositol polyphosphate 6-/3-kinase is a nuclear protein that complements a yeast mutant lacking a functional ArgR-Mcm1 transcription complex. Plant Cell 15: 449–463.

Xia HJ, Yang G (2005) Inositol 1,4,5-trisphosphate 3-kinases: Functions and regulations. Cell Res 15: 83–91.

Apomixis

Genetic Analyses of Aposporous Embryo Sac Formation in Sorghum

John G. Carman[1,2], Michelle S. Jamison[2], Jayasree Pattanayak[2], Jeff Lacey[3], Jeong-Soon Kim[4], Estella G. Elliott[2], Patricia Klein[4], Tom Ulrich[3], and Krishna Dwivedi[2]

[1]Department of Plants, Soils and Biometeorology, Utah State University, Logan, UT 84322-4820, USA.
[2]Caisson Laboratories, Inc., 1740 Research Park Way, North Logan, UT 84341, USA.
[3]Idaho National Laboratory, Idaho Falls, ID 83415, USA.
[4]Institute for Plant Genomics and Biotechnology, Texas A&M University, College Station, TX 77843, USA.

Aposporous embryo sac (AES) formation has been documented embryologically in a few genotypes of *Sorghum bicolor* L. Moench where it occurs in <25% of ovules (Hanna et al., 1970; Tang et al., 1980; Rana et al., 1981; Schertz, 1992; Carman, 2007). Parthenogenesis and the formation of apomictically derived seed have not been definitively demonstrated in any line. To the contrary, detailed embryological studies in several AES forming lines failed to reveal parthenogenic embryo formation (Carman, 2007). Nevertheless, formation of AES in an otherwise sexual plant is important in that it may provide opportunities to map and clone genes responsible for this initial element of apomixis.

To understand the genetics of AES formation in sorghum, we produced a mapping population consisting of 302 F_2, which segregated for aposporous initial (AI) and AES formation (Carman, 2007). Many F_2 did not produce AI or AES, and the frequency of ovules containing one or more AI and/or AES in the best F_2 was

Z. Xu et al. (eds.), Biotechnology and Sustainable Agriculture 2006 and Beyond, 305–307.

only 23%. In this mapping population, frequency distributions for AI and AES were highly skewed toward 0%. The data suggest this skewedness may be caused by a threshold where a minimum combination of appropriate quantitative trait loci (QTL) alleles must occur together for AI or AES to form. Accordingly, F_2 with high percentages of ovules containing one or more AES would require rare combinations of QTL alleles. Many F_2 contained AI but no AES, which suggests important QTL for AI formation differ from important QTL for AES formation (Carman, 2007). Of the 302 F_2, 179 were selected for mapping. Preliminary results suggest frequencies AI and AES formation are quantitative traits, but confirmation of putative QTL will require additional markers.

Alternatively, AI and AES formation may be unlinked simply inherited traits with penetrance affected by modifiers. If this is correct, AES would form in an F_2 only if *i*) an AI had previously formed, i.e. the AI locus would epistatically control the AES locus, and *ii*) appropriate allele(s) are present at the AES locus. According to this two gene epistasis model, 227 F_2 (75% of 302) should have contained AI, and 170 F_2 (9/16th of 302) should have contained AES. Observed values were 243 (significantly higher, Chi square, $P < 0.05$) and 166 (NS, Chi square = 0.20), respectively. Before rejecting this model, it should be noted that while AES are easily and accurately identified microscopically, large nucellar cells can be easily mistaken as AI. Hence, some F_2, especially with AI frequencies near zero, may have been misclassified as AI-formers. This could have caused the significant chi square for AI (Carman, 2007).

To further evaluate the inheritance of AI and AES formation in sorghum, we designed and are conducting expression profiling experiments. Publicly available sorghum EST, mostly from pistils and other flower parts, but also from other plant tissues, were obtained, and 12 K arrays were produced (CombiMatrix CustomArrays, Mukilteo, WA). Two sibs from our F_2 mapping population were then selected, one in which percentage ovules with AI or AES was zero (sexual) and the other where 23% of ovules contained either AI and/or AES (aposporous). Ovules were then dissect from each sib at early (approximately meiosis I to meiosis II), mid (early sexual ES formation), and late (early egg formation) ovule development stages (about 15,000 total). Total RNA was extracted and used to synthesize cDNA labeled with Cy3 or Cy5 (Ambion, Austin, TX; GE Healthcare, Piscataway, NJ). A saturated loop design was used to compare the six samples to each other (15 arrays total). Groups of EST uniquely expressed only in sexual samples and only in aposporous samples were identified, and these are currently being subjected to bioinformatics analyses.

Most differentially expressed genes in ovules of sexual and aposporous lines probably belong to gene cascades induced by few controlling genes. To identify controlling genes, we plan to locate candidates, identified by expression profiling, to chromosomes of the *Sorghum* genome once the first iteration sequence of the genome is completed (work in progress) and then determine if any of these locations match those identified by mapping. Genes located to the same chromosome regions by both expression profiling and mapping will be considered strong candidates and will be further evaluated for their possible involvement in conferring apomixis.

References

Carman JG (2007) Do duplicate genes cause apomixis? In: Hrandl E, Grossniklaus U, Van Dijk P, Sharbel T (eds) Apomixis: Evolution, Mechanisms and Perspectives, pp. 63–91. Regnum Vegetabile 147, International Association for Plant Taxonomy and Koeltz Scientific Books, Vienna, 424 p.

Hanna WW, Schertz KF, Bashaw EC (1970) Apospory in Sorghum bicolor (L.) Moench. Science 170: 338–339.

Rana BS, Reddy CS, Rao VJM, Rao NGP (1981) Apomixis in grain sorghums: analysis of seed set and effects of selection. Indian J Genet Plant Breed 41: 118–123.

Schertz KF (1992) Apomixis in sorghum. In: Elgin JH, Miksche JP (eds) Proceedings of the Apomixis Workshop, February 11–12. USDA-ARS, ARS-104, Atlanta, Georgia.

Tang CY, Schertz KF, Bashaw EC (1980) Apomixis in sorghum lines and their F_1 progenies. Bot Gaz 141: 294–299.

Epigenetic Regulation of Seed Development

Claudia Köhler

Department of Plant Developmental Biology, Institute of Plant Sciences, ETH Zürich, CH8092 Zürich.
Email: koehlerc@ethz.ch

The correct timing of developmental decisions is ensured by Polycomb group (PcG) proteins, which are evolutionary conserved master regulatory proteins present in plants as well as in animals but not in yeast. Mutations in members of the *FERTILIZATION INDEPENDENT SEED* (*FIS*) PcG genes from Arabidopsis cause the formation of seed like structures without fertilization, establishing the role of FIS proteins in blocking precocious cell divisions of egg cell and central within the female gametophyte (Ohad et al., 1996; Chaudhury et al., 1997). After fertilization *fis* mutations cause the formation of an overproliferated endosperm and a developmentally arrested embryo. The identified *FIS* genes in plants are *MEDEA* (*MEA*; Grossniklaus et al., 1998; Kiyosue et al., 1999), *FERTILIZATION INDEPENDENT ENDOSPERM* (*FIE*) (Ohad et al., 1999), *FIS2* (Luo et al., 1999) and *MULTI COPY SUPPRESSOR OF IRA1* (*MSI1*) (Köhler et al., 2003a; Guitton et al., 2004). The corresponding FIS homologs in Drosophila and mammals are *E*(*z*) (*Ezh1*, *Ezh2*), *Esc* (*Eed*), *Su*(*z*)*12* and *Nurf55* (*RbAp46/48*), respectively (reviewed in Cao and Zhang, 2004). The FIS proteins MEA, FIE and MSI1 interact and form a protein complex of ~650 kDa (Köhler et al., 2003a) that most likely contains FIS2 as well (Chanvivattana et al., 2004). Similarly, PcG proteins in animals interact and are localized within large protein complexes that change their composition during development (reviewed in Lund and Lohuizen, 2004). One of the best

Z. Xu et al. (eds.), Biotechnology and Sustainable Agriculture 2006 and Beyond, 309–311.

characterized PcG complexes in animals is the Polycomb repressive complex 2 (PRC2), which is similar in size and composition to the plant FIS complex (reviewed in Lund and Lohuizen, 2004). Insect and mammalian PRC2 complexes have intrinsic histone methyltransferase activity targeting lysine 27 of histone H3, suggesting that the maintenance of cellular memory involves methylation of histones (reviewed in Cao and Zhang, 2004). Histone methylation is mediated by the PRC2-subunit E(Z) and its homologs, which all contain an evolutionary conserved 130- residue SET domain (for SU[VAR]3-9, E(Z), TRX).

Microarray expression profiling of *fis* mutants and Chromatin Immunoprecipitation identified the type I MADS-box gene *PHERES1* (*PHE1*) as a direct target gene of the FIS PcG complex (Köhler et al., 2003b). *PHE1* expression is regulated by genomic imprinting (Köhler et al., 2005) and repression of the maternal *PHE1* allele is likely to be established in the female gametophyte by the FIS mediated application of repressive histone methylation (Makarevich et al., 2006). *PHE1* is not only under FIS mediated repression, but also CURLY LEAF (CLF) and SWINGER (SWN) containing PcG complexes repress *PHE1* expression. Also *FUSCA3* (*FUS3*) expression is under control of PcG complexes containing MEA, CLF and SWN, suggesting that different PcG complexes share a common set of target genes (Makarevich et al., 2006). *FUS3* encodes a B3-type transcription factor and loss of *FUS3* function causes heterochronic developmental aberrations, leading to precocious onset of the germination pathway (Bäumlein et al., 1994; Keith et al., 1994; Meinke et al., 1994). Whether FIS mediated repression of *FUS3* is established by similar mechanisms as *PHE1* repression is currently investigated.

References

Bäumlein H, Miséra S, Luerßen H, Kölle K, Horstmann C, Wobus U, Müller AJ (1994) The *FUS3* gene of *Arabidopsis thaliana* is a regulator of gene expression during late embryogenesis. Plant J 6: 379–387.

Cao R, Zhang Y (2004) The functions of E(Z)/EZH2-mediated methylation of lysine 27 in histone H3. Curr Opin Genet Dev 14: 155–160.

Chanvivattana Y, Bishopp A, Schubert D, Stock C, Moon YH, Sung ZR, Goodrich J (2004) Interaction of Polycomb-group proteins controlling flowering in Arabidopsis. Development 131: 5263–5276.

Chaudhury AM, Ming L, Miller C, Craig S, Dennis ES, Peacock WJ (1997) Fertilization-independent seed development in Arabidopsis thaliana. Proc Natl Acad Sci USA 94: 4223–4228.

Grossniklaus U, Vielle-Calzada JP, Hoeppner MA, Gagliano WB (1998) Maternal control of embryogenesis by MEDEA a Polycomb group gene in Arabidopsis. Science 280: 446–450.

Guitton AE, Page DR, Chambrier P, Lionnet C, Faure JE, Grossniklaus U, Berger F (2004) Identification of new members of FERTILIZATION INDEPENDENT SEED Polycomb group pathway involved in the control of seed development in Arabidopsis thaliana. Development 131: 2971–2981.

Keith K, Kraml M, Dengler NG, McCourt P (1994) fusca3: A heterochronic mutation affecting late embryo development in Arabidopsis. Plant Cell 6: 589–600.

Kiyosue T, Ohad N, Yadegari R, Hannon M, Dinneny J, Wells D, Katz A, Margossian L, Harada JJ, Goldberg RB, Fischer RL (1999) Control of fertilization-independent endosperm development by the MEDEA Polycomb gene in Arabidopsis. Proc Natl Acad Sci USA 96: 4186–4191.

Köhler C, Hennig L, Bouveret R, Gheyselinck J, Grossniklaus U, Gruissem W (2003a) Arabidopsis MSI1 is a component of the MEA/FIE Polycomb group complex and required for seed development. EMBO J 22: 4804–4814.

Köhler C, Hennig L, Spillane C, Pien S, Gruissem W, Grossniklaus U (2003b) The Polycomb-group protein MEDEA regulates seed development by controlling expression of the MADS-box gene PHERES1. Genes Dev 17: 1540–1553.

Köhler C, Page DR, Gagliardini V, Grossniklaus U (2005) The Arabidopsis thaliana MEDEA Polycomb group protein controls expression of PHERES1 by parental imprinting. Nat Genet 37: 28–30.

Lund AH, Lohuizen MV (2004) Polycomb complexes and silencing mechanisms. Curr Opin Cell Biol 16: 239–246.

Luo M, Bilodeau P, Koltunow A, Dennis ES, Peacock WJ, Chaudhury AM (1999) Genes controlling fertilization-independent seed development in Arabidopsis thaliana. Proc Natl Acad Sci USA 96: 296–301.

Makarevich G, Leroy O, Akinci U, Schubert D, Clarenz O, Goodrich J, Grossniklaus U, Köhler C (2006) Different Polycomb group complexes regulate common target genes in Arabidopsis. EMBO Rep 7(9): 947–952.

Meinke DW, Franzmann LH, Nickle TC, Yeung EC (1994) Leafy cotyledon mutants of Arabidopsis. Plant Cell 6: 1049–1064.

Ohad N, Margossian L, Hsu Y-C, Williams C, Fischer P, Repetti RL (1996) A mutation that allows endosperm development without fertilization. Proc Natl Acad Sci USA 93: 5319–5324.

Ohad N, Yadegari R, Margossian L, Hannon M, Michaeli D, Harada JJ, Goldberg RB, Fischer RL (1999) Mutations in FIE, a WD Polycomb group gene, allow endosperm development without fertilization. Plant Cell 11: 407–416.

Biotechnology in Cereals

Optimisation of Tissue Culture, Regeneration and *Agrobacterium*-Mediated Transformation Paramaters in Winter Wheat Cultivars (*T.durum* cv. Kızıltan-91 and *T.aestivum* cv. Bezostaja-01)

M. Kavas[*], H.A. Öktem, and M. Yücel

Department of Biological Sciences, Middle East Techical University, 06531, Ankara,Turkey.
[*]Email: mkavas@metu.edu.tr

Introduction

Wheat, like other cereal species, shows a common recalcitrance and strong genotypic variation in tissue culture and *Agrobacterium*-mediated transformation. A variety of explant sources have been used for obtaining wheat plants from *in vitro* cultures and *Agrobacterium*-mediated transformation. These are: isolated microspores (Mejza et al., 1993), mature embryos (Özgen et al., 1998; Delporte et al., 2001), immature embryos (Ozias-Akins and Vasil, 1982; Vasil et al., 1990; Pellegrineschi et al., 2004), and immature inflorescences (Ozias-Akins and Vasil ,1982; He and Lazzeri, 2001). Also there are different target tissues for *Agrobacterium*-mediated transformation for wheat. Immature inflorescences have been reported as an alternative source of explant for stable transformation of durum wheat (He and Lazzeri, 1998) and bread wheat (Rasco-Gaunt and Barcelo, 1999). This explant has also shown good expression of T-DNA delivered genes after *Agrobacterium* co-cultivation (Amoah et al., 2001). Advantages of the immature inflorescence system are

Z. Xu et al. (eds.), Biotechnology and Sustainable Agriculture 2006 and Beyond, 315–318.

that they are easier to isolate and are harvested from much younger plants allowing more efficient use of space in donor-plant growth facilities.

The objective of the present study was to investigate the influence of vernalization period of donor plant and callus age on regeneration capacity of calli, number of tillers, number of seeds and *Agrobacterium*-mediated transformation efficiency produced from immature inflorescence cultures of two winter wheat cultivars.

Material and methods

Two winter wheat cultivars were used; *Triticum durum* cv. Kızıltan-91 and bread wheat *Triticum aestivum* cv. Bezostaja-01. On 7th day of growth plants were subjected to vernalization for 1, 2, 3, 4, 5 weeks at 4°C with a 12 h photoperiod. In order to obtain immature inflorescence in the desired stage, tillers containing immature inflorescence were collected prior to the emergence of flag leaf which corresponds to 50–60 days for Kızıltan-91 and 80–90 days for Bezostaja-01. Tillers obtained from plants that were subjected to different vernalization periods were collected at the same time when inflorescence were generally between 0.5 and 1.0 cm in length. To evaluate the effects of vernalization period and callus age on regeneration capacity, calli produced from 3 to 5 weeks vernalized explants were used. For the purpose of determination of effect of transformation medium, callus age and vernalization time on transformation efficiency, two different transformation protocols were used. Inflorescence based calli of both cultivars were transformed with *Agrobacterium tumefaciens* AGL I strain.

Results and discussion

The highest inflorescence formation (78% and 73% in average) was observed in 4 and 5 weeks vernalized Kızıltan-91 and Bezostaja-01 plants, respectively. Genotype and medium composition were found to have a significant ($p < 0.05$) effect on regeneration capacity of calli.

In Kızıltan-91 cultures, vernalisation period and callus age have very important effects on regeneration capacity. When callus ages were examined in terms of regeneration capacity, 6 weeks old calli had higher regeneration capacity than the others. Increased callus age decrease the regeneration capacity, therefore 15 weeks old calli had the lowest regeneration capacity (27.4%) in Kızıltan cultures. Calli originated from 3 weeks vernalized explants had lower regeneration capacity in Kızıltan cultures. When vernalization period and callus age were examined for their influence on regeneration capacity of Bezostaja cultures, no difference was observed in regeneration capacity. Effects of vernalization time of explants and age of calli on number of tillers per regenerated callus were evaluated. No difference was observed between calli that belong to different vernalization period in each cultivar in terms of number of tillers per regenerating callus. In contrast to vernalization period, callus age has significant influence on number of tillers in both cultivars. In Kızıltan cultures, 15 weeks old calli have significantly ($p < 0.001$) higher number tiller (mean 9) than the others. However, 6 weeks old calli have higher number of tillers (mean 7.4) than the others in Bezostaja cultures. When we compared two transformation protocols (MG/L and MMA based) in terms of transient GUS expression, MG/L based transformation protocol was more efficient than MMA based one for two cultivars. When we evaluated the transformation results in terms of vernalisation time, there were no significant differences between inflorescence based calli in both cultivars. As a consequence, vernalisation based optimization of regeneration and transformation protocols can be used for improvement of local winter wheat varieties.

References

Amoah BK, Wu H, Sparks C, Jones HD (2001) Factors influencing *Agrobacterium*-mediated transient expression of uidA in wheat inflorescence tissue. J Exp Bot 52: 1135–1142.

Delporte F, Mostade O, Jacquemin JM (2001) Plant regeneration through callus initiation from thin mature embryo fragments of wheat. Plant Cell Tissue Organ Cult 67: 73–80.

He GY, Lazzeri PA (1998) Analysis and optimisation of DNA delivery into wheat scutellum and tritordeum inflorescence explants by tissue electroporation. Plant Cell Rep 18: 64–70.

He GY, Lazzeri PA (2001) Improvement of somatic embryogenesis and plant regeneration from durum wheat (*Triticum turgidum* var. durum Desf.) scutellum and inflorescence cultures. Euphytica 119: 369–376.

Mejza SJ, Morgant V, Di Bona DE, Wong JR (1993) Plant regeneration from isolated microspores of *Triticum aestivum*. Plant Cell Rep 12: 149–153.

Özgen M, Türet M, Altınok S, Sancak C (1998) Efficient callus culture induction and plant regeneration from mature embryo culture of winter wheat (*Triticum aestivum* L) genotypes. Plant Cell Rep 18: 331–335.

Ozias-Akins P, Vasil IK (1982) Plant regeneration from cultured immature embryos and inflorescences of *Triticum aestivum* (L. Wheat): Evidence for somatic embryogenesis. Protoplasma 110: 95–105.

Pellegrineschi A, Brito RM, McLean S, Hoisington D (2004) Effect of 2,4-dichlorophenoxyacetic acid and NaCl on the establishment of callus and plant regeneration in durum and bread wheat. Plant Cell Tissue Organ Cult 77: 245–250.

Rasco-Gaunt S, Barcelo P (1999) 'Immature inflorescence culture of cereals: A highly responsive system for regeneration and transformation', In Hall R (ed) Methods in Molecular Biology – Plant Cell Culture Protocols. Humana Press Inc., Totowa, NJ, pp. 71–81.

Vasil V, Redway F, Vasil IK (1990) Regeneration of plants from embryogenic suspension culture protoplasts of wheat (*Triticum aestivum* L). Bio/Technol 8: 429–434.

Enhancement of the Brassinosteroid Biosynthesis Pathway Improves Grain Yield in Rice

Chuan-Yin Wu[1], Shing Kwok[1], Sam Harris[1], Anthony Trieu[1], Parthiban Radhakrishnan[1], Andres Salazar[1], Ke Zhang[1], Jiulin Wang[2], Jianmin Wan[2], Shozo Fujioka[3], Ken Feldmann[1], and Roger Pennell[1]

[1]Ceres, Inc., 1535 Rancho Conejo Boulevard, Thousand Oaks, CA 91320, USA.
[2]Institute of Crop Sciences, Chinese Academy of Agricultural Sciences, Beijing 100081, China.
[3]Plant Functions Laboratory, RIKEN, Wako-shi, Saitama 351-0198, Japan.

Many of the semi-dwarf but high-yielding crop varieties that were developed during Green Revolution are defective in gibberellin biosynthesis or unresponsive to gibberellin (Peng et al., 1999). Of the other kinds of plant hormones, brassinosteroids (BRs) seem to be among the most useful for controlling crop productivity – BR mutants can also be dwarfs and over-expression lines can be high yielding (Choe et al., 2001; Mori et al., 2002; Sakamoto et al., 2006).

Like all hormones, BRs affect a broad range of plant processes including those that control size, branch number and leaf angle, and manipulation of BR levels in different parts of crop plants is one way to increase commercial grain yield (Sakamoto and Matsuoka, 2004; Morinaka et al., 2006). While the specific manipulations that would be optimal for crop productivity are difficult to define, the *osdwf4-1* mutations that result in smaller plants with reduced leaf area for light and CO_2 capture and with slight negative effects on harvest index in rice (Sakamoto et al., 2006) suggest that procedures resulting in reduced BR levels could be used to improve planting

Z. Xu et al. (eds.), Biotechnology and Sustainable Agriculture 2006 and Beyond, 319–322.

density but not intrinsic yield. But while over-expression of BR pathway genes can increase seed yield it can also make leaves more horizontal and more overlapping and render plants taller and prone to lodging, suggesting that over-expression procedures need to be targeted to certain plant parts or to specific stages in development for optimization of light harvesting, plant density and seed productivity.

We have been trying to manipulate BR levels in rice plants in order to improve commercial grain yields. One of the things we have been trying to do is to affect the flow of assimilate from the leaves to the seeds – BRs have been implicated in the control of photosynthesis (Yu et al., 2004; Sakamoto et al., 2006) – and we have been exploring the possibility that the manipulation of BR levels in the green parts of rice plants might be a useful way to enhance unloading of assimilates in seeds and the overall weight of the grain. We have been using *DWF4* and *DWF11* genes from *Arabidopsis*, rice and corn and a promoter that is broadly active only in the stems, leaves and roots of rice plants to do this. *AtDWF4*, *OsDWF4* and *ZmDWF11* encode cytochrome P450 proteins necessary for the rate-limiting hydroxylation of campestanol (CN) to 6-deoxocathasterone (6-DeoxoCT) in BR biosynthesis and are associated with growth, development and seed yield traits in rice (Yamamuro et al., 2000; Tanabe et al., 2005), and a 1953 nt part of an *Arabidopsis* S-adenosylmethionine synthase promoter (pAs) is active in vegetative tissues, but not active in the flowers other than in the filaments of the stamens or in the seeds other than in the epidermis of the seed coat.

Expression of *DWF4* and *DWF11* genes under the control of pAs resulted in plants bigger than the wild type control (Figure 1). GC-MS analysis of transgenic seedlings carrying either *DWF4* genes or *DWF11* showed that the production of 6-DeoxoCT was increased by two to five folds (Figure 2). As a result, grain yield of transgenic plants was significantly increased in both greenhouse and field trials (Figures 3 and 4). The expression pattern of pAs, as well as microarray and photosynthetic efficiency analysis of leaves and seeds (data not shown), suggest that BR levels in the photosynthetic tissues are able to affect the flow of assimilates to the ovules and seeds. These results suggest that promoters that are active in specific parts of rice plants together with genes that regulate the levels of BRs can be used to optimize the grain yield of rice.

Figure 1. Transgenic vs wt plants. T, transgenic; wt, wild type control.

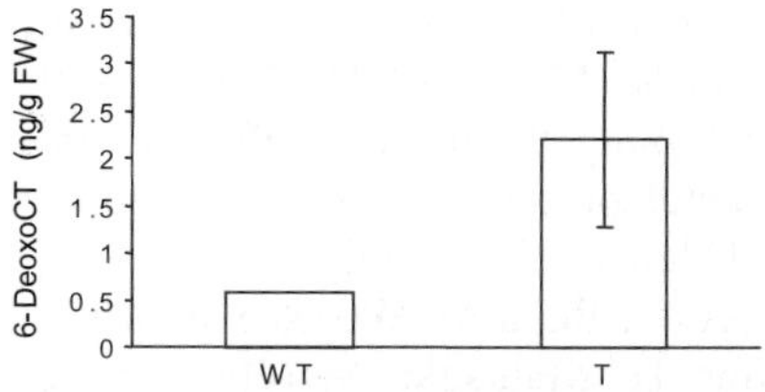

Figure 2. Content of 6-DeoxoCT in seedlings.

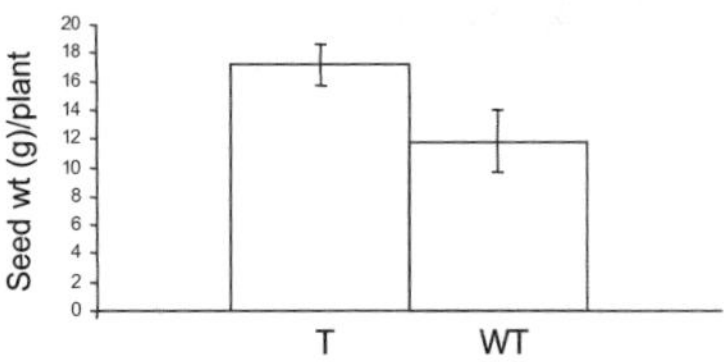

Figure 3. Grain yield of greenhouse plants.

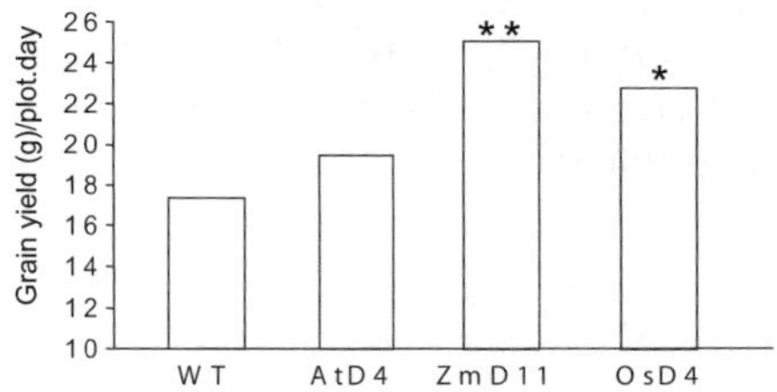

Figure 4. Grain yield of field plants.
* Significant at 5%; ** at 1% by LSD test.

References

Choe S, Fujioka S, Noguchi T, Takatsuto S, Yoshda S, Feldmann KA (2001) Overexpression of DWARF4 in the brassinosteroid biosynthetic pathway results in increased vegetative growth and seed yield in Arabidopsis. Plant J 26: 573–582.

Mori M, Nomura T, Ooka H, Ishizaka M, Yokota T, Sugimoto K, Okabe K, Kajiwara H, Satoh K, Yamamoto K, Hirochika H, Kikuchi S (2002) Isolation and characterization of a rice dwarf mutant with a defect in brassinosteroid biosynthesis. Plant Physiol 130: 1152–1161.

Morinaka Y, Sakamoto T, Inukai Y, Agetsuma M, Kitano H, Ashikari M, Matsuoka M (2006) Morphological alteration caused by brassinosteroid insensitivity increases the biomass and grain production of rice. Plant Physiol 141: 924–931.

Peng J, Richards DE, Hartley NM, Murphy GP, Devos KM, Flintham JE, Beales J, Fish LJ, Worland AJ, Pelica F, Sudhakar D, Christou P, Snape JW, Gale MD, Harberd NP (1999) Green revolution genes encode mutant gibberellin response modulators. Nature 400: 256–261.

Sakamoto T, Morinaka Y, Ohnishi T (2006) Erect leaves caused by brassinosteroid deficiency increase biomass production and grain yield in rice. Nat Biotechnol 24: 105–109.

Sakamoto T, Matsuoka M (2004) Generating high-yielding varieties by genetic manipulation of plant architecture. Curr Opin Biotechnol 15: 144–147.

Tanabe S, Ashikari M, Fujioka S, Takatsuto S, Yoshida S, Yano M, Yoshimura A, Kitano H, Matsuoka M, Fujisawa Y, Kato H, Iwasaki Y (2005) A novel cytochrome P450 is implicated in brassinosteroid biosynthesis via the characterization of a rice dwarf mutant, dwarf11, with reduced seed length. Plant Cell 17: 776–790.

Yamamuro C, Ihara Y, Wu X, Noguchi T, Fujioka S, Takatsuto S, Ashikari M, Kitano H, Matsuoka M (2000) Loss of function of a rice brassinosteroid insensitivel homolog prevents internode elongation and bending of the lamina joint. Plant Cell 12: 1591–1605.

Yu JQ, Huang LF, Hu WH, Zhou YH, Mao WH, Ye SF, Nogues S (2004) A role for brassinosteroids in the regulation of photosynthesis in Cucumis sativus. J Exp Bot 55: 1135–1143.

Functional Genomics of Hexaploid Wheat

Yasunari Ogihara

Kihara Institute for Biological Research, Yokohama City University, Yokohama 244-0813, Japan.

Compared to rice, wheat exhibits characteristic growth habits and contains complex genome constituents. In order to assess global changes in gene expression patterns, we conducted large scale analysis of expressed sequence tags (ESTs). At first, ten tissues during the life cycle of common wheat were used to construct cDNA libraries. Several thousands colonies were randomly selected from each of these ten cDNA libraries and sequenced from both 5' and 3' ends. These ESTs were able to be grouped into representative contigs which correspond to each of three homologues. By computing abundantly expressed ESTs, correlated expression patterns of genes across the tissues and relationships of gene expression profiles among the ten wheat tissues were identified (Ogihara et al., 2003). Furthermore, homologous gene expression in hexaploid wheat can be discriminated by SNPs found among homoloci by using nullisomics-tetrasomics series of Chinese Spring wheat (Mochida et al., 2003). This technique was applied for gene expression profiles in 21 stress-induced tissues, such as callus, as well as liquid cultures and seedlings under abiotic stress conditions (temperature treatment, desiccation, photoperiod, moisture and ABA). These EST data bases enable us to infer the relationships between gene expression profiles among the stress-induced tissues. Multi-dimensional analysis of EST data is analogous to microarray experiments. As in the case of microarray, genes specifically induced and/or suppressed by stresses in comparison to the control expression level were selected from the virtual display *in silico*. The selected genes were annotated with the

Z. Xu et al. (eds.), Biotechnology and Sustainable Agriculture 2006 and Beyond, 323–324.

BLAST search. *In silico* selection of screened genes from virtual display should provide a powerful tool for functional plant genomics (Mochida et al., 2006).

In total, 335,014 sequences were pooled in our data base. 73,483 contigs were grouped with the CAP3, and these contigs were classified into 35,004 gene clusters, which cover more than 90% of rice-predicted genes. Our current activity is shown in Figure 1.

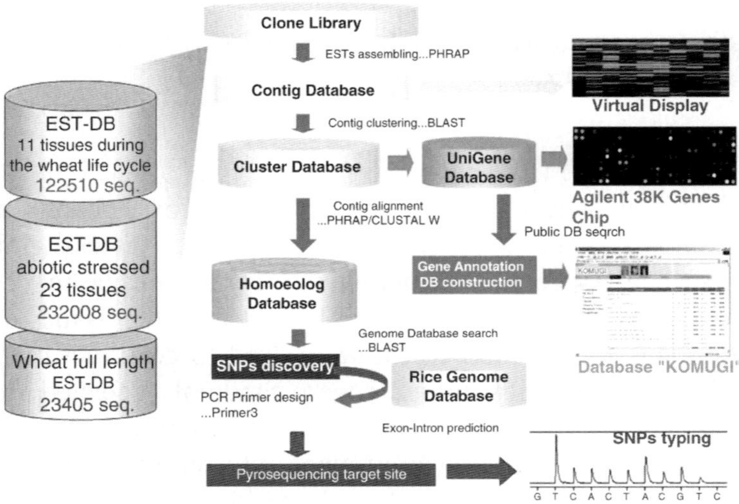

Figure 1. Wheat ESTs clustering pipeline: WECPI

References

Mochida K, Kawaura K, Shimosaka E, Kawakami N, Shin-I T, Kohara Y, Yamazaki Y, Ogihara Y (2006) Tissue expression map of a large number of expressed sequences and its application to in silico screening of stress response genes in common wheat. Mol Genet Genomics (E-pub).

Mochida K, Yamazaki Y, Ogihara Y (2003) Discrimination of homoeologous gene expression in hexaploid wheatby SNP analysis of contigs grouped from a large number of expressed sequence tabs. Mol Gen Genomics 270: 371–377.

Ogihara Y, Mochida K, Nemoto Y, Murai K, Yamazaki Y, Shin-I T, Kohara Y (2003) Correlated clustering and virtual display of gene expression patterns in the wheat life cycle by a large-scale statistical analyses of expressed sequence tags. Plant J 33: 1001–1011.

Biotechnology in Economic Crops

Isolation and Characterization of Senescence-Associated Ethylene Genes from *Dendrobium* Orchids

Zuliana Razali[1], S. Chandran[1], Lee Ai Ling[1], A.N. Boyce[1], and Helen Nair[2]

[1]Institute of Biological Sciences, Faculty of Science, University of Malaya, Kuala Lumpur, Malaysia.
[2]Department of Biotechnology, Faculty of Applied Sciences, Asian Institute of Medicine, Science and Technology, Amanjaya Campus, Amanjaya, 08000 Sungai Petani, Kedah, Malaysia.
Email: helen_nair@aimst.edu.my

Abstract

Pollination hastens senescence in most flowers via the regulation of the ethylene pathway. This phenomenon impacts on the lucrative orchid cut flower industry when attractive hybrids that are ethylene-sensitive undergo rapid post-pollination induced senescence. Manipulation of ethylene-associated genes could overcome the problem in transgenic orchids if ethylene biosynthesis is suppressed or if the signal transduction pathway is confounded. While ACCO oxidase and ACC synthase have been acknowledged as key enzymes in the ethylene biosynthesis pathway, knowledge about ethylene receptors in orchids is still quite limited and often confusing. In our studies with *Dendrobium* Pompadour orchids, we have isolated the genes for ACC oxidase and ACC synthase as well as those for two ethylene receptors first characterized in *Arabidopsis*, namely ETR1and ERS1. Physiological studies have shown that within 24 h, pollinated *Dendrobium* flowers display an increase in ethylene production after

Z. Xu et al. (eds.), Biotechnology and Sustainable Agriculture 2006 and Beyond, 327–332.

pollination, alongside distinct visual changes. The presence of ACC oxidase, ACC synthase, ETR1 and ERS1 genes in pollinated and unpollinated flowers was detected using RT-PCR. Detailed analysis, with Scan Prosite, showed that the nucleotide sequence for the RT-PCR product of ACC oxidase (925 bp) displays more than 60% homology with known ACC oxidases from other orchids, viz. a *Phalaenopsis* cv SM9108, *Doritaenopsis* sp. and *Dendrobium crumenatum* as well as carnation and rose. The ETR1 RT-PCR product (744 bp) shares more than 60% nucleotide sequence homology with the orchid, *Phalaenopsis* as well as tobacco, rice, petunia and carnation. Sequence results for ACCS1 and ERS1 as analysed and characterized using bioinformatics tools are also presented.

Introduction

The regulation of pollination-induced senescence in flowers by ethylene has long been an accepted dogma (Arditti, 1979). *Dendrobium* hybrid flowers experience an increase in ethylene production and senesce within 24 h after pollination (Nair et al., 1994), thus shortening their life span and decreasing their value to the cut flower industry. ACC (1-aminocyclopropane-1-carboxylic acid) is the immediate precursor of ethylene while ACC synthase and ACC oxidase play a central role in ethylene biosynthesis, converting S-adenosyl methionine into ACC and ACC into ethylene respectively (see Srivastava, 2002). One of the most successful manipulations of ACCO was the production of transgenic carnations with antisense ACCO which showed a significant delay in petal senescence concomitant with low ethylene production (Savin et al., 1995). Later, the discovery of a family of membrane-associated ethylene receptors in *Arabidopsis* started to unravel the mechanism(s) behind ethylene signalling. The first ethylene receptor to ever be identified was ETR1 in *Arabidopsis* (Bleecker et al., 1988) and later in other plants, including the orchid (Do et al., 1999). The mutant *etr1-1* gene of *Arabidopsis* was shown to significantly delay fruit ripening, flower senescence and flower abscission when expressed in petunia and tomato plants (Wilkinson et al., 1997). Subsequently, other ethylene receptors, ETR2, ERS1, ERS2 and EIN4 were also identified (see Srivastava, 2002). In this paper, we report the cloning and characterization of both *ACCO and ACCS* and two ethylene receptors, ETR1 and ERS1 from a popular tropical orchid genus, the *Dendrobium,* as part of initial steps towards understanding the regulation of ethylene-sensitivity in pollinated and non-pollinated monocot flowers.

Materials and methods

RNA extraction and RT-PCR analysis

Total RNA from *Dendrobium* flowers (pollinated/unpollinated) was extracted at specific time intervals, using an RNA extraction procedure, modified from Liu et al. (1998). RNA samples were verified for purity by visualizing fractionated and ethidium bromide-stained samples, prior to RT-PCR procedures with the AccessQuick RT-PCR System (Promega). Primers for ACCO were designed from the sequence for ACC oxidase gene for *Phalaepnopsis* cv SM9108 (Nadeau et al., 1993) and for ETR1 gene from *Arabidopsis* (Chang et al., 1993). Multiple alignments provided primers for ACC synthase 1 and ERS1.

Sequencing

First Base Laboratories, Kuala Lumpur sequenced the RT-PCR product. The CLUSTAL W (1.82) program was employed for the sequence analysis and multiple sequence alignments. A detailed analysis of protein structure was also conducted in order to identify the specific regions characteristic to ACC oxidase, ACC synthase, ETR1 and ERS1 proteins.

Results and discussion

The ACC oxidase primers produced a product of about 1kb in length and more than 65% nucleotide sequence homology with known ACC oxidase nucleotide sequences of *Doritaenopsis* sp. (Nadeau and O'Neill, 1995), *Dendrobium crumenatum* (Yang et al., 1999),

Phalaenopsis cv SM9108 (Nadeau et al., 1993) and carnation (Wang and Woodson, 1991). The sequence of about 965 bp predicts a protein of approximately 322 amino acids. A preliminary phylogenetic tree by multiple sequence alignments with CLUSTAL W (1.8) shows that the ACC oxidase from *Dendrobium* Pompadour flowers is more closely related to carnation than to *Doritaenopsis* sp. or *Dendrobium crumenatum.* A detailed analysis of protein structure with Scan Prosite shows that the ACC oxidase sequence from *Dendrobium* Pompadour has four different groups characteristic to the ACC oxidase protein family, viz. one N-glycosylation site, one protein kinase C phosphorylation site, two casein kinase II phosphorylation sites and seven N-myristoylation sites. The ACC synthase primers produced a product of 1.3 kb in length with more than 80% homology with ACC synthase nucleotide sequences reported in *Arabidopsis* (Q06402), *Pelargonium hortorum* (Q43810) and soybean (Q27ZJ5). Analysis with Scan Prosite showed the presence of two N-glycosylation sites, one cAMP- and cGMP-dependent protein kinase phosphorylation sites, four protein kinase C phosphorylation sites, four protein kinase C phosphorylation sites, two amidation sites, ten casein kinase II phosphorylation sites and five N-myristoylation sites.

With the *ETR1* primers, the RT-PCR product gave a nucleotide sequence of 744 bp with more than 65% homology to tobacco (Knoester et al., 1997), rice (Yau and Yip, 1997), petunia (Lai and Shaw, 2001), *Phalaenopsis* (Do et al., 1999) and carnation (Shibuya et al., 1998). The RT-PCR product predicted a sequence of about 248 amino acids. The phylogenetic tree, different from that for ACCO, showed that the *ETR1* sequence from *Dendrobium* Pompadour is more closely related to petunia than *Phalaenopsis* .The Scan Prosite results identified regions characteristic to the ETR1 protein family, viz. one N-glycosylation site, three protein kinase C phosphorylation sites at different residues, and two N-myristoylation sites. ERS 1 primers produced a bigger product of 1.2 kb, compared to that of ETR1 with more than 80% homology with delphinium (AB201245) petunia hybrid (AF145972) and *Oncidium* Gower Ramsey (AF276235). Analysis with Scan Prosite showed the presence of one N-glycosylation site, seven protein kinase C phosphorylation sites, three casein kinase II phosphorylation sites and six N-myristoylation sites.

In this study we have isolated and cloned good-size ethylene-associated genes of *Dendrobium* Pompadour that are linked to post-pollination induced senescence.

References

Arditti J (1979) Aspects of the physiology of orchids. Adv Bot Res 7: 421–655.

Bleecker AB, Estelle MA, Somerville C, Kende H (1988) Insensitivity to ethylene conferred by a dominant mutation in *Arabidopsis thaliana*. Science 24a1:1086–1089.

Chang C, Kwok SF, Bleecker AB, Meyerowitz EM (1993) Arabidopsis ethylene-response gene ETR1: Similarity of product to two-component regulators. Science 262: 539–544.

Do YY, Chen YC, Huang PL (1999) Molecular analysis of a cDNA encoding ethylene receptor that expresses in *Phalaenopsis* petals (Accession Number AF055894) (PGR99-047). Plant Physiol 119: 1567.

Knoester M, Hennig J, van Loon LC, Bol JF, Linthorst HJM (1997) Isolation and characterization of a tobacco cDNA encoding an ETR1 homolog (Accession No. AF022727) (PGR97-188). Plant Physiol 115: 1731.

Lai JP, Shaw JF (2001) *Petunia multiflora* hybrid putative ethylene receptor (PETR3) mRNA. (Accession No. AF 145975). Unpublished.

Liu JJ, Goh CJ, Loh CS, Liu P, Pua EC (1998) A method for isolation of total RNA from fruit tissues of banana. Plant Mol Bio Reporter: 1–6.

Nadeau JA, O'Neill SD (1995) Nucleotide sequence of a cDNA encoding 1-aminocyclopropane-1carboxylate oxidase from senescing orchid petals. Plant Physiol 108: 833–834.

Nadeau JA, Zhang XS, Nair H, O'Neill SD (1993) Temporal and spatial regulation of 1-aminocyclopropane-1-carboxylate oxidase in the pollination-induced senescence of orchid flowers. Plant Physiol 103: 31–39.

Nair H, Lim TH, Jamilah Suhaimi (1994) Pollination-induced senescence in *Dendrobium* orchid flowers. Trans Mal Soc Plant Physiol 5: 111–114.

Savin KW, Baudinette SC, Graham MW, Micheal MZ, Nugent GD, Lu CY, Chandler SF, Cornish EC (1995) Antisense ACC oxidase RNA delays carnation petals senescence. HortScience 30(5): 970–972.

Shibuya K, Satoh S, Yoshioka T (1998) A cDNA encoding a putative ethylene receptor related to petal senescence in carnation (*Dianthus caryophyllus L.*) flowers (Accession No. AF034770) (PGR98-019). Plant Physiol 116: 867.

Srivastava LM (2002) In: Plant Growth and Development – Hormones and Environment. Academic Press.

Wang H, Woodson WR (1991) A flower senescence-related mRNA from carnation shares sequence similarity with fruit ripening-related mRNAs involved in ethylene biosynthesis. Plant Physiol 96: 1000–1001.

Wilkinson JQ, Lanahan MB, Clark DG, Bleecker AB, Chang C, Meyerowitz EM, Klee HJ (1997) A dominant mutant receptor from Arabidopsis confers ethylene insensitivity in heterologous plants. Nat Biotechnol. 15(5): 444–447.

Yang XH, Pua EC, Goh CJ (1999) Isolation of a cDNA clone encoding 1-aminocyclopropane-1-carboxylate oxidase from *Dendrobium crumenatum*. Plant Physiol 119(2): 805–805.

Yau CP, Yip WK (1997) Nucleotide sequence of a full-length cDNA encoding an ethylene receptor from rice (Accession No. AF013979)(PGR97-134). Plant Physiol 115: 315.

Biotechnological Improvement of Forage Crops

Zeng-Yu Wang*, Jeremey Bell, Xiaofei Cheng, Yaxin Ge, Kun Jun Han, Xuefeng Ma, Elane Wright, Yajun Xi, Xirong Xiao, Ji-Yi Zhang, Andrew Hopkins, and Joseph Bouton

Forage Improvement Division, The Samuel Roberts Noble Foundation, 2510 Sam Noble Parkway, Ardmore, OK 73401, USA.
*Email: zywang@noble.org

Forages play a unique role in agriculture and the environment; they contribute through animals to our food supply. Genetic improvement is one of the most effective ways to improve forage productivity. The potential of biotechnology in the development of improved forage grass cultivars has been recognized in recent years. Transgenic approaches are expected to complement or accelerate conventional breeding, since they offer the opportunity to generate unique genetic variation that would otherwise be absent or low in heritability.

We have been interested in developing biotechnological methods to improve a number of important forage grass and legume species, such as tall fescue (*Festuca arundinacea*), bermudagrass (*Cynodon dactylon*), switchgrass (*Panicum virgatum*), alfalfa (*Medicago sativa*) and white clover (*Trifolium repens*). This report briefly summarizes our work in the following aspects: establishment of efficient genetic transformation systems for different forage species, cloning of potentially useful agronomical genes and promoters, and generation of transgenic forage plants with improved agronomic characteristics.

Efficient plant regeneration and genetic transformation systems have been established for all the relevant grass and legume species in which we are interested. Transgenic grass plants were first obtained

Z. Xu et al. (eds.), Biotechnology and Sustainable Agriculture 2006 and Beyond, 333–338.

by biolistic transformation of embryogenic cell cultures (Chen et al., 2003, 2004; Wang et al., 2003, 2004a). Later, new protocols based on *Agrobacterium*-mediated transformation were developed (Wang and Ge, 2005a, b, 2006). It is known that *Agrobacterium*-mediated transformation generally results in a lower copy number and an improved stability of gene expression than the free DNA delivery methods.

For *Agrobacterium*-mediated transformation of grasses, a chimeric hygromycin phosphotransferase (*hph*) gene, which renders transformed cells resistant to hygromycin, was used as a selectable marker. To produce transgenic tall fescue, switchgrass and darnel ryegrass (*Lolium temulentum*) plants, embryogenic calluses induced from mature embryos/seeds were co-cultivated with *Agrobacterium* strain EHA105 carrying binary vectors. Hygromycin resistant calluses were obtained after selection and transgenic plants were subsequently regenerated from the resistant calluses (Wang and Ge, 2005a). For bermudagrass, creeping bentgrass (*Agrostis stolonifera*) and zoysiagrass (*Zoysia japonica*) transformation, stolon nodes were co-cultivated with *Agrobacteria* and transgenic plants were obtained directly from the infected stolon (Wang and Ge, 2005b; Ge et al., 2006). The procedure allows for the rapid production of transgenics bypassing the callus phase (Wang and Ge, 2006). Transgenic nature of the regenerated plants was demonstrated by Southern hybridization analysis. Expression of the transgenes was confirmed by northern hybridization analysis, GUS staining and detection of GFP signals (Wang and Ge, 2005a, b; Ge et al., 2006).

Compared with grasses, *Agrobacterium*-mediated transformation is relatively easier in legume species. Large numbers of transgenic alfalfa, white clover and *Medicago truncatula* plants have been routinely produced (Zhang et al., 2005; Crane et al., 2006; Wright et al., 2006; Xie et al., 2006). Leaves and cotyledons have been used as explants for alfalfa and white clover transformation, respectively. Cotyledons, roots and leaves have been successfully used for *M. truncatula* transformation (Crane et al., 2006; Wright et al., 2006; Xie et al., 2006).

The development of transformation technologies in grasses and legumes provides opportunities to generate value-added materials by the direct introduction of chimeric genes. Projects aimed at improving agronomic traits of forage grasses and legumes are in progress. The target traits are improved forage quality, drought tolerance, increased phosphate uptake and flowering time control.

Forage quality, particularly digestibility, is a limiting factor for animal productivity. It is known that lignification of plant cell walls is largely responsible for lowering digestibility of forage tissues. We analyzed lignin deposition at different development stages of tall fescue and cloned cDNAs of two enzymes involved in lignin biosynthesis: cinnamyl alcohol dehydrogenase (CAD) and caffeic acid *O*-methyltransferase (COMT) (Chen et al., 2002). Transgenic tall fescue plants were produced using sense and antisense CAD and COMT gene constructs. Severely reduced mRNA levels and significantly decreased enzymatic activities were found in some transgenic lines. These transgenic tall fescue plants had reduced lignin content, altered lignin composition and increased *in vitro* dry matter digestibility (Chen et al., 2003, 2004). Lignin also negatively affects enzymatic hydrolysis and utilization of plant structural polysaccharides for ethanol production. Currently, we are using a similar genetic engineering strategy to reduce lignin in switchgrass, a biomass crop that has huge potential for biofuel production.

Drought tolerance is an important target for improvement in perennial forages. We characterized a novel AP2 domain-containing transcription factor gene (*WXP1*) from the model legume *Medicago truncatula* (Zhang et al., 2005). Overexpression of *WXP1* under the control of CaMV35S promoter led to a significant increase in cuticular wax loading on leaves of transgenic alfalfa. The increase in wax accumulation was mainly contributed by C30 primary alcohol. Transgenic leaves showed decreased water loss and reduced chlorophyll leaching. Transgenic alfalfa plants with increased cuticular waxes showed enhanced drought tolerance (Zhang et al., 2005).

Phosphorus is immobile and often deficient in pasture soils. Improving phosphate uptake in plants is an economic way to increase forage production. In addition, improved phosphate uptake could help to remove excessive phosphate in areas where large amounts of poultry litter have been applied. We cloned and characterized a constitutive promoter, two root-specific promoters, a novel phytase gene and a purple acid phosphatase gene from *M. truncatula* (Xiao et al., 2005a, b, 2006a, b). Transgenic expression of the phytase gene or the purple acid phosphatase gene in *Arabidopsis* and white clover led to significant improvement in organic phosphorus uptake and plant growth. The results showed that the transgenic approach has great potential for improving plant organic phosphorus utilization and for phytoremediation (Xiao et al., 2005a, 2006a).

Genetic manipulation of flowering time in forages can be beneficial for forage quality and transgene containment. We found that certain *CONSTANS-LIKE* genes play opposite roles to the *CONSTANS* gene in flowering time control (Cheng and Wang, 2005). *L. temulentum* has been proposed as a model species for study of floral genes because it is an inbred, diploid grass species with a short life cycle. We have developed tissue culture responsive genotypes and an efficient transformation system for *L. temulentum* (Wang et al., 2002, 2005).

Information regarding gene flow in wind-pollinated outcrossing forage crops is essential for regulation and releases of transgenic cultivars (Wang et al., 2004c). Pollen viability and longevity were similar between transgenic and control plants (Wang et al., 2004b). Relatively large-scale experiments on gene flow in transgenic tall fescue and Italian ryegrass are being carried out in the field, which will give us insights into the distance and frequency of pollen flow, and the probability/frequency of cross hybridization between several grass species.

References

Chen L, Auh C, Chen F, Cheng XF, Aljoe H, Dixon RA, Wang Z-Y (2002) Lignin deposition and associated changes in anatomy, enzyme activity, gene expression and ruminal degradability in stems of tall fescue at different developmental stages. J Agric Food Chem 50: 5558–5565.

Chen L, Auh C, Dowling P, Bell J, Chen F, Hopkins A, Dixon RA, Wang Z-Y (2003) Improved forage digestibility of tall fescue (*Festuca arundinacea*) by transgenic down-regulation of cinnamyl alcohol dehydrogenase. Plant Biotechnol J 1: 437–449.

Chen L, Auh C, Dowling P, Bell J, Lehmann D, Wang Z-Y (2004) Transgenic down-regulation of caffeic acid *O*-methyltransferase (COMT) led to improved digestibility in tall fescue (*Festuca arundinacea*). Functional Plant Biol 31: 235–245.

Cheng X-F, Wang Z-Y (2005) Overexpression of *COL9*, a *CONSTANS-LIKE* gene, delays flowering by reducing *CO* and *FT* expression in *Arabidopsis thaliana*. Plant J 43: 758–768.

Crane C, Wright E, Dixon RA, Wang Z-Y (2006) Transgenic *Medicago truncatula* plants obtained from *Agrobacterium tumefaciens*-transformed roots and *Agrobacterium rhizogenes*-transformed hairy roots. Planta 223: 1344–1354.

Ge Y, Norton T, Wang Z-Y (2006) Transgenic zoysiagrass (*Zoysia japonica*) plants obtained by *Agrobacterium*-mediated transformation. Plant Cell Rep 25: 792–798.

Wang Z-Y, Bell J, Ge YX, Lehmann D (2003) Inheritance of transgenes in transgenic tall fescue (*Festuca arundinacea* Schreb.). In Vitro Cell Dev Biol Plant 39: 277–282.

Wang Z-Y, Bell J, Lehmann D (2004a) Transgenic Russian wildrye (*Psathyrostachys juncea*) plants obtained by biolistic transformation of embryogenic suspension cells. Plant Cell Rep 22: 903–909.

Wang Z-Y, Ge Y (2005a) *Agrobacterium*-mediated high efficiency transformation of tall fescue (*Festuca arundinacea* Schreb.). J Plant Physiol 162: 103–113.

Wang Z-Y, Ge Y (2005b) Rapid and efficient production of transgenic bermudagrass and creeping bentgrass bypassing the callus formation phase. Functional Plant Biol 32: 769–776.

Wang Z-Y, Ge Y (2006) Recent advances in genetic transformation of forage and turf grasses. In Vitro Cell Dev Biol – Plant 42: 1–18.

Wang Z-Y, Ge Y, Mian R, Baker J (2005) Development of highly tissue culture responsive lines of *Lolium temulentum* by anther culture. Plant Sci 168: 203–211.

Wang Z-Y, Ge YX, Scott M, Spangenberg G (2004b) Viability and longevity of pollen from transgenic and non-transgenic tall fescue (*Festuca arundinacea*) (*Poaceae*) plants. Am J Bot 91: 523–530.

Wang Z-Y, Hopkins A, Scott M (2002) Plant regeneration from embryogenic cell suspension cultures of *Lolium temulentum*. In Vitro Cell Dev Biol Plant 38: 446–450.

Wang Z-Y, Lawrence R, Hopkins A, Bell J, Scott M (2004c) Pollen-mediated transgene flow in the wind-pollinated grass species tall fescue (*Festuca arundinacea* Schreb.). Mol Breed 14: 47–60.

Wright E, Dixon RA, Wang Z-Y (2006) *Medicago truncatula* transformation using cotyledon explants. In: Wang K (ed) *Agrobacterium* Protocols (2nd edition). Humana Press, Totowa, NJ, pp. 129–135.

Xiao K, Harrison M, Wang Z-Y (2005a) Transgenic expression of a novel *M. truncatula* phytase gene results in improved acquisition of organic phosphorus by *Arabidopsis*. Planta 222: 27–36.

Xiao K, Katagi H, Harrison M, Wang Z-Y (2006a) Improved phosphorus acquisition and biomass production in Arabidopsis by transgenic expression of a purple acid phosphatase gene from *M. truncatula*. Plant Sci 170: 191–202.

Xiao K, Liu J, Dewbre G, Harrison M, Wang Z-Y (2006b) Isolation and characterization of root-specific phosphate transporter promoters from *Medicago truncatula*. Plant Biol 8: 439–449.

Xiao K, Zhang C, Harrison M, Wang Z-Y (2005b) Isolation and characterization of a novel plant promoter that directs strong constitutive expression of transgenes in plants. Mol Breed 15: 221–231.

Xie D-Y, Sharma SB, Wright E, Wang Z-Y, Dixon RA (2006) Metabolic engineering of proanthocyanidins through co-expression of anthocyanidin reductase and the PAP1 MYB transcription factor. Plant J 45: 895–907.

Zhang J-Y, Broeckling CD, Blancaflor EB, Sledge M, Sumner LW, Wang Z-Y (2005) Overexpression of *WXP1*, a putative *Medicago truncatula* AP2 domain-containing transcription factor gene, increases cuticular wax accumulation and enhances drought tolerance in transgenic alfalfa (*Medicago sativa*). Plant J 42: 689–707.

Biotechnology in Medicinal, Woody and Horticultural Plants

Lignan and Norlignan Biosynthesis and Biotechnology

Toshiaki Umezawa

Research Institute for Sustainable Humanosphere, Kyoto University, Uji, Kyoto 611-0011, Japan.

Lignans and norlignans are two major classes of plant phenylpropanoid compounds and their biosynthesis has been receiving widespread interests for their importance in heartwood formation and pharmaceutical applications (Umezawa, 2003, 2005).

Initial steps of the lignan biosynthetic pathway, that is, the pathway from coniferyl alcohol to matairesinol, have been well-established, and several cDNAs encoding enzymes catalyzing these steps have been cloned (Umezawa, 2003, 2005). However, further conversion of matairesinol to many other lignans remained to be elucidated. For example, in spite of frequent occurrence of methoxyl groups in lignan molecules, little is known about *O*-methyltransferases (OMTs) involved in *O*-methylation of lignans including matairesinol.

Arctigenin is one of the simplest lignans with a methoxyl group introduced by lignan-OMT-catalyzed methylation. Because the lignan occurs in *Taiwania cryptomeriodes* heartwood (Chang et al., 2000), matairesinol OMT (MROMT) which methylates regioselectively matairesinol to afford arctigenin plays an important role in heartwood formation.

Recently, we detected a MROMT activity from *Carthamus tinctorius* (safflower) seeds, which catalyzed regioselective methylation of matairesinol to yield arctigenin, but not isoarctigenin (Umezawa

Z. Xu et al. (eds.), Biotechnology and Sustainable Agriculture 2006 and Beyond, 341–343.

et al., 2004). In addition, we isolated a novel *Carthamus tinctorius* gene encoding the OMT by use of a PCR-guided strategy (Umezawa et al., 2004). The recombinant *Ct*MROMT expressed in *E. coli* exhibited the same regioselectivity as the plant protein, whereas the recombinant enzyme did not catalyze *O*-methylation of phenylpropanoid monomers. A flavonoid (apigenin) was found to be a poor substrate. Thus, this is the first molecular cloning of a cDNA that encodes a plant OMT specific to lignan methylation.

As for norlignan biosynthesis, hinokiresinol is the simplest norlignan and, therefore, a good target for norlignan biosynthetic studies (Umezawa, 2003, 2005). This norlignan has two geometrical isomers, (*Z*)- and (*E*)-hinokiresinols. *Asparagus officinalis* and *Cryptomeria japonica* give the (*Z*)- and (*E*)-isomers, respectively. From these plants, we detected for the first time enzyme activities catalyzing formation of (*Z*)-hinokiresinol (Suzuki et al., 2002) and (*E*)-hinokiresinol (Suzuki et al., 2004), respectively.

Then, we purified (*Z*)-hinokiresonol synthase from *A. officinalis*, and isolated cDNAs encoding the enzyme. It was found that the enzyme is composed of two subunits. When each recombinant subunit was incubated individually with 4-coumaryl 4-coumarate, (*E*)-hinokiresonol, but not (*Z*)-isomer, was formed. Interestingly, however, incubating the same substrate with a mixture of the two subunits gave rise to (*Z*)-hinokiresinol that is the isomer occurring in *A. officinalis*. Thus, the subunit composition can control the stereochemistry of the product.

In addition to these two topics, current status of lignan and norlignan biosynthesis will be discussed.

References

Chang S-T, Wang DS-Y, Wu C-L, Shiah S-G, Kuo Y-H, Chang C-J (2000) Cytotoxicity of extractives from *Taiwania cryptomerioides* heartwood. Phyotchemistry 55: 227.

Suzuki S, Nakatsubo T, Umezawa T, Shimada M (2002) First in vitro norlignan formation with *Asparagus officinalis* enzyme preparation. Chem Commun, 1088–1089.

Suzuki S, Yamamura M, Nakatsubo T, Hattori T, Shimada M, Umezawa T (2004) A heartwood norlignan, (*E*)-hinokiresinol, is formed from 4-coumaryl 4-coumarate by a *Cryptomeria japonica* enzyme preparation. Chem Commun, 2838–2839.

Umezawa T (2003) Diversity in lignan biosynthesis. Phytochem Rev 2: 371–390.

Umezawa T (2005) Biosynthesis of lignans, lignins, and norlignans. Kagaku to Seibutsu 43: 461–467.

Umezawa T, Li L, Suzuki S, Sakakibara N, Nakatsubo T, Chiang VL (2004) A novel O-methyltransferase catalyzing a regioselective methylation of lignan Proceedings 49th Lignin Symposium. Tsukuba, Japan, pp. 33–36.

Polyketide Metabolism in *Hypericum perforatum* and Related Species

Ludger Beerhues* and Benye Liu

Institute of Pharmaceutical Biology, Technical University of Braunschweig, Mendelssohnstraße 1, D-38106 Braunschweig, Germany; Key Laboratory of Photosynthesis and Environmental Molecular Physiology, Institute of Botany, Chinese Academy of Sciences, Nanxincun 20, Haidian District, Beijing 100093, China.
*Email: l.beerhues@tu-bs.de

Extracts from St. John's wort (*Hypericum perforatum* L., Clusiaceae) are widely used for the treatment of mild to moderate depression. The major antidepressant constituent is hyperforin, a polyprenylated bicyclic polyketide derivative. This broad-band neurotransmitter re-uptake inhibitor uses a novel mechanism of action which affects the sodium gradient across the neuronal membrane. In addition, hyperforin exhibits antibacterial and antitumoral activities. Clusiaceae (Guttiferae) are rich in polyprenylated phloroglucinol derivatives. Bridged polycyclic compounds with caged skeletons, such as sampsoniones and garcinol, result from cyclizations of the prenyl side chains. No total synthesis has been reported for any of these complex molecules. The combination of challenging chemical structure and intriguing pharmacological activity makes these metabolites attractive compounds for biotechnological research.

The nuclei of acyl- and benzoylphloroglucinols are formed by type III polyketide synthases (PKSs) that use isobutyryl-CoA and benzoyl-CoA, respectively, as starter substrates. In cell cultures of *H. calycinum*, three PKSs, isobutyrophenone synthase (BUS),

Z. Xu et al. (eds.), Biotechnology and Sustainable Agriculture 2006 and Beyond, 345–348.

benzophenone synthase (BPS), and chalcone synthase (CHS), were separated by anion exchange chromatography. cDNAs encoding BPS and CHS were cloned and the recombinant proteins functionally expressed in *E. coli*. BPS prefers benzoyl-CoA as starter substrate and catalyzes the stepwise condensation with three molecules of malonyl-CoA to give a tetraketide intermediate which then is cyclized by intramolecular Claisen condensation into 2,4,6-trihydroxybenzophenone. Among the starter substrates used by PKSs, benzoyl-CoA is a rare starter molecule. BPS is inactive with CoA-linked cinnamic acids such as 4-coumaroyl-CoA. This metabolite, however, is the preferred starter molecule for CHS. Active site residues of the two PKSs were altered by site-directed mutagenesis. A CHS triple mutant preferred benzoyl-CoA over 4-coumaroyl-CoA and formed phlorbenzophenone. A BPS single mutant limited chain extension to two condensations with malonyl-CoA, leading to the formation of phenylpyrone instead of phlorbenzophenone. These mutagenesis experiments were guided by homology modeling, based on the crystal structure of alfalfa CHS.

The aromatic nucleus of hyperforin, phlorisobutyrophenone, is stepwise prenylated. The first prenylation step is catalyzed by a soluble and Fe^{2+}-dependent dimethylallyl transferase. This enzyme is the first plant prenyltransferase to prefer Fe^{2+}, other enzymes are most active with Mg^{2+} or Mn^{2+}. The majority of aromatic prenyltransferases studied are integral membrane proteins. In cell cultures of *H. calycinum*, the formation of hyperforin and its homologue adhyperforin was preceded by an increase in dimethylallyltransferase activity.

In benzophenone metabolism, simple benzophenones are either polyprenylated or intramolecularly cyclized. 2,3',4,6-Tetrahydroxybenzophenone which forms under catalysis of benzophenone 3′-hydroxylase undergoes regioselective ring closures to give either 1,3,5- or 1,3,7-trihydroxyxanthones. These metabolites are the precursors of the majority of plant xanthones. The oxidative phenol couplings occurring either *ortho* or *para* to the 3'-hydroxy group are catalyzed by cytochrome P450 enzymes. The reactions strictly require NADPH and O_2. P450 enzymes are also responsible for subsequent hydroxylation steps, with xanthone 6-hydroxylation being the first modification. NADPH-cytochrome P450 reductase was purified and its cDNA cloned.

Another benzoic acid-specific type III PKS is biphenyl synthase (BIS) which was detected in yeast-extract-treated cell cultures of *Sorbus aucuparia* (Rosaceae). The recombinant enzyme shares 54% amino acid sequence identity with BPS. Both PKSs catalyze the iterative condensation of benzoyl-CoA with three acetyl units from malonyl-CoA and form identical linear tetraketides. While BPS cyclizes this intermediate via an intramolecular C6→C1 Claisen condensation, BIS catalyzes an intramolecular C2→C7 aldol condensation and decarboxylative elimination of the terminal carboxyl group to give 3,5-dihydroxybiphenyl. Biphenyls and dibenzofurans are the phytoalexins of the Maloideae. This subfamily includes economically important species such as apple and pear. In cell cultures of *S. aucuparia*, BIS expression was rapidly, strongly and transiently induced by yeast extract treatment, whereas CHS expression was not.

In a phylogenetic tree, the PKSs from angiosperms fall into two clusters. One cluster comprises CHSs, the other cluster contains the functionally divergent enzymes including BIS and BPS. The two groups originate from an ancient duplication of an ancestral gene. While one gene retained the CHS function which is indispensable due to the important roles of flavonoids, the other gene underwent impressive functional diversification in various taxa. BIS and BPS grouped together closely, indicating their origin from a relatively recent functional divergence of a common ancestor.

A precursor of benzophenones and biphenyls is benzoic acid which is formed from cinnamic acid *via* a CoA-dependent and non-*ß*-oxidative route in cell cultures of *H. androsaemum*. Three CoA ligases preferring either cinnamic acid or benzoic acid and 4-coumaric acid were separated. Cinnamate: CoA ligase channels cinnamic acid from the general phenylpropanoid pathway into the benzoic acid biosynthetic route. The central step is catalyzed by cinnamoyl-CoA hydratase/lyase which adds water to the cinnamoyl double bond and cleaves the resulting 3-hydroxy-3-phenylpropionyl-CoA. In cell cultures of *Centaurium erythraea* (Gentianaceae), 3-hydroxybenzoic acid is derived directly from the shikimate pathway in the presence of ATP. Furthermore, centaury BPS prefers 3-hydroxybenzoyl-CoA.

This starter substrate is supplied by 3-hydroxybenzoate: CoA ligase which was purified to apparent homogeneity.

For metabolic engineering, a transformation protocol for *H. perforatum* is being established. Since the *Agrobacterium tumefaciens*-mediated method failed, gene transfer *via* particle bombardment is under work. A protocol for efficient regeneration is available.

In Vitro Selection and Molecular Markers for Early Screening of Fusarium Head Blight Resistance Wheat

F. Eudes*, Ana Badea, A. Laroche, D. Gaudet, R. Graf, and S. Sadasivaiah

Bioproducts and Bioprocesses, Lethbridge Research Centre, Agriculture and Agri-Food Canada, 5403 – 1st Avenue South, Lethbridge, AB, Canada, T1J 4B1.
*Email: eudesf@agr.gc.ca

The breeding of FHB resistant wheat cultivars has been a slow and frustrating process due to the multigenic nature of resistance, numerous modes of resistance, and difficulties in obtaining reliable results from screening nurseries. The result has been that only a few moderately resistant cultivars have been registered in the United States (e.g., Alsen, Ernie, Glenn, Truman) and Canada (e.g., FT Action, FT Wonder, 5602PR) over the past decade. Two methods proposed for accelerating the rate of introgression from unadapted germplasm, and ultimately increasing the frequency of potential wheat cultivars, are molecular marker assisted selection (Somers et al., 2005) and *in vitro* selection (Eudes et al., 2007).

In vitro selected doubled haploid (DH) lines were produced from crosses among five Canadian wheat classes (Superb (CWRS), AC Crystal (CPS-red), AC2000 (CPS-white), Radiant (CWRW), and AC Reed (CWSWS)) and candidate sources of resistance from China, Brazil, CIMMYT, and Ukraine: CIMMYT-1 (Sha3/Catbird) and CIMMYT-11 (80456/Yangmai 5); CM82036 (Sumai 3/Thornbird) derived from Sumai 3; Chinese line 894013 (somaclonal variant of Ningmai 3); Brazilian lines BR32 and BR38; doubled haploids FE1

Z. Xu et al. (eds.), Biotechnology and Sustainable Agriculture 2006 and Beyond, 349–352.

and FE17 (Norseman/Frontana), FE9 and FE30 (Norseman/Katepwa), and FE39 and FE41 (Norseman/Nyu-Bay). Anther culture was conduced on induction medium supplemented with 0.23 mg L^{-1} deoxynivalenol, 0.23 mg L^{-1} 15-oxy-acetyldeoxynivalenol, 0.47 mg L^{-1} nivalenol and 0.7 mg L^{-1} T2 toxin. Doubled haploid lines were produced and evaluated in the greenhouse using point inoculation and in the field from 2003 to 2005.

In vitro selection leading to improved FHB resistance was observed for most crosses, especially using source of moderate resistance CIMMYT 1 (Sha3/Catbird), CIMMYT 11 (80456/Yangmai 5) and Frontana. Differences in ratings were observed between both phenotypic evaluation methods. Only eight DH lines were shown to have resistance equal to or better than the resistant parents in both phenotypic evaluation methods. However, under field conditions, 29 lines out of 170 were statistically equal to or more resistant than their respective resistant parents. The field evaluation likely tested all mechanisms of FHB resistance that the plant might have, including resistance or tolerance to trichothecenes. A higher proportion of DH lines resistant to FHB in the field were regenerated from in vitro selection of backcross lines utilizing a parent developed by the selection method and verified for FHB resistance. Transgressive segregation for improved FHB resistance was also identified. This process appears to offer greater value for any new candidate source of resistance.

Ten of 17 SSR primers located on chromosomes 3A, 3BS and 5A (QTLs identified for Sumai3 and Frontana) revealed polymorphism among the parental wheat lines. Based on the polymorphism revealed by parental lines, molecular screening was performed for the descendents from the following crosses: CM82036 with AC2000, AC Crystal and AC Superb for 3Bs and 5A QTLs; FE1 with Radiant; FE17 with AC Superb and Radiant for 3A and 5A QTLs. In addition, descendents from crosses between FE1 with AC Crystal for 5A QTL, BR38 with AC2000 and Radiant for 3A QTL; CIMMYT11, FE9, and FE41 with AC Reed for 3A QTL were tested, too. A better efficiency of *in vitro* selection was observed in case of type I resistance (47% for Sumai3 derivates, and 58% for Frontana derivates) than type II resistance (33% for Sumai3 derivates) based on phenotypic results. Similar *in vitro* selection efficiency (47%, 55% and 33%, respectively) was revealed by genotypic analysis, too. Genotypes with marker alleles from resistant parent had a lower FHB rating than those with marker alleles from susceptible cultivar. However, in case of type I resistance one line of Sumai3 derivatives and seven lines of Frontana with the resistant alleles for tested markers were found to have a phenotypic response like susceptible parent, and conversely two, respectively six DH lines with the susceptible alleles had a response as resistant parent. The variation within genotypic classes might be generated by the genetic effects of other QTLs, environmental variation and the variability in phenotypic evaluation.

The usefulness of MAS as an early screening method was shown by the proportion of genotypes which were retained in common in both genotypic and phenotypic selection. Among the DH lines with similar or better response than the parent source of resistance for Frontana derivates 68% carried the resistant parent alleles for type I resistance. A better percentage showed Sumai3 derivatives with 71% for type I and 100% for type II resistance. Within the DH lines with similar response as susceptible parent for Frontana derivatives 50% carried the susceptible parent alleles for type I resistance. Meanwhile for Sumai3 derivatives 75% for type I and 100% for type II resistance were observed.

Efficacy of in vitro selection observed during phenotypic evaluation was confirmed by QTLs analysis. The in vitro selection efficiency was observed in case of type I, in field condition, but not as much in case of type II resistance, using point inoculation. However, marker assisted selection was very accurate when selection was carried out for type II resistance. Using other sources of resistance than Sumai3 and Frontana, *in vitro* selection enabled to accelerate breeding of more FHB resistant lines and reduced the size of the DH population for further phenotypic evaluation of FHB resistance.

According to our results effective screening for FHB resistance in early generations is possible by *in vitro* and molecular means. As the relationships among the FHB resistance mechanisms are not obvious, *in vitro* selection for type III (resistance expressed in the developing kernel) and type IV (reduced accumulation of mycotoxins) resistance, in combination with marker assisted selection (if available), followed by phenotypic evaluation, is a valuable refinement to the phenotypic evaluation and selection.

References

Eudes F, Comeau A, Rioux S, Colin J (2007) Can. J. Plant Pathol. Accepted.
Somers DJ, Thomas J, DePauw R, Fox S, Humphreys G, Fedak G (2005) Theor Appl Genet 111: 1623–1631.

The Geo-Model for Cell Wall Formation: A Design Tool for Plants

Carolina Cifuentes and Anne Mie C. Emons*

Laboratory of Plant Cell Biology, Wageningen University, 6703 BD Wageningen, the Netherlands.
*Email: annemie.emons@wur.nl

Biological importance of cellulose as main cell wall component and its value for industrial applications

Cell walls are highly organized composites of polysaccharides, proteins, and aromatic compounds that undergo dynamical changes during cell division, cell expansion and differentiation (Carpita and Gibeaut, 1993). Cellulose microfibrils form a complex network with xyloglucans which in primary walls contributes to about 70% of their total strength. Since cellulose corresponds to about 40% of the dry weight of mature plants, it can certainly be considered to be a major contributor to the overall plant strength (Hayashi, 1989; Delmer and Amor, 1995). Besides providing structural support to plant cells, plant cell walls fulfill other important functions such as establishment of cell shape and size, interaction with environmental factors including those required for nutrition, response to abiotic stresses and biological attack by other organisms (Carpita and McCann, 2000).

The primary cell wall combines unique properties of strength and rigidity but also of plasticity and viscosity (Carpita and Gibeaut, 1993). Among the different wall materials cellulose microfibrils represent the most important component in the control of anisotropic expansion. Inhibition of cellulose biosynthesis with compounds such

Z. Xu et al. (eds.), Biotechnology and Sustainable Agriculture 2006 and Beyond, 353–360.

as DCB, isoxaben or thaxtomin (Scheible et al., 2003) result in rapid loss of growth anisotropy; this effect has also been observed in cellulose deficient mutants of *Arabidopsis thaliana* (Arioli et al., 1998). Defects in the secondary cell wall are characterized by a collapse of xylem vessels that are unable to withstand the negative pressure generated during water transport through the xylem (Turner and Somerville, 1997).

Cellulose constitutes the most abundant renewable polymer resource available today worldwide. It occupies a very important place in industry, it is extensively used as a construction material mainly in the form of intact wood (in which cellulose constitutes 40–50%) or textile fibers, such as cotton (90% cellulose) and flax (80% cellulose), or in the form of paper and board. Cellulose is used as starting material for chemical conversions that are applied in the production of artificial cellulose-based threads and films as well as a variety of stable cellulose derivatives (Klemm et al., 1998). To exploit plant cell wall texture for the future production of (ligno)cellulosic (nano)materials with unique multifunctional properties is a major challenge in nanotechnology.

Cell wall composition

Plant cells are surrounded by a heterogeneous and dynamic extracellular matrix called the cell wall; primary cell walls are deposited in growing cells. Therefore they must be mechanically stable and sufficiently extensible to allow cell expansion while supporting turgor pressure. Secondary walls are deposited after the cessation of cell growth, conferring mechanical strength upon specialized cell types such as xylem elements and sclerenchyma cells.

The composition of plant cell walls varies significantly from one cell type to another, but basically they are biphasic structures: cellulose microfibrils embedded in a matrix of non-cellulosic polysaccharides (hemicelluloses) and pectins (Carpita and Gibeaut, 1993).

Cell wall matrix

The cell wall matrix consists of hemicelluloses, proteins, glycoproteins and phenolic substances; the polysaccharides enter the cell wall by exocytosis, a process in which the post Golgi vesicle membrane inserts into the plasma membrane and delivers the vesicle content to the existing cell wall. Hemicelluloses are branched polysaccharides with a backbone composed of 1,4-linked ß–D-pyranosyl residues. Within this group are found the xyloglucans, glucurono-arabinoxylans and mannose-containing polysaccharides. Xyloglucans constitute the major hemicellulosic polysaccharides of primary cell walls of non-graminaceous monocots and dicotyledonous plants. They bind tightly to cellulose microfibrils via hydrogen bonds forming a cross-linked network that is considered the major load-bearing structure in the cell wall. Therefore, xyloglucans contribute to the formation of macro-molecular architecture, and to the determination of mechanical properties of the cell wall (Carpita and Gibeaut, 1993; O'Neil and York, 2003)

Pectins form cross linked three-dimensional hydrated networks that may regulate cell wall porosity, hydration, and pH, in turn regulating the access or activities of cell wall modifying enzymes to either the pectic network or to the cellulose cross-linking glycan networks (Sugimoto et al., 2004).

Apart from polysaccharides, cell walls contain several proteins and enzymes which are involved in cell wall metabolism, as well as in the formation and disruption of cross-links between the different components, therefore influencing the mechanical properties of the walls. Expansins for instance have been proven to increase extensibility of isolated walls *in vitro* under mechanical stress conditions, and to induce cell expansion *in vivo* under acidic conditions (McQueen-Manson et al., 1992; Cosgrove, 2000). Later in life, cell walls may become impregnated with for instance lignin (Rogers and Campbell, 2004).

Cellulose and callose

Cellulose and callose constitute 15–30% of the dry mass of all primary cell walls and an even larger percentage in secondary cell walls (McCann et al., 2004), both are made by enzyme complexes embedded in the plasma membrane.

Cellulose consists of crystalline aggregates of linear polymers of (1,4) ß-D glucan chains (Bacic and Stone, 1988; Reiter, 1998), where every other glucose residue is rotated approximately 180° to its neighbor. This orientation allows the chain to adopt a flat ribbon-like structure, in which the glucan chains are bound to each other by intra and inter-molecular hydrogen bonds, as well as Van der Waals forces to form cellulose microfibrils (CMFs).

Cellulose is synthesized by a large complex of proteins that is localized in the plasma membrane. In vascular plants this complex appears as a rosette with a six-fold symmetry with a diameter of 25–30 nm. Based on measurements of CMFs, it has been inferred that each of the six subunits of a rosette synthesizes approximately 36 glucan chains that are extruded into the extracellular space (reviewed by Somerville et al., 2004). Higher plant cellulose microfibrils have a width of approx 3.5 nm (Emons, 1988; Jarvis, 2003). Callose is a polymer with (1,3) ß-D glucan chains that form helical duplexes and triplexes rather than crystalline microfibrils. It is a natural component of the initial cell plate of dividing cells, of the pollen tube wall, the sieve plate pores of phloem elements, and of plasmadesmatal channels. In addition, callose is deposited after plants are exposed to a range of abiotic and biotic stresses, including wounding or pathogen attack (Stone and Clarke, 1992). Callose is synthesized by a different group of enzymes from those that synthesize cellulose (McCann et al., 2004). Callose has anti aids virus activity, and is being used as a formulation aid, processing aid, stabilizer, and thickener or texture modifier for food use (Mikio et al., 1995; Spicer et al., 1999).

Cellulose microfibrils arrange in one microfibril thick lamellae that together form the cell wall texture. The orientation between CMFs within a lamella is on the whole constant but it may vary between lamellae. The classification of cell wall textures is based on the distribution of CMF orientation both between lamellae and with respect to the cell axis. Basic types of cell wall texture are axial, transverse, helical, helicoidal, crossed polylamellate and random. Although cortical microtubules appear to orient the native cellulose microfibrils (Paredez et al., 2006), cellulose synthases move in ordered patterns in their absence (Paredez et al., 2006, Figure 4i), and regular wall textures are formed in microtubule mutants (Sugimoto et al., 2003; Himmelspach et al., 2003), cells treated with microtubule drugs (Emons et al., 1990; Sugimoto et al., 2003) and in cells depositing secondary cell wall for instance in root hairs with helicoidal cell walls (Emons et al., 1992) and other secondary walls in several algae (Itoh, 1989).

The Geo-model

A bottleneck in the development of novel germplasm with new wall properties is our lack of understanding of the impact of specific genes on wall texture and the impact of wall texture on industrial properties. One would like to intelligently select and manipulate candidate cell wall genes to rationally design plants with the required properties. The systems biology approach of iterative modeling and experimentation is increasingly being applied to improving human health, but has as much potential for agricultural-based applications. The geo-model (Emons and Mulder, 1998) is a space-filling geometrical, mathematical model for the creation of wall texture. The model quantitatively relates the deposition angle of cellulose microfibrils to (1) the density of active synthases in the plasma membrane, (2) the distance between individual microfibrils within a wall lamella, and (3) the geometry of the cell (Eq. 1).

$$\text{Sine of CMF angle } (\alpha) = \frac{\text{Number of microfibrils } (N) \times \text{width of fibril + adherent matrix } (d)}{\text{Circumference of cell } (\pi D)} \qquad (1)$$

The Geo-model explains wall texture as a space-time process, mainly dependent on the number of active cellulose synthases at the location of cellulose microfibril deposition. It predicts wall properties, in particular microfibril angle. The ‘Ansatz’ of the model is that the activation of the synthases occurs in domains, which move, or are activated, along the plasma membrane. Since the distance between produced cellulose microfibrils, i.e., the ratio between them and the whole wall matrix, influences microfibril angle, plant properties can be predictably manipulated by up- or down-regulating a limited number of candidate cellulose- and matrix-associated genes. Thus, apart from the constituents of the cellulose synthase itself, every enzyme in the production cascade leading to a matrix wall polysaccharide or glycoprotein is a potential candidate to rationally alter the cell wall and consequently the plant design.

Importance of the geo-model for industrial plant design

Microfibril orientation underlies several technological properties such as length, strength and fineness of natural fibres. However, the major genetic determinants remain unknown. Thus, knowledge about how a cell determines cell wall texture, the mechanical consequences thereof, and the genes that can be used to modify wall texture is required for the rational engineering of industrial plant traits. This knowledge will also lead to novel strategies for determining the orientation of fibrils in composite materials, which is a major challenge. Understanding Nature's strategy in designing plant cell walls by embedding a texture of stiff cellulose microfibrils in a soft matrix of mainly hemicellulose and pectin polymers, eventually impregnated with lignin, provides a key for technical applications in nanotechnology.

Acknowledgements

This work was financially supported by the Netherlands Organization for Scientific Research (NWO), Physical Biology Program 2, project number 805.47.065.

References

Arioli T, Peng L, Betzner AS, Burn J, Wittke W, Herth W, Camilleri C, Höfte H, Plazinski J, Birch R, Cork A, Glover J, Redmond J, Williamson RE (1998) Molecular analysis of cellulose biosynthesis in Arabidopsis. Science 279: 717–720.

Bacic A, Harris PJ, Stone BA (1988) Structure and function of plant cell walls. In: The Biochemistry of Plants, Academic Press, London, Vol. 14, pp. 297–371.

Carpita NC, Gibeaut DM (1993) Structural models of primary-cell walls in flowering plants – consistency of molecular structures with the physical properties of the wall during growth. Plant J 3: 1–30.

Carpita NC, McCann M (2000) The cell wall. In: Buchanan B, Gruissem W, Jones RL (eds) Biochemistry and Molecular Biology of Plants. Rockville, Maryland: American Society of Plant Physiologists, pp. 52–108.

Cosgrove DJ (2000) Loosening of Plant cell walls by expansins. Nature 407: 321–326.

Delmer DP, Amor Y (1995) Cellulose biosynthesis. Plant Cell 7: 987–1000.

Emons AMC (1988) Methods for visualizing cell wall architecture. Acta Bot Neerl 37: 31–38.

Emons AMC, Derksen J, Sassen MMA (1992) Do microtubules orient plant cell wall microfibrils? Physiol Plant 84: 486–493.

Emons AMC, Mulder BM (1998) The making of the architecture of the plant cell wall: How cells exploit geometry. Proc Natl Acad Sci USA 95: 7215–7219.

Emons AMC, Wolters-Arts AMC, Traas JA, Derksen J (1990) The effect of colchicine on microtubules and microfibrils in root hairs. Acta Bot Neerl 39: 19–27.

Hayashi T (1989) Measuring β-glucan deposition in plant cell walls. In: Linskens HF, Jackson JF (eds) Modern Methods of Plant Analysis. Springer Verlag, Berlin, Vol. 10, pp. 138–160.

Himmelspach R, Williamson RE, Wateneys RO (2003) Cellulose microfibril alignment recovers from DCB-induced disruption, despite microtubule disorganization. Plant J 36: 565–575.

Itoh T (1989) Biogenesis of cellulose microfibrils and the role of microtubules in green algae. In: Lewis NG, Paice MG (eds) Plant Cell Wall Polymers: Biogenesis and Biodegradation. ACS Symposium Series, Washington DC Vol. 399, pp. 257–277.

Jarvis M (2003) Cellulose stacks up. Nature 426: 611–612.

Klemm D, Philipp B, Heinze T, Heinze U, Wagenknecht W (1998) Comprehensive Cellulose Chemistry. Willey-VCH , Weinheim.

McQueen-Mason S, Durachko DM, Cosgrove DJ (1992) Two endogenous proteins that induce cell wall extension in plants. The Plant Cell 4: 1425–1433.

Mikio K, Yoshiro O, Hirotomo O, Yoshikazu K, Hiroshi I, Hisashi M, Takeshi K (1995) Water soluble ß-1,3-glucan derivative and antiviral agent containing the derivative (Japanese patent 07228601).

O'Neill MA, York WS (2003) The composition and structure of plant primary cell walls. In: Jocelyn KCR (ed) The Plant Cell Wall. Blackwell, USA, Vol. 8, pp. 1–54.

Paredez AR, Somerville CR, Ehrhardt DW (2006) Visualization of cellulose synthase demonstrates functional association with microtubules. Science 312: 1491–1495.

Reiter W (1998) The molecular analysis of cell wall components. Trends Plant Sci 3: 27–32.

Rogers LA, Campbell MM (2004) The genetic control of lignin deposition during plant growth and development. New Physiologist 164: 17–30.

Scheible WR, Fry B, Kochevenko A, Schindelasch D, Zimmerli L, Somerville S, Loria R, Somerville CR (2003) An Arabidopsis mutant resistant to thaxtomin, a cellulose synthesis inhibitor from Streptomyces species. Plant Cell 15: 1781–1794.

Somerville C, Bauer S, Brininstool G, Facette M, Hamann T, Milne J, Osborne E, Paredez A, Persson S, Raab T, Vorwerk S, Youngs H (2004) Toward a systems approach to understand plant cell walls. Science 306: 2206–2211.

Spicer EJF, Goldenthal EI, Ikeda T (1999) A toxicological assessment of curdlan. Food Chem Toxicol 37: 455–479.

Stone BA, Clarke AE (1992) Chemistry and Biology of (1,3)- ß-glucans. La Trobe University Press, Melbourne, Australia.

Sugimoto K, Carpita NC, McCann MC (2004) The cell wall: a sensory panel for signal transduction. In: Hussey PJ (ed) The Plant Cytoskeleton in Cell Differentiation and Development. CRC Press, USA/Canada, Vol. 10, pp. 176–203.

Sugimoto K, Carpita NC, McCann MC (2004) The cell wall: a sensory panel for signal transduction. In: Hussey PJ (ed) The Plant Cytoskeleton In Cell Differentiation and Development. Blackwell, UK, Vol. 10, pp. 176–203.

Sugimoto K, Himmelspach R, Williamson RE, Wasteneys GO (2003). Mutation or drug-dependent microtubule disruption causes radial swelling without altering parallel cellulose microfibril deposition in *Arabidopsis*. Plant Physiol 124: 1493–1506.

Turner SR, Somerville CR (1997) Collapsed xylem phenotype of Arabidopsis identifies mutants deficient in cellulose deposition in the secondary cell wall. Plant Cell 9: 689–701.

Tissue Culture in Biodiversity Preservation

Integrating In Vitro Methods for Propagating and Preserving Endangered Plants

V.C. Pence*, S.M. Charls, B.L. Plair, M.A. Jaskowiak, G.D. Winget, and L.L. Cleveland

Center for Conservation and Research of Endangered Wildlife, Cincinnati Zoo and Botanical Garden, 3400 Vine Street, Cincinnati, OH 45220, USA.
*Email: valerie.pence@cincinnatizoo.org

Introduction

In vitro methods provide a variety of tools to supplement traditional methods for collecting, propagating and preserving endangered plant species (Henshaw and Blakesley, 1996; Pence, 1999; Sugii and Lamoureux, 2004). These methods are in use by researchers in several programs around the world, including at the Center for Conservation and Research of Endangered Wildlife (CREW) at the Cincinnati Zoo & Botanical Garden, where they are used to address reproductive challenges in a variety of United States endangered species. *In vitro* methods also provide a supplemental approach for addressing Target 8 of the Global Strategy for Plant Conservation (Wyse Jackson, 2004), when traditional methods of ex situ conservation are not adequate.

Established in 1875, the Cincinnati Zoo & Botanical Garden (CZBG) is the second oldest zoo in the United States, and officially became a botanical garden in 1987. The conservation research program, known as CREW, was established in 1980 to apply scientific methods to reproductive questions in endangered animals. In 1987, the Plant Research Division was added, with a major focus on using

Z. Xu et al. (eds.), Biotechnology and Sustainable Agriculture 2006 and Beyond, 363–373.

in vitro methods for the propagation and preservation of endangered plants. This became known as the Endangered Plant Propagation Program (EPPP).

CREW's work in endangered plant propagation has been done with a number of collaborators, including the Center for Plant Conservation (CPC, St. Louis, MO). The CPC is a network of more than 30 botanical gardens and arboreta dedicated to preventing the extinction of plants native to the United States. Gardens within the CPC network have nominated species for in vitro work at CREW, and development of in vitro propagation protocols for these species has been supported by several grants from the Institute of Museum and Library Services (IMLS). Once propagation protocols have been established, some species are moved into outplanting projects, and grants from the U.S. Fish & Wildlife Service have supported a portion of that work. Currently the Plant Research Division is targeting over 40 species of vascular plants, both seed plants and ferns. This review will provide a brief overview of this work and the various in vitro and other procedures used to collect, propagate, preserve and utilize endangered plant species.

Culture initiation

Tissue culture lines are initiated either from seeds/spores or from shoot or leaf tissues from adult plants, depending on the species and the availability of seeds or spores. When seeds are available, smaller seeds are surface sterilized and germinated *in vitro* to produce aseptic seedlings from which shoot cultures are initiated. Larger seeds are germinated in soil, and shoot tips from the seedlings are surface sterilized and initiated into culture. Seeds of some species may be dormant and require stratification, scarification, or other dormancy breaking treatments. Seeds of *Rorippa subumbellata* (a shoreline species from Lake Tahoe, NV and CA), for example, could be germinated *in vitro* only if cytokinin was added to the medium (Pence et al., 2006), whereas seeds of *Ranunculus aestivalis* (a very rare species from wet meadows in UT) were stratified in vitro to stimulate germination.

Spores of pteridophytes can also be germinated in vitro, as was done in this lab with *Asplenium heterosiliens* (a fern of limestone outcrops of the southeastern United States), producing cultures of gametophytes, which later produced sporophytes. Alternatively, spores can be germinated in soil and the resulting gametophytes surface sterilized to initiate *in vitro* cultures. When this is done, much of the gametophyte tissue will likely be damaged by the sterilant, but surviving fragments will regenerate new gametophytes to establish culture lines.

Shoot propagating cultures are most easily initiated from seeds or spores, but for some endangered species, seeds or spores may not be produced, they may be dormant, or they may not be available at the time they are needed in the laboratory. In these cases, cultures may be initiated from young, growing shoot tips or, in some cases, through regeneration of shoots from young leaf tissues. In some situations, these tissues can be sent overnight to the laboratory for surface sterilization and transfer to medium within 24 h of collecting. For this to be successful, tissues must not be exposed to extremes of temperature or undergo deterioration in transit. In some cases, however, stable transport cannot be guaranteed, or overnight delivery to the laboratory may not be possible due to distance, remote collecting areas, or the need to cross international borders.

When overnight shipment of shoot tips is not possible or is inefficient, *in vitro* collecting (IVC) may also be used to initiate cultures. IVC is a method for starting tissue cultures in the field that has been used by a number of researchers for collecting tissues from agricultural and other plants (Pence et al., 2002). It has also been tested by CREW researchers extensively in Trinidad, Costa Rica, and at various locations around the United States (Pence, 2005, unpublished). Small pieces of shoot tips or leaves are taken from the plant and surface sterilized with either 70% alcohol or a 5–10% solution of commercial bleach with a water rinse and quickly transferred to small vials of sterile media. The procedures are conducted in the open air and ambient contamination is less than 5%. A fungicide and antibiotic solution are used in the initial collection to reduce contamination from endogenous sources, in some cases reducing contamination levels from 95% to less than 30% (Pence, 2005).

In some situations, IVC can provide unique advantages over conventional collecting methods. With IVC, only small amounts of tissues are removed, leaving most of the endangered plant undisturbed. Tissues are provided with a growth medium and hormones immediately after excision and sterilization, and thus, they can initiate growth rapidly and do not deteriorate as they might in transit without medium. Furthermore, the sterilized bits of tissue are often easier to transport across international boundaries than are samples that might harbor insects or soil. IVC is a method that has been used for a number of species, including agricultural species, and is very flexible in the types of tissues, media, containers, sterilization procedures, etc. that can be used (Withers, 1987, 1995; Pence et al., 2002).

IVC has been used at CREW particularly when a large amount of genetic diversity is required for reintroduction projects or for cryopreservation of a particular species. With *Asimina tetramera* (a small, rare tree from the sandpine scrub of southeast FL), IVC collections have been made from over 100 individual plants, from nine of the 17 known populations of the species, providing a source for a genetically diverse population of tissue culture propagated plants for outplanting. Because the seeds of this species are recalcitrant, this collection is also providing shoot tips for long-term germplasm storage in liquid nitrogen (see below).

Propagating shoot cultures

Once clean shoot cultures are established, they are multiplied to provide material for rooting and acclimation, as well as for cryopreservation. Although standard tissue culture methods can often be used, different species require modifications of procedures for growth *in vitro*. Species under study at CREW include those from a wide variety of habitats (temperate forests, sandy scrublands, rocky soils, wetlands, deserts) and a wide variety of life-forms (woody, herbaceous, monocots, dicots, pteridophytes, hemi-parasites), and their response in culture is likely related to their growth in the wild.

Some species multiply relatively rapidly in culture, such as *R. subumbellata*, an herbaceous creeping perennial from the moist shores of Lake Tahoe, in Nevada and California, while others are much slower to form new shoots, such as *Asimina tetramera*, a woody species from the sandy scrub soils of southern Florida. Species from cooler habitats, such as *Aconitum noveboracensis,* an herbaceous plant from cool shaded areas of the temperate forest, prefer growth at 21°C rather than at 26°C. Similarly, this species, as well as others from shaded areas, such as *Arenaria cumberlandensis* (native to sandy shaded soils beneath sandstone outcrops) prefer lower light levels. In some cases media are adjusted depending on the future disposition of the cultures. Shoots designated for cryopreservation are generally grown on agar-gelled media, even though other gelling agents can be used for standard propagation. Ferns offer a unique opportunity for propagation in that either gametophyte or sporophyte cultures can be maintained, thus providing two methods for multiplying tissues of pteridophyte species *in vitro*.

The goal of conservation is to maintain maximum genetic diversity, but *in vitro* shoot propagation is a clonal process, producing many genetically identical individuals. At CREW, an attempt is made to maintain any genetic diversity that is received. If cultures are initiated from tissues from different individuals or from seeds, cultures from each individual or seed will be maintained as separate genetic lines. For example, in the case of *Ranunculus aestivalis* and *Crotalaria avonensis* (a small legume from the central Florida scrub), special projects are underway, and there are approximately 50 lines from seeds and 150 lines initiated from shoots in culture, respectively. Only a few individuals of each genetic line are being maintained, in order to produce genetically diverse populations of rooted shoots for outplanting projects.

Rooting

Micropropagated shoots are rooted in order to produce plants for reintroduction, research and display. In some cases, shoots may spontaneously root on propagation media, as in the case of *Rorippa subumbellata.* In other cases, rooting is much less frequent, and there is some suggestion that plants from areas with dry or very well-drained soils are more difficult to root than species from wet areas. Attempts were made to root shoot cultures of *Asimina tetramera* (from sandy soils in Florida) on indolebutyric acid (IBA) at standard levels (0.5 or 1.0 mg L^{-1}) without success. However, when silver thiosulfate (STS), an ethylene inhibitor, was added to the medium, rooting occurred at a rate of approximately 25% (Charls et al., 2003), suggesting an inhibitory role for ethylene in this system. Further research on the rooting medium has shown that treating the shoots with very high levels of auxin followed by culture on charcoal can produce rooting rates of 40–60% in some lines of *A. tetramera* (Jaskowiak et al., 2006). Although the role of ethylene in this system is still unclear (Jaskowiak et al., 2005), in practical terms, this improved rate of root initiation will allow a more efficient production of plants for outplantings into sites in Florida.

Cryopreservation of shoot tips

A portion of the propagated shoots, or (in the case of some ferns) gametophytes, are directed toward cryopreservation protocols, in order to preserve the lines in liquid nitrogen in CREW's 'Frozen Garden' for future use. Both encapsulation dehydration (Fabre and Dereuddre, 1990) and encapsulation vitrification (Tannoury et al., 1991) are used, with the method of choice depending on the response of a particular species. For example, *Asimina tetramera* shoot tips respond best to encapsulation vitrification (Plair and Pence, 2003), whereas encapsulation dehydration has provided good survival for other species, such as *Lobelia boykinii* (a wetland species from the eastern temperate forest), and for gametophytes of several ferns (Pence, 2000; Plair, unpublished).

Although for most species in the Endangered Plant Propagation Program, cryopreservation is used to maintain the micropropagating lines, in other cases, cryostorage is used as a primary method for a species' long-term germplasm storage. With some endangered species, traditional seed storage is not possible because the seeds are recalcitrant, or there are few or no seeds produced. In the case of *A. tetramera,* the recalcitrant seeds of this species are large and embryo isolation is difficult, but in vitro propagation protocols have been developed. Work at CREW has been directed at collecting as many genetic lines of this species as possible by IVC (Pence and Charls, 2003), in order to establish multiple genetic lines and to cryopreserve the shoot tips as a method for germplasm storage (Plair and Pence, 2003). Protocols for cryopreservation have also been worked out for *Hedeoma todsenii* (Plair, unpublished) (a rare herbaceous perennial of rocky soils in the mountains of New Mexico), a species that produces few or no seeds in the wild. Cryopreservation of multiple genetic lines of this species would similarly provide a method for long-term storage, where none currently exists. Although cryopreserving shoot tips for germplasm storage is more labor intensive than seed banking, for those species with no seeds or with recalcitrant seeds, it provides a method for long-term storage of genetic diversity that is more cost-effective than maintaining populations of individuals ex situ.

Genetic analysis

Genetic analysis of tissue culture lines is providing information on the diversity of the collections, as a basis for documenting the diversity both of plants produced for outplanting and of cryopreserved tissue lines. Using PCR, genetic analyses have been done on multiple lines of *Asimina tetramera* that are being cryopreserved for long-term storage. Thus far, 20 lines have been analyzed, showing differences in all of the lines and indicating that storage of this germplasm will capture some of the genetic diversity of this species. The ten-mers that have been used to successfully analyze *A. tetramera* have also worked well with two closely related species, *Deeringothamnus pulchellus* and *D. rugelii* (two small, rare shrubs found in flatwoods in Florida), which, like *A. tetramera*, have recalcitrant seeds and are also being cryopreserved as shoot tips. Similarly, lines of *Crotalaria avonensis* are being analyzed genetically and cryopreserved, since this species produces very little seed, and multiple lines are being propagated for a large outplanting project. In the case of *Aconitum noveboracense,* genetic analysis has helped define the relationship of the two remaining Ohio populations, both of which are being propagated at CREW.

Examples of acclimation and utilization of micropropagated plants

Once shoots are rooted in vitro, they are moved into acclimation protocols, in order to produce plants for outplanting or reintroduction projects, research, education or display. In some cases, as with *Ranunculus aestivalis*, the plants are sent to collaborators as rooted cultures for acclimation. *R. aestivalis* is an herbaceous perennial, found in moist meadows in two small populations in Utah. At least 300 plants produced from approximately 50 different genetic lines of *R. aestivalis* will be part of a reintroduction project funded by a grant from the U.S. Fish and Wildlife Service. Rooted cultures from CREW are being sent to collaborators at The Arboretum at Flagstaff, in Flagstaff, Arizona, where the cultures will be acclimated. The plants will then be transported to the Sevier Valley Preserve of The Nature Conservancy in Utah where they will be planted and monitored for at least 5 years by researchers from Utah Valley State College. This preserve is home to one of the two remaining populations of *R. aestivalis*, with less than 30 plants remaining, and the tissue culture propagated plants will help to increase numbers in the declining population.

Asimina tetramera, is a small tree found in the sand pine scrub habitat in two counties in southeastern Florida. This difficult-to-root species, has seen the rate of rooting of shoots in vitro increase from 0%, on lower levels of IBA, to 25% on IBA plus STS, to over 50% with a pulse of a high concentration of IBA. This improvement in rooting has made a significant difference in the production of plants for acclimation. Some of these are being acclimated at CREW and some by collaborators at Historic Bok Sanctuary in central Florida. Whereas survival through acclimation is still below 25%, growth of surviving plants has been good, with the young plants producing significant root growth in the soil, just as normally occurs in the wild.

In the case of *Arenaria cumberlandensis,* acclimation is done entirely at CREW. This small perennial species is found only in southern Kentucky and northern Tennessee in the Daniel Boone National Forest. Its habitat is the cool, shaded, sandy soil beneath sandstone outcroppings known as ‘rockhouses’. Plants acclimated in sand with less than 5% organic material/clay showed good survival. Seventy-seven plants from ten genetic lines of *A. cumberlandensis* micropropagated at CREW were planted at a site in the Daniel Boone National Forest in Kentucky in September, 2005, by collaborators from the U.S. Forest Service. One month later, 65% of the plants were still alive. The following spring, almost all of these plants were still surviving and some were producing flowers.

Some plants are produced for projects other than outplantings. In the case of *Houstonia montana* (a rare, herbaceous perennial native to a few mountaintops in the Appalachian Mountains), 100 plants were produced to support research on this species that could not be done in situ (Euliss, 2004). In other cases, plants are produced for display at the CZBG or at other sites, as with *Clematis socialis* (a rare perennial of open grass-sedge areas of Alabama and Georgia), which was sent to the Birmingham Botanical Garden for display. Additionally, some plants are being sent to the Molecular Targets Development Program of the Natural Products Branch of the National Cancer for screening for anti-cancer activity.

Summary

Over 40 species have been targeted, thus far, in the Endangered Plant Propagation Program at the Cincinnati Zoo & Botanical Garden. *In vitro* techniques have been applied to species for which traditional propagation and preservation methods are not adequate, and different methods or combinations of in vitro methods are used, depending on the species. *In vitro* techniques provide a variety of possible approaches, and adapting and applying these methods to individual species can overcome reproductive challenges and produce plants for outplanting projects and germplasm storage.

Acknowledgements

This research was supported in part by grants from the Institute of Museum and Library Services and the U.S. Fish & Wildlife Service.

References

Charls SM, Clark JR, Pence VC (2003) Rooting in cultures of two endangered Florida pawpaws, *Asimina tetramera* and *Deeringothamnus rugelii.* In Vitro Cell Dev Biol 39: 50A.

Euliss AC (2004) Growth and resource allocation of *Houstonia montana*, a rare endemic plant in the southern Appalachians. M.S. Thesis, Appalachian State University, Boone, NC, p. 66.

Fabre J, Dereuddre J (1990) Encapsulation dehydration: A new approach to cryopreservation of *Solanum* shoot tips. Cryo-Letters 11: 413–426.

Henshaw GG, Blakesley D (1996) The scope of in vitro technology in conservation of biological diversity. In: Normah MN, Narimah MK, Clyde MM (eds) In Vitro Conservation of Plant Genetic Resources. Universitii Kebangsaan Malaysia, Kuala Lumpur, Malaysia, pp. 217–227.

Jaskowiak MA, Pence VC, Charls SM (2005) The effects of ethylene inhibitors on rooting of three endangered Florida pawpaws *in vitro*. *In Vitro* Cell Dev Biol 41: 54A.

Jaskowiak MA, Pence VC, Charls SM (2006) Improvements in in vitro rooting of *Asimina tetramera*. In Vitro Cell Dev Biol – Animal, 42.

Pence VC (1999) The application of biotechnology for the conservation of endangered plants. In: Benson EE (ed) Plant Conservation Biotechnology. Taylor & Francis, Ltd, London, pp. 227–250.

Pence VC (2000) Cryopreservation of *in vitro* grown fern gametophytes. Am Fern J 90: 16–23.

Pence VC (2005) In vitro collecting (IVC) I. The effect of media and collection method on contamination in temperate and tropical collections. In Vitro Cell Dev Biol – Plant 41: 324–332.

Pence VC, Charls SM (2003) *In vitro* collecting and establishment of tissue culture lines of three endangered Florida pawpaws. In Vitro Cell Dev Biol 39: 19A.

Pence VC, Guerrant EO, Raven AN (2006) Cytokinin stimulation of seed germination in *Rorippa subumbellata* Rollins. Seed Sci Technol 34: 241–245.

Pence VC, Sandoval JA, Villalobos VM, Engelmann F (eds) (2002) In Vitro Collecting Techniques for Germplasm Conservation. IPGRI Technical Bulletin No. 7. International Plant Genetic Resources Institute, Rome, Italy, p. 100, ISBN 92-9043-534-1.

Plair BL, Pence VC (2003) Cryopreservation of shoot tips of the endangered *Asmina tetramera* by encapsulatiion-vitrification. *In Vitro* Cell Dev Biol 39: 19A.

Sugii N, Lamoureux C (2004) Tissue culture as a conservation method: An empirical view from Hawaii. In: Guerrant E, Havens K, Maunder M (eds) Ex Situ Plant Conservation – Supporting Species Survival in the Wild. Island Press, Covelo, CA, pp. 189–205.

Tannoury M, Ralambosoa J, Kaminski M, Dereuddre J (1991) Cryopreservation by vitrification of coated shoot-tips of carnation (*Dianthus caryophyllus* L.)

cultured *in vitro.* Comptes Rendus de l'Academie des Sciences, Paris, Ser. III, 313: 633–638.

Withers LA (1987) *In vitro* methods for collecting germplasm in the field. FAO/IBPGR Plant Genet Resour Newsl 69: 2–6.

Withers LA (1995) Collecting *in vitro* for genetic resources conservation. In: Guarino L, Rmanatha Rao V, Reid R (eds) Collecting Plant Genetic Diversity. CAB International, Wellingford, UK, pp. 511–526.

Wyse Jackson P (2004) Developing international targets for botanic gardens in conservation. B Gjournal 1(1), http://www.bgci.org.uk/worldwide/article/0226/.

Micropropagation and Germplasm Conservation of Staple Food Crops and Traditional Medicinal Plants in Micronesia

Virendra Mohan Verma

Micronesia Plant Propagation Research Center, Cooperative Research and Extension (USDA Land Grant Program), College of Micronesia-FSM, Kosrae FM 96944, Micronesia.
Email: vmv_vmv@hotmail.com

Introduction

The state of Kosrae is a single volcanic island at the south-eastern end of the Caroline Islands, located at 5°19'N of the equator and 163°E of the Greenwich Meridian. It is one of the four states in the Federated States of Micronesia and has a land area of 43.2 square miles. A narrow coastal strip circles a steep and rugged interior. About 15% is foot slope, alluvial fans and bottom land and freshwater swamp. Another 15% is mangrove swamp and coastal strands. Mountains account for about 70% of the land area.

Agriculture is an important industry and it could help greatly in the economic development and growth of Kosrae state. However, current agricultural programs are mostly on a subsistence level and economic development is largely dependent on the outside world.

Staple food crops like bananas, citrus, soft taro, sweet potato, yam, and traditional medicinal plants like noni and kava are placed on high agricultural priority by the state government and local producers but

Z. Xu et al. (eds.), Biotechnology and Sustainable Agriculture 2006 and Beyond, 375–379.

limitations in availability of disease-free and elite seedlings is a major bottleneck in quality production. Kosrae is a small island state, therefore, quarantine measures are very strict and entry of any planting material is strictly prohibited. The only way to multiply the planting material of vegetatively propagated crops in bulk is to use tissue culture through which we can produce seedlings throughout the year. In the tropics, it is especially difficult to maintain disease-free parental stocks and meristem culture is increasingly being appreciated as a potential means of germplasm preservation and for the production of virus-free planting materials on mass scale.

In vitro propagation of banana provides excellent advantages over traditional propagation, including a high multiplication rate, physiological uniformity, the availability of disease-free material throughout the year, safe and rapid dissemination of new plant materials throughout the world, uniformity of shoots, short harvest interval in comparison with conventional plants, and faster growth in the early growing stages compared to conventional materials (Vuylsteke, 1989; Arias, 1992). Tissue culture also plays a vital role in the distribution of germplasm, conservation, safe exchange of internal planting material and rapid propagation of newly selected hybrid cultivars. Traditional breeding methods of taro are time consuming and, therefore, must be finely tuned with the use of biotechnological approaches. *In vitro* culture is of great advantage for mass propagation of vegetatively propagated crop taro in which, because of low propagation ratio, seed corm supply has become a major bottleneck in large-scale cultivation. Sweet potato ranks seventh in the world among food crops with respect to annual production (Jabsson and Raman, 1991) and is the fifth most important food crop in developing countries. Sweet potato is a hardy crop and can be grown in poor soils with little fertilizer. However, conventional breeding programs based on sexual hybridization of sweet potato remain poorly developed because of its sterility and cross incompatibility. To overcome these limitations, there is a need for incorporation of novel approaches such as micropropagation into conventional breeding programs.

Figure 1. *In vitro* multiplication and conservation. Banana micropropagation, taro micropropagation, citrus *in vitro* conservation, tissue cultured banana plants in nursery, noni elite seedlings, tissue cultured taro plants in nursery, taro somatic embryogenesis, and sweet potato micropropagation (from top left to clockwise).

Biotechnology is a powerful tool and appropriate and skillful use of biotechnological approach like tissue culture multiplication through micropropagation, regeneration and somatic embryogenesis is helping in successful multiplication of the elite plantlets for raising food and medicinal crops for sustenance as well providing opportunities for the conservation of germplasm in an inexpensive way. The extended use of *in vitro* conservation is now being considered as a more secure strategy for safeguarding genetic resources. Similar attempts to develop the use of biotechnology for propagation and conservation purposes are also ongoing in the south pacific region (Taylor, 1998).

The Micronesia Plant Propagation Research Center is the only north-pacific regional source for multiplying the planting material through tissue culture in such a bulk quantity and is actively engaged in plant biotechnology research. The research center also plays a vital role in germplasm conservation of economically important staple food and traditional medicinal crops. The research center is engaged in *in vitro* multiplication, conservation, and distribution of different varieties of banana, taro, sweet potato, yam, citrus, noni and kava (Figure 1). Recently, a new research project, focused on multiplication of staple food crops through micropropagation and *in vitro* germplasm conservation in Kosrae state was developed and has been approved from U. S. Department of Agriculture (USDA). The research center facilitates the coordination between the national and international agencies for research and extension activities related with staple food crops in the pacific region.

Acknowledgements

The author thanks the U. S. Department of Agriculture, the Western Sustainable Agriculture Research and Education Program and Kosrae state for financial support.

References

Arias O (1992) Commercial micropropagation of banana. In: Biotechnology Applications for Banana and Plantain Improvement. Inibap, San Jose, Costa Rica, pp. 139–142.

Jabsson RK, Raman KV (1991) Sweet potato pest management: A global overview. In: Jabsson RK, Raman KV (eds) Sweet Potato Pest Management: A Global Perspective. Westview Press, Boulder, USA, pp. 1–12.

Taylor M (1998) Biotechnology in the south pacific island region. Acta Horticulturae (ISHS) 461: 55–64.

Vuylsteke D (1989) Shoot-tip culture for the propagation, conservation and exchange of Musa germplasm. IBPGR, Rome.

Efficient In Vitro Bulblet Regeneration from Immature Embryos of Endemic and Endangered Geophyte Species in *Sternbergia, Muscari* and *Fritillaria* Genera

Sebahattin Özcan*, İskender Parmaksız[1], Semra Mirici[2], Satı Çöçü, Serkan Uranbey[3], Arif İpek, Cengiz Sancak, Ercüment O. Sarıhan[4], Bilal Gürbüz, Cafer S. Sevimay, and Neşet Arslan

Department of Field Crops, Faculty of Agriculture, University of Ankara, 06110, Dişkapi, Ankara, Turkey.
*Email: ozcan@agri.ankara.edu.tr
[1]Department of Biology, Faculty of Science and Arts, University of Gaziosmanpaşa,Tokat, Turkey.
[2]Faculty of Education, University of Akdeniz, Antalya, Turkey.
[3]General Directorate of Agricultural Research, Ministry of Agriculture and Rural Affairs, Ankara, Turkey.
[4]Department of Field Crops, Faculty of Agriculture, University of Mustafa Kemal, Hatay, Turkey.

Introduction

Flora of Turkey possesses approximately 9000 plant species, from which 3000 are endemic. Flowerbulbs (geophytes) are the important part of this rich biodiversity and contain many important endemic and endangered species such as *Sternbergia candida*, *Muscari muscarimi*, *Fritillaria alburyana* and *F. whittallii*. Besides use as ornamental plants because of their attractive flowers, these species have also a great potential in perfume and pharmaceutical industry. However, the natural propagation rate of most geophytes is relatively low (Arslan et al., 2002) and this often hampers the large-scale field

Z. Xu et al. (eds.), Biotechnology and Sustainable Agriculture 2006 and Beyond, 381–383.

cultivation of these plants. Mass propagation of these plants is therefore very important for germplasm conservation and commercial production. In the present paper we described a system for efficient *in vitro* bulblet regeneration from immature embryo explants of *S. candida*, *M. muscarimi*, *F. alburyana* and *F. whittallii* species.

Materials and methods

Bulbs and immature fruits of *Sternbergia candida*, *Muscari muscarimi*, *Fritillaria alburyana* and *F. whittallii* were collected from wild flora or experimental field of Field Crops Department of Ankara University and surface-sterilized by treatment in commercial bleach. Bulb explants with two scale segments and immature embryos were isolated and cultured on MS basal medium supplemented with various concentrations of BAP and NAA. Immature embryos were also cultured on 2 mg L^{-1} 2,4-D containing N6 medium using a four step culture method as described earlier (Mirici et al., 2005) and subcultured every 1 or 1.5 month to allow bulblets to develop further. Regenerated bulblets were kept at 5°C for 5 weeks and then transplanted to a potting mixture of peat moss, vermiculite and perlite (1:1:1) as described by Paek and Murthy (2002).

Results and discussion

Bacterial and fungal contaminations were always observed on cultured bulb scale explants. After 3 weeks in culture, small buds were visible between the scales. These buds subsequently developed into the bulblets approximately 5–10 mm in diameter after 5–6 months of culture initiation. Bulblet formation occurred on most of the media tested. However, the average number of bulblet per explant was low and ranged from one to four.

Unlike bulb scale explants, no contamination was seen on cultured zygotic embryos and large numbers of shoots or embryos were visible on callus in both MS medium supplemented with different combinations of NAA and BAP or N6 medium containing 2 mgL^{-1} 2,4-D using a four step culture method within 5 months of culture initiation.

Callus tissue carrying shoots or embryos were divided into small pieces and subcultured onto fresh medium every 1 or 1.5 month to allow bulblets to develop further. At the end of 12–14 months, large numbers of well-developed bulblets were produced on both NAA and BAP or 2,4-D containing media. The highest number of bulblets (over 100 bulblets/explants) were achieved from immature embryos of the species studied on MS medium supplemented with 4 mgL^{-1} BAP and 0.25 $mg\ L^{-1}$ NAA or 2 $mg\ L^{-1}$ 2,4-D. Regenerated bulblets were kept at 5°C for 5 weeks and then transplanted to a potting mixture of peat moss.

Bulb scales are most commonly used as explants for *in vitro* culture of geophytes. However, the use of bulbs as a source of explant is often associated with heavy bacterial and fungal contamination (Ziv and Lilien-Kipnis, 2000) and the frequency of bulblet development remained low. In the present study, we produced large number of bulblets from immature embryos of many endemic and endangered geophyte species without having any contamination problem. The procedure described here may help the micropropagation and conservation of endangered geophytes.

References

Arslan N, Gurbuz B, Gumuşçu A, Ozcan S, Mirici S, Khawar KM (2002) Cultivation of *Sternbergia fischeriana* (Herbert) Rupr. and a study on its morphological characteristics. Pak J Bot 34: 411–418.

Mirici S, Parmaksiz I, Ozcan S, Sancak C, Uranbey S, Sarihan EO, Gumuscu A, Gurbuz B, Arslan N (2005) Efficient *in vitro* bulblet regeneration from immature embryos of endangered *Sternbergia fischeriana*. Plant Cell Tissue Organ Cult 80: 239–246.

Paek KY, Murthy HN (2002) High frequency of bulblet regeneration from bulb scale sections of *Fritillaria thunbergii*. Plant Cell Tissue Organ Cult 68: 247–252.

Ziv M, Lilien-Kipnis H (2000) Bud regeneration from inflorescence explants for rapid propagation of geophytes *in vitro*. Plant Cell Rep 19: 845–850.

In Vitro Strategies for Conservation of Plant Diversity in India

Pramod Tandon

Plant Biotechnology Laboratory, Department of Botany,
North-Eastern Hill University, Shillong 793 022, India.

The Indian Sub-continent has been identified as one of the world's 'mega-diversity' regions with two 'hotspots', namely, the Western Ghats/Sri Lanka and the Indo–Burma region (covering the Eastern Himalayas) (Myers et al., 2000). About 33% of the Indian flora is endemic to the country and is confined to the Western Ghats, Northeast, Northwest Himalayas, and the Andaman and Nicobar islands. However, in recent years, many a plant species including several unique and irreplaceable varieties has been under the threat of extinction.

Conservation of plant genetic resources can be accomplished both by *in situ* and *ex situ* methods. Biotechnology has emerged as an important link between conservation and sustainable utilization of genetic diversity. The molecular marker technology and molecular diagnostics, *in vitro* technologies and cryopreservation techniques have been applied for germplasm conservation. *In vitro* propagation methods have led to re-establishment of several endangered plant species in their natural habitat (Tandon and Kumaria, 2005). The use of *in vitro* propagation methods for conservation of endangered plants has been extensively reviewed (Pence, 1999). Realizing the importance of conservation of endangered Indian flora, the Department of Biotechnology, Government of India has initiated a program on species recovery. Besides several research institutions and Universities, a National Bureau of Plant Genetic Resources (NBPGR) at New Delhi has been developed as a national facility for collection, evaluation, conservation and exchange of germplasm.

Z. Xu et al. (eds.), Biotechnology and Sustainable Agriculture 2006 and Beyond, 385–386.

Several plants, viz. *Nepenthes khasiana* Hook f. (Nepenthaceae), *Nymphaea tetragona* Linn. (Nympheaceae), *Ilex khasiana* Hook f. (Aquifoliaceae) and *Mantisia spathulata* Schult (Zingiberaceae) that are endemic to Northeast India have been listed as endangered plants, for which micropropagation protocols have been developed in our laboratory. *In vitro* propagation of *N. khasiana* (the Indian insectivorous pitcher plant) was accomplished using seed germination and nodal explants and a couple of thousand plants have been re-introduced in its natural habitat. *N. tetragona*, a very dwarf aquatic herb restricted to a single pond, was successfully multiplied using immature embryos. For *M. spathulata*, an extremely threatened perennial herb, new emerging shoot sprouts were used; about one thousand hardened plants have been re-established in nature. A large number of plantlets of *I. khasiana*, a small forest tree, restricted to certain pocket areas of Khasi Hills, were produced *in vitro* through axillary buds.

We have extensively used the in-vitro seed germination for multiplication of rare orchid species. Seeds are preferred for multiplication of rare and endangered species, as this would ensure genetic diversity. In certain cases, explants from the *in-vitro* raised seedlings are used to initiate cultures for micropropagation (Kumaria and Tandon, 2001). In view of the ever-increasing threat to biodiversity, efforts are on towards recovery of several more of the endangered plant species of the region.

References

Kumaria S, Tandon P (2001) Orchids: The world's most wondrous plants. In: Pathak P, Sehgal RN, Shekhar N, Sharma M, Sood A (eds) Orchids: Science and Commerce. Bishen Singh Mahendra Pal Singh, India, pp. 17–28.

Myers N, Mittermeier RA, Mittermeier CG, da Fonseca GAB, Kent J (2000) Biodiversity hotspots for conservation priorities. Nature 403: 853–858.

Pence VC (1999) The application of biotechnology for the conservation of endangered plants. In: Bensen EE (ed) Plant Conservation Biotechnology. Taylor and Francis, London, pp. 227–250.

Tandon P, Kumaria S (2005) Prospects of plant conservation biotechnology in India with special reference to Northeast region. In: Tandon P, Sharma M, Swarup R (eds) Biodiversity: Status and Prospects. Narosa Publishing House, New Delhi, India, pp. 79–92.

Plants as Bioreacter/Molecular Pharming

Molecular Farming in Plants: An Update

Elisabetta Lupotto[1] and Maria Rosaria Stile[2]

[1]CRA – Istituto sperimentale per la Cerealicoltura, Vercelli, Italy.
[2]ISTA S.p.A – Parco Tecnologico Padano, Lodi, Italy.

With the terms 'plant molecular farming' or 'biopharming' an increasingly important area of plant science is indicated, namely referring to the application of biotechnology to agricultural plants in order to use them as bioreactors for the production of valuable products. These products belong to different categories, but the terms stem mainly for Plant Made Pharmaceuticals (PMPs) or Plant Made Biologics (PMBs) of protein origin, such as recombinant therapeutic proteins, enzymes, growth factors and vaccines. Biopharmaceuticals actually represent almost 25% of the new molecular entities coming to the market. Plants have the potential to synthesize animal and human proteins by means of genetic engineering. The expression of the foreign protein can be targeted to a specific plant tissue, thus facilitating harvest and storage, obtaining safe and consistent expression. The simple investment of cropping areas relative to demand, and lower costs for mass production make the perspective of using plants a very attractive hypothesis for a wide number of biological molecules. Advances in applied biotechnology offer the possibility to overcome the traditional idea of crops as a source of food and feed toward the recognition of the great economic potential of using plants as bioreactors for the synthesis of therapeutic molecules. PMPs and biopharming is an important emerging industry and Pharma Crops cultivation an interesting field of investigation relative to various aspects of the production chain from field to shelf, regulatory issues, biosafety and environmental implications.

Z. Xu et al. (eds.), Biotechnology and Sustainable Agriculture 2006 and Beyond, 389–396.

In the recent years, molecular farming in plants has been widely reviewed and discussed (e.g., Horn et al., 2004; Gomord et al., 2005; Howard, 2005; Ma et al., 2005; Walsh, 2005). The current scientific activity is continuously updated and a wide variety of PMPs are on the market and on the pipeline for registration. The present note does not pretend to make an exhaustive overview of the field but aims at focusing on some aspects of the basic tools available in plant biopharming, relative to the most recent achievements in the field.

Types of molecules and recent examples

Higher plants can be used as biofactories with the highest economic benefit respect to transgenic animals, animal cells, yeast, fungi and bacteria, to produce foreign proteins of economical value. A fundamental advantage of biopharming in plants is the extreme versatility of the system, being the plant cell able to potentially produce a wide range of recombinant molecules. They include small peptides, polypeptides, complex multimeric proteins not always obtainable in microbial systems (Ma et al., 2003).

Although one can talk about molecular farming since the first transgenic plant was produced (Fraley et al., 1983), the first reports exploiting the potential of plants as producers of PMPs date back to late eighties when first recombinant plant-derived protein (human antibodies) were produced in tobacco (During, 1988; Hiatt et al., 1989), and the human serum albumin was produced in tobacco and potato plants (Sijmons et al., 1990). Commercial production of avidin, an egg protein with important properties primarily used as diagnostic reagent, in transgenic maize was the first example of real plant molecular farming (Hood et al., 1999). In maize it was also farmed the proteolytic enzyme trypsin (TrypZean, Sigma) which stems as the first large-scale protein product from plant biopharming (Woodard et al., 2003). Respect to microbial production systems, pharma crops have a number of further advantages in terms of safety and ease of scale-up (Daniell et al., 2001). Being eukaryotes, plant cells display a number of metabolic processes much more similar to mammalian cells, such as the presence of chaperones in the production of antibodies, which are important in the control of the efficiency in protein assembly and degradation, or the presence of N-glycosylation and other post-translational modifications essential in the downstream processing (Gomord et al., 2004).

Today, a wide array of plant-derived pharmaceutical proteins (PDPs) are on the market including antibodies, cytokines, growth factors, hormones, recombinant enzymes, human and veterinary vaccines (Twyman et al., 2005). There are more than twenty biotech organizations specialized in PMPs: they have focused in one or more crop of choice as a platform for therapeutic production, most of them developed as spin-off of research centres and universities (Elbehri, 2005). Besides the PMPs already on the market, an increasing number are in the pipeline and some are close to be commercialized, such as the recombinant bovine aprotinin produced in transgenic corn seed (Zhong et al., 1999; Azzoni et al., 2002) or human lactoferrin produced in rice and maize seed (Nandi et al., 2002; Legrand, 2003), edible vaccines and monoclonal antibodies (Daniell et al., 2001).

The production systems and tools

The recent years have seen technological developments at many levels, including transformation methods, control of gene expression, protein targeting, processing and accumulation, the use of different crop plants as production platforms (Twyman et al., 2003).

The production of proteins in plants depends upon the availability of DNA sequence coding and on the construction of a gene expression cassette suitable for plant transformation. The strategies for a plant expression cassette construction depend on the desired goal. A recombinant protein level can be enhanced, in a particular expression system, by choosing appropriate regulatory elements in the vector. This can result in higher protein accumulation by improving the efficiency of transcription and translation, by intracellular sorting and targeting or by increasing mRNA and protein stability (Stoger et al., 2002).

Most research in the area of plant-based protein production has relied on expression of transgenes stably integrated into the plant genome. This system requires a suitable plant transformation vector, and a method for transferring the foreign genes into the plant cells, usually using *Agrobacterium tumenfaciens* or particle bombardment or other standard transformation techniques. *Nicotiana tabacum* is widely used as a model expression system, but other plants are also considered, including *Nicotiana benthamiana*, *Arabidopsis thaliana*, tomato, banana, turnip, black-eyed bean, oilseed rape, Ethiopian mustard, potato, soybean alfalfa, rice, wheat, and maize. Field testing of pharma plants occurs since early 1990; in recent years, the place and number of the pharma-field trials have accelerated. In the United States, according to APHIS data, more than 325 sites of field trials were approved for the period 1991–2004, including pharmaceutical, novel protein, and industrial enzymes (Elbehri, 2005). Maize dominated as pharma-plant in most cases but the recent developments tend to use non food plants for biopharming, due to regulatory issues and concern about open-field cultivation of pharma-crops.

However, greenhouse and other containment-grown cultures are also suitable for biopharming. Transgenic plants containing a gene coding for the target protein can be also grown hydroponically in a way that allows release of the desired product as part of the root exudates into the hydroponic medium. In this system the purification of the desired product is considerably easier since no tissue disruption is needed, and the quantity of contamination protein is low (Horn et al., 2004). However, the system still needs improvement because to date the production of large quantities of any protein product is a limiting factor. Plant suspension cultured cells have also been used to produce over 20 different recombinant proteins. Plant cell transformation and protein expression follow the same criteria of those described for transgenic plant transformation. Production systems that use whole plants lack several of the intrinsic benefits of *in vitro* cultured cells, including the precise control over growth conditions, batch-to-batch product consistency, a high level of containment and the ability to produce recombinant protein in compliance with Good Manufacturing Practice (Miele, 1997). On the other hand, the most important challenge associated with suspension cells is the relatively low protein yields (0.0064–4% of total soluble protein or from 0.5 $\mu g\ L^{-1}$ to 200 $mg\ L^{-1}$ based on the cultured volume) compared with plant storage organs. Besides this, it has to be underlined that long term cell cultures are subjected to a range of genetic modifications through the mechanism of somaclonal variation, and this may lead to undesirable effects on the production system (Hellwig et al., 2004).

The choice of an appropriate promoter for constitutive or tissue-specific expression is determinant for each plant-derived product and can strongly influence the transgene expression level. The expression of recombinant proteins in transgenic plants is mostly controlled by a strong constitutive promoter such as the 35S CaMV promoter. This however can result in low yields of product, in the range of 0.1–0.5% of total soluble protein (Gleba et al., 2005). This yield may be improved by the use of the enhanced 35S CaMV promoter, as well documented in tobacco, rice and potato. On the other side, the maize ubiquitin-1 promoter with the first intron is the preferred choice for cereal plants engineering (discussed in Stoger et al., 2002). Many cases use cereal seeds as preferred expression organs, also because of the easiness of storage. Restricting expression to the seeds can lead to higher yields up to 3% of total soluble protein (Vandekerckhove et al., 1989). The development of inducible promoters offers an interesting tool to control and modulate the synthesis of the engineered product, separating for example the biomass production in the field from the protein synthesis step. This is the case reported for the patented post-harvest expression system, in which plant tissues harvested from the field can be subsequently induced to produce the foreign protein during storage (Cramer and Weissenborn, 1997).

Cytoplasmically synthesized proteins are often retained in the endoplasmic reticulum where they generally accumulate to a much higher proportion of total soluble protein (1–5%) and where various choices for post-translational modifications are provided. However, recombinant proteins can also be targeted to different subcellular compartments, such as chloroplasts, chromoplasts and amyloplasts, which may result more suitable than the cytoplasm for some enzymatic reactions, for modifications, processing and so on.

The gene encoding the protein of interest can be integrated on plastid chromosome (Bogorad, 2000). Chloroplast transformation uses two flanking sequences that, through homologous recombination, insert foreign DNA into the spacer region between the functional genes of chloroplast genome. The expression levels of transgenes incorporated in the plastid chromosome can reach 47% of total soluble protein (De Cosa et al., 2001) and the chloroplasts can process eukaryotic proteins, including correct folding and formation of disulfide bridges. The expressed protein is however confined to the plastid, an organelle that does not provide many desired post-translational modifications such as glycosylation and, like in bacteria, some proteins form inclusion bodies that have to be refolded during the purification process. Like nuclear transformation, plastid transformation is also a slow process, and is unable to support rapid production of sub-gram quantities of recombinant protein within 12–18 months (Gleba et al., 2005). Moreover, the current limitation to frequent use of chloroplast transformation is that – although routine in tobacco and in the green alga *Chlamydomonas reinhardtii* – it is difficult and still requiring optimisation in other edible plant species.

When considering the time needed to engineer a plant for biopharming, usually 3–9 months are needed to generate stably transformed plants available for testing the function and characteristics of the expressed protein. The most advanced transformation system is now the 'systemic transfection' which implies the use of RNA viral vectors as DNA precursor and transient full expression resulting in gene amplification in mature leaf tissues (Marillonnet et al., 2005). Transient expression of genes delivered by *Agrobacterium* and viral vector-based transient amplification in plants are also two processes that are being routinely used to produce protein. In both cases, expression relies on the delivery of transgene by a vector (bacterium or virus), and there is no need for stable transgene integration. Transient gene expression is rapid compared to stable plant transformation and gives results in days: 3–4 days necessary for full expression in leaves infiltrated with Agrobacteria, and 10–14 days for maximum expression from viral vectors. The transfection at the whole plant level is termed 'magnifection' and stems as the most promising and updated way to produce heterologous proteins in plants (Marillonnet et al., 2005).

References

Azzoni AR, Kusnadi AR, Miranda EA, Nikolov ZL (2002) Recombinant aprotinin produced in transgenic corn seed: Extraction and purification studies. Biotechnol Bioeng 80: 268–276.

Bogorad L (2000) Engineering chloroplast: An alternative site for foreign genes, proteins, reactions and products. Trends Biotechnol 18: 257–263.

Cramer CL and Weissenborn DL (1997) HMG2 promoter expression system and post-harvest production of gene products in plants and plant cell cultures. US Patent, 5, 670 349.

Daniell H, Streatfield SJ, Wycoff K (2001) Medical molecular farming: production of antibodies, biopharmaceuticals and edible vaccines in plants. Trends Plant Sci 6: 219–226.

De Cosa B, Moar W, Lee SB, Miller M, Daniell H (2001) Hyper-expression of Bt Cry2Aa2 operon in chloroplast leads to formation of insecticidalk crystals. Nat Biotechnol 19: 71–74.

During K (1988) Wound-inducible expression and secretion of T4 lysozyme and monoclonal antibodies in *Nicotiana tabacum*. Ph.D Thesis, University Koln, Germany.

Elbehri A (2005) Biopharming and the Food System: examining the potential benefits and risks. AgBioForum 8: 18–25.

Fraley RT, Rogers SG, Horsch RB, Sanders PR, Flick JS, Adams SP, Bittner ML, Brand LA, Fink CL, Fry JS, Galluppi GR, Goldberg SB, Hoffman NL, Woo SC (1983) Expression of bacterial genes in plant cells. Proc Natl Acad Sci USA 80: 4803–4807.

Gleba Y, Klimyuk V, Marillonet S (2005) Magnification – a new platform for expression vaccines in plants. Vaccine 23: 2042–2048.

Gomord V, Chamberlain P, Jefferis R, Faye L (2005) Biopharmaceutical production in plants: problems, solutions and opportunities. Trends Biotechnol 23: 559–565.

Gomord V, Sourrouille C, Fitchette A-C, Bardor M, Pagny S, Lerouge P, Faye L (2004) Production and glycosylation of plant-made pharmaceuticals: The antibodies as a challenge. Plant Biotechnol J 2: 83–100.

Hellwig S, Drossard J, Twyman RM, Fischer R (2004) Plant cell cultures for the production of recombinant proteins. Nat Biotechnol 22: 1415–1422.

Hiatt A, Cafferkey R, Bowdish K (1989) Production of antibodies in transgenic plants. Nature 342: 76–78.

Hood EE, Kusnadi A, Nikolov Z, Howard JA (1999) Molecular farming of industrial proteins from transgenic maize. In: Shahidi F, Kolodziejczyk P, Whitaker JR, Munguia AL Fuller G (eds) Chemicals via Higher Plant Bioengineering. Kluwer/Plenum, New York, pp. 127–147.

Horn ME, Woodard SL, Howard JA (2004) Plant molecular farming: systems and products. Plant Cell Rep 22: 711–720.

Howard JA (2005) Commercialization of biopharmaceutical and bioindustrial proteins from plants. Crop Sci 45: 468–472.

Legrand D, Salmon V, Spik G, Gruber V, Bournat P, Merot B (2003) Recombinant lactoferrin, methods of production from plants and uses. US Patent 6: 569–831.

Ma J K-C, Chikwamba R, Sparrow P, Fischer R, Mahoney R, Twyman RM (2005) Plant-derived pharmaceuticals – the road forward. Trends Plant Sci 10: 580–585.

Ma J K-C, Drake PMW, Christou P (2003) The production of recombinant pharmaceutical proteins in plants. Nat Rev Genet 4: 794–805.

Marillonnet S, Thoeringer C, Kandzia R, Klimyuk V, Gleba Y (2005) Systemic *Agrobacterium tumefaciens*-mediated transfection of viral replicons for efficient transient expression in plants. Nat Biotechnol 23: 718–723.

Miele L (1997) Plants as bioreactors for biopharmaceuticals: regulatory considerations. Trends Biotechnol 15: 45–50.

Nandi S, Suzuki YA, Huang J, Yalda D, Pham P, Wu L, Bartley G, Huang N, Loennerdal B (2002) Expression of human lactoferrin in transgenic rice grains for the application in infant formula. Plant Sci 163: 713–722.

Sijmons PC, Dekker BM, Schrammeijer B, Verwoerd TC, Van den Elzen PJ, Hoekema A (1990) Production of correctly processed human serum albumin in transgenic plants. Biotechnology 8: 217–221.

Stoger E, Sack M, Perrin Y, Vaquero C, Torres E, Twyman RM, Christou P, Fischer R (2002) Practical considerations for pharmaceutical antibody production in different crop systems. Mol Breed 9: 149–158.

Twyman RM, Schillberg S, Fischer R (2005) Trangenic plants in the biopharmaceutical market. Expert Opin Emergent Drugs 10: 185–218.

Twyman RM, Stoger E, Schillberg S, Christou P, Fischer R (2003) Molecular farming in plants: Host systems and expression technology. Trends Biotechnol 21: 570–578.

Vandekerckhove J, Van Damme J, Van Lijsebettens M, Botterman J, Leemans J, Van Montagu M, Krebbers E (1989) Enkephalins produced in transgenic plants using modified storage proteins. Bio/Technol 7: 929–932.

Walsh G (2005) Biopharmaceuticals: recent approvals and likely directions. Trends Biotechnol 23: 553–558.

Woodard SL, Mayor JM, Bailey MR, Barker DK, Love RT, Lane JR, Delaney DE, McComas-Wagner JM, Mallubhotla HD, Hood EE, Dangott LJ, Tichy SE, Howard JA (2003) Maize-derived bovine trypsin: characterization of the first large-scale, commercial protein product from transgenic plants. Biotechnol Appl Biochem 38: 123–130.

Zhong G-Y, Peterson D, Delaney DE, Bailey M, Witcher DR, register JC III, Bond D, Li C-P, Marshall L, Kulisek E, Ritland D, Meyer T, Hood EE, Howard JA (1999) Commercial production of aprotinin in transgenic maize seeds. Mol Breed 5: 345–356.

In Vivo and In Vitro Studies on Huperzine A Producing Ferns

Somika Bhatnagar, Yuan Qing Deborah Hong, Guang Yuan Wang, and Yan Hong*

Plant Biotechnology Group, Temasek Life Sciences Laboratory, 1 Research Link, National University of Singapore, Singapore 117604.
*Email: hongy@tll.org.sg

The alkaloid Huperzine A (Hup A) is the active component extracted from fern allies of family Lycopodiaceae, of which many members have also been used in Traditional Chinese Medicine. Its popularity as a treatment for Alzheimer's disease is depleting the limited natural resources at a very fast pace while the chemical synthesis remains prohibitively expensive. In our studies, plants from tropical and subtropical regions were collected and grown under controlled conditions. Molecular systematic study was conducted with fluorescent Amplified Fragment Length Polymorphism (AFLP) and DNA sequencing. Hup A content was determined by High Performance Liquid Chromatography (HPLC). An *in vitro* propagation system was established for one *Phlegmariurus* species starting from cell division from spores, formation of callus, gametophyte development followed by fertilization and growth of sporophytes. *In vivo* propagation was also successful for some other species through manipulating usual dichotomous branching followed by rooting. A subtractive cDNA library and a general cDNA library was constructed and >3000 cDNA contigs were sequenced. After co-cultivation of spores with *Agrobacterium tumefaciens* harboring a green fluorescent protein (GFP) gene, fluroscence of GFP could be observed

Z. Xu et al. (eds.), Biotechnology and Sustainable Agriculture 2006 and Beyond, 397–398.

under UV illumination in the germinating spores. This suggested the possibility of genetic transformation. With *in vivo* propagation, *in vitro* regeneration and genetic transformation systems and characterization of genes involved in Hup A biosynthesis, we aim to have better understanding of plant secondary metabolism and to ensure sustainable supply of this alkaloid of medical importance.

Production of Recombinant Antibodies in Pea Seeds and Their Oral Application in Piglets

Isolde Saalbach[1,2], Marcus Riehl[1], Martin Giersberg[1], Jochen Kumlehn[2], and Dieter Falkenburg[1,*]

[1] Novoplant GmbH, Am Schwabeplan 1b, 06466 Gatersleben, Germany.
[2] Leibniz Institute of Plant Genetics and Crop Plant Research (IPK), Corrensstr. 3, 06466 Gatersleben, Germany.
[*]Email: dieter.falkenburg@novoplant.de

Introduction

Transgenic plants offer an effective and economically interesting alternative for the heterologous production of biomolecules for medical and veterinary purposes. Plants are free of human and animal pathogens and offer virtually unlimited scale-up potential with low initial investment. Especially the large seeds of cereal and legume crop plants represent attractive bioreactors (Giersberg et al., 2004).

Here we report on the development of feed pea as a production system for single-chain antibodies (scFvs) for animal health applications. Feed pea is a minor field crop with established low-cost processes for planting, harvesting, storage and processing. As a GMO peas provide a very high safety level as they are self-pollinating with a extremely low outcrossing risk (Polowic et al., 2002), show sexual incompatibility to most of the wild pea species and produce a low number of volunteers. We use seed-specific expression of scFvs with the goal to apply the milled seed containing the antibody as part of the regular feed to animals for the prevention of gastrointestinal

Z. Xu et al. (eds.), Biotechnology and Sustainable Agriculture 2006 and Beyond, 399–402.

diseases. Expensive purification issues and purification losses are thus avoided, application is made easy and no extra handling costs are required for the farmer.

ETEC infections in piglets

Enterotoxigenic Escherichia coli (ETEC) is a major cause of diarrhea in neonatal and postweaning piglets (Nataro and Kaper, 1998). Bacterial surface structures (fimbriae) attach to the microvilli of the small intestine and produce enterotoxins that act locally on enterocytes. This causes hypersecretion of water and electrolytes and reduced absorption resulting in diarrhea. E. coli bearing F4 (K88) fimbriae are responsible for ca. 50% of the 10 million piglet deaths/year. We have isolated various scFv antibodies from a phage display library which specifically bind to F4 fimbriae. Some of them are capable to neutralize the attachment of ETEC to enterocytes in *in vitro* assays. The one with the strongest neutralizing effect (designated BA11) was expressed in bacteria and yeasts as well as in tobacco and pea seed.

Antibody stability

A homozygous, marker free and single insert feed pea line was developed. Using the seed specific USP promoter (Zakharov et al., 2004) yield of the scFv BA11 is about 2 g/kg seed. Under greenhouse conditions plants are now grown in the 7th generation, stably expressing the functional antibody. To date a first field trial is performed with this line in the United States. Ongoing stability tests with pea seeds containing BA11 have so far showed no loss in quantity and functionality of the antibody after an observation period of 18 months at room temperature. Furthermore we could demonstrate that pea seeds containing BA11 can be integrated into established feed processing procedures (pelletizing) without significant loss of function. These data indicate that pea seed provides a highly protective environment for heterologous biomolecules and is thus ideally suited as an industrial production process.

Animal studies

BA11 expressed either in yeast or in pea seed was orally applied with the feed to experimentally infected piglets. Preliminary data with small number of animals indicate positive effects upon feeding of the antibody, as demonstrated by a lower rate of clinical symptoms and faster elimination of the applied E. coli strain. Microbiological examination was based on rectal swabs, furthermore fecal dry matter, chloride, sodium and pH were recorded. After the trial all animals were sacrificed and organ samples were taken during necropsy. No antibody could be detected either in blood, liver, spleen, muscle, heart, lymph nodes, tonsils, kidneys or in urine. These data show that BA11 is not taken up systemically. No waiting periods for clearance of antibody residues from the body would have to be observed as is the case with antibiotics and other chemicals.

Conclusions

Feed pea (*Pisum sativum* L.) has been developed as a tool for the production of single-chain antibodies (scFvs) using a seed-specific expression system. This system has demonstrated genetic stability over many generations, high yield of the heterologous protein and an excellent environmental safety profile. Our data with an antibody (BA11) directed against F4 fimbriae of Enterotoxigenic *E.coli* (ETEC) indicate that pea offers a stabilizing environment for the heterologous proteins. Both long-term storage at room temperature and integration into established feed processing procedures are possible. First animal studies indicate protective effects upon feeding of BA11 to experimentally infected piglets. BA11 is solely confined to the GI tract and not taken up systemically or excreted. The inclusion of antibody containing pea seeds into the diet of piglets is easy to achieve and might be an interesting alternative to established antimicrobial treatments for the prevention of *E. coli* related diseases in weaned piglets.

References

Giersberg M, Bäumlein H, Saalbach I (2004) Gene farming in pea under field conditions. In: Fisher R, Schillberg S (eds) Molecular Farming. Wiley-VCH-Verlag GmbH & Co., KgaA, Weinheim, pp. 1–8.

Nataro JP, Kaper JB (1998) Diarrheagenic Escherichia coli. Clin Microbiol Rev 11: 142–201.

Polowic PL, Vandenberg A, Mahon JD (2002) Field assessment of outcrossing from transgenic pea (Pisum sativum L.) plants. Transgenic Res 11: 515–519.

Zakharov A, Giersberg M, Hosein F, Melzer M, Muntz K, Saalbach I (2004) Seed-specific promoters direct gene expression in non-seed tissue. J Exp Bot 55: 1463–1471.

Molecular Breeding

Breeding for Improved and Novel Starch Characteristics in Potato

Farhad Nazarian Firouzabadi, Geraldine Kok-Jacon, Ji Qin, Jean-Paul Vincken, Luc Suurs, and Richard G.F. Visser

Laboratory of Plant Breeding, Graduate School Experimental Plant Sciences, Wageningen University, P.O. Box 386, 6700 AJ Wageningen, the Netherlands.

Starch is the major form in which carbohydrates are stored and is present in almost all plant organs at one or another time during development. Storage starch is formed in amyloplasts as dense granules ranging in size from 1 to over 100 μM. They are composed of an essential linear glucose polymer and a branched glucose polymer termed amylose and amylopectin respectively (Ball et al 1998).

Knowledge about starch biosynthesis is required to be able to modify starches for industrial or food uses. This knowledge has been acquired by studies in a broad range of plants using both mutants and reverse genetic approaches. In this way the function and impact of a number of the essential genes in the starch biosynthetic pathway has been assessed.

Modification of starch biosynthesis pathways holds an enormous potential for tailoring granules or polymers with new functionality's. Until recently a lot of effort was put into investigating the individual components in the starch biosynthetic pathway. Both enzymes involved in synthesizing and degrading carbohydrate polymers have been down-regulated using different inhibition approaches. In most cases this were single gene knock out attempts although in some cases two or even three genes were addressed at once. Examples of

Z. Xu et al. (eds.), Biotechnology and Sustainable Agriculture 2006 and Beyond, 405–408.

this are the production of amylose-free starch (Kuipers et al 1994), high amylopectin starch and starch with an increased degree of branching. This has lead to a number of interesting new starches, which have helped to shape the scientific knowledge about starch biosynthesis in general (Visser et al 1997).

By altering the starch structure, granular starches might be obtained with novel physical properties and a potentially unlimited range of new industrial applications. This has proven to be more difficult although altered morphology of starch granules has been obtained in potato plants in which the activity of two specific starch synthases has been inhibited or in plants in which in starch granules co-polymers of starch and fructan were produced.

By adding novel enzyme activities to the starch biosynthetic machinery of plants other storage sugars might be produced (Kok Jacon et al 2003). We have embarked on introducing new linkage types into potato starch by overexpressing bacterial genes from *Leuconostoc* and *Streptococcus* strains. These genes are responsible for the production of dextrans (water-soluble polymers with α-1,6-linked glucose residues) and mutans (water-insoluble polymers with α-1,3-linked glucose residues). In addition, these polymers present different degrees of branching, generating an extensive structural variation. The physico-chemical properties of these starches were severely affected. Analysis of these transgenic starches with scanning electron microscopy demonstrates a severely altered granule morphology. In addition, the granule size distribution is significantly modified (Qin et al 2004). Further biochemical characterizations of these transgenic starches are in progress. Finally the ability to introduce and anchor into the starch granule novel activities which would render certain chemical- and enzymatic-modification procedures obsolete is a final step in producing in planta or postharvest tailor made starches.

Various transgenic approaches have been explored for modifying the amylose content of potato starch. In earlier research, we were able to show that a tandem repeat of two SBDs had a ten-fold higher affinity towards starch granules than a single SBD domain (Qin et al 2003). Hence we hypothesized that engineering of a series of multiple high-affinity repeats of family 20 starch binding domain might be able to out compete native granule-bound starch synthase I (GBSSI) from the starch granules. Therefore, a series of constructs of SBDs (SBD_2–SBD_5) were introduced in an amylose-containing genetic background (Nazarian Firouzabadi et al 2007a). Our results showed that none of the multiple repeat SBDs was able to displace GBSSI from the starch granules. Neither the amount of GBSSI protein nor the apparent amylose content of any of the transformants was changed. However, it was shown that SBDn expression can affect physical processes underlying granule assembly, in both amylose-containing and amylose-free potato genetic backgrounds, without altering the primary structure of the constituent starch polymers and the granule melting temperature. The amount of SBD accumulated in starch granules increased progressively from SBD to SBD3; however, it seemed as if less SBD4 and SBD5 were accumulated inside the starch granules.

Starch acetylation is required because native starch does not have the desired characteristics demanded by industry. Improved stability against retrogradation, lower gelatinization temperature, more transparency and improved elasticity are some of the desired properties of acetylated starches. Starch acetylation is achieved through chemical reactions using some environmentally pollutant procedures. In an attempt to acetylate potato tuber starch *in planta*, *E. coli* maltose-acetyltransferase (MAT) was targeted to the starch granules either alone or fused to SBD. A significant higher degree of acetylation was obtained in an amylose-containing genetic background. However, this did not seem to be enough to see altered physicochemical properties of starch. Interestingly, MAT maintained its activity inside the starch granules when provided with acetyl-CoA and different acceptor substrates. Our results evidenced for the first time that MAT has affinity for starch granules on its own. In comparison to the SBD2 trasformants, fusion proteins comprising MAT were accumulated in a larger amount inside the starch granules. MAT trimer formation on the one hand and fusion to SBD on the other hand might explain its affinity for starch granules as well as its high accumulation in the starch granules (Nazarian Firouzabadi et al 2007b).

References

Ball SG, van de Wal MHBJ, Visser RGF (1998) Progress in understanding the biosynthesis of amylose. Trends Plant Sci 3: 462–467.

Qin J, Vincken JP, Suurs LCM, Visser RGF (2003) Microbial starch binding domains as a tool for targeting proteins to starch granules during starch biosynthesis. Plant Mol Biol 51: 789–801.

Qin J, Oomen R, Vincken JP, Bolam DN, Gilbert H, Suurs LCM, Visser RGF (2004) Reduction of starch granule size by expression of an engineered tandem starch binding domain in potato plants. Plant Biotechnology Journal 2: 251–260.

Kok-Jacon G, Qin J, Vincken JP, Visser RGF (2003). Towards a more versatile alpha-glucan biosynthesis in plants. J Plant Physiol 160: 765–778.

Kuipers GJ, Jacobsen E, Visser RGF (1994) Formation and deposition of amylose in the potato tuber starch granule are affected by the reduction of granule-bound starch synthase expression. Plant Cell 6: 43–52.

Nazarian Firouzabadi F, Vincken JP, Qin J, Suurs LCJM, Buléon A, Visser RGF (2007a). Accumulation of multiple-repeat starch-binding domains (SBD2-SBD5) does not reduce amylose content of potato starch granules. Planta 225: 919–933.

Nazarian Firouzabadi F, Vincken JP, Qin J, Suurs LCJM, Visser RGF (2007b). Expression of an engineered granule-bound *E. coli* maltose acetyl transferase in wild-type and *amf* potato plants. Plant Biotechnology Journal 5: 134–145.

Visser RGF, Suurs LCJM, Bruinenberg PM, Jacobsen E (1997) Comparison between amylose-free and amylose containing potato starches. Starch/Staerke 49: 438–443.

Biofortification of Cassava Using Molecular Breeding

Kirsten Jørgensen[1], Charlotte Sørensen[1], Susanne Jensen[1], Marc Morant[1], Martin Fregene[2], and Birger Lindberg Møller[1]

[1]The Royal Veterinary and Agricultural University, Department of Plant Biology, Plant Biochemistry Laboratory and Center for Molecular Plant Physiology, 40 Thorvaldsensvej, DK-1871 Frederiksberg C, Denmark.
[2]Centro Internacional de Agricultura Tropical (CIAT), KM 17 Recta Cali-Palmira, AA6713 Cali, Columbia.

Cassava, *Manihot esculenta,* is the most important root crop and the third most important staple food in the Tropics. Cassava has gained such a high importance because (1) cassava produces well on poor soils and may be grown with low resource input; (2) cassava is drought tolerant and the tubers may be left in the soil as a storage of carbohydrates for extended periods; (3) in periods of food shortage, farmers are not required to save a part of the edible cassava crop for replanting because cassava is propagated vegetatively via stem cuttings ('stakes'). The tubers are a rich source of starch and afford a high percentage of the calorie intake in Sub-Saharan Africa, in South America and in South east Asia. Cassava leaves have high protein content but are not routinely used as a food source. The importance of cassava has been increasing for the last decade due to its agricultural performance and the high yield of starch.

Major drawbacks of the cassava crop are (1) a low protein content in the tubers, (2) a fast post-harvest tuber perishability and (3) a high content of the two cyanogenic glucosides linamarin and lotaustralin in all tissues of the cassava plant. Upon disruption of the cassava tissue e.g., during food preparation, the cyanogenic glucosides

Z. Xu et al. (eds.), Biotechnology and Sustainable Agriculture 2006 and Beyond, 409–411.

are brought in contact with degrading enzymes (β-glucosidase and α-hydoxy nitrilase) which results in release of hydrogen cyanide and ketones.

In regions where cassava serves as the major staple food, careful processing to remove the released hydrogen cyanide is required to avoid acute or chronic cyanide intoxication. Unfortunately, processing to remove hydrogen cyanide is typically accompanied by loss of protein, minerals, and vitamins.

The long term aim of our research programme is to improve the nutritional value of the cassava tubers from African elite cultivars by blocking the accumulation of cyanogenic glucosides; increasing the pro-vitamin A content and enhancing the protein content.

The cyanogenic glucosides linamarin and lotaustralin are synthesized from the amino acids valine and isoleucine, respectively. The biosynthetic pathway is catalyzed by two membrane bound multifunctional cytochrome P450s and a soluble family 1 UGT glucosyltransferase. The initial part of the pathway is conversion of the parent amino acid to the corresponding oxime. This multistep reaction is catalyzed by the CYP79 paralogs, CYP79D1 and CYP79D2. The amino acid sequence of the two proteins is 59% identical. The two enzymes show identical catalytic properties and are found in the same tissues. The presence of paralogous enzymes is ascribed to the allotetraploid nature of the cassava plant. Using RNA interference technology, the expression of the two genes encoding CYP79D1 and CYP79D2 was blocked and more than 300 independent transgenic cassava lines produced. In some lines, the content of cyanogenic glucoisdes in leaves and tubers was reduced to 0.3% and 8%, respectively, in wild-type plants at the same developmental stage. When grown on agar with a reduced mineral content, the transgenic lines with the lowest content of cyanogenic glucosides showed a distinct phenotype characterized by elongated stems with few leaves that withered quickly. A wild-type phenotype was restored by transfer of the transgenic plants to a medium with increased mineral content and upon transfer of the plants to soil. The characteristic phenotype enables selection of low cyanide releasing lines without the use of an antibiotic resistance gene (Jørgensen et al., 2005). It has not been possible using traditional breeding to achieve acyanogenic cassava cultivars. This may partly be ascribed to the low seed set obtained with cassava and its alloploidicity.

The protein level in the current elite cassava cultivars is very low. This may partly be ascribed to the continued breeding of cassava for high starch content since a high starch content is inversely correlated to a high protein content. Breeding for high protein content would be at least partly in vain when the major part of the protein is lost anyway during processing to provide a safe food source. Fortunately, some wild cassava varieties has been found to contain quite high levels of protein. An investigation of these lines shows that the composition of the storage proteins in these lines vary considerably. The wild cassava cultivars were provided by Dr. Martin Fregene, CIAT, Colombia. These varieties exhibited great quantitative and qualitative differences in the amount of protein bound to isolated starch grains (cassava flour). These storage proteins are now being characterized to identify predominant proteins that have an optimal content of essential amino acids. This work is carried out in collaboration with Dr. Markus Piotrowski at Universität Bochum, Germany. Using molecular breeding, genes encoding desired storage proteins will be combined with the RNAi technology decreasing the content of cyanogenic glucosides in the African elite cassava lines.

Cassava lines with a high content of carotenoids have been obtained by classical breeding and have been made available to us by Dr. Alfred Dixon, IITA, Ibadan, Nigeria. We have succeeded in the transformation of these cassava lines and have by the use of RNAi technology managed to downregulate the synthesis of linamarin and lotaustralin. Again, in the absence of cyanogenic glucosides, thorough processing is rendered superfluous thus offering the possibility to develop a cassava plant that produces tubers with increased protein and pro-vitamin A content that will be preserved in the derived food products since these will not need excessive processing as normally dictated by the presence of high amounts of cyanogenic glucosides.

Reference

Jørgensen et al. (2005) Plant Physiol 139: 363–374.

Evaluation of GM Poplars Expressing Relevant Traits for Herbicide Tolerance, Disease Resistance and Production of Pharmaceutics: Biochemical, Molecular and Microbiological Studies on Plants and Detection of Transgene Sequences in Soil

S. Zelasco[1,*], D. Carbonera[2], A. Giorcelli[1], M. Confalonieri[3], F. Mattivi[4], M. Bonadei[2], M. Gennaro[1], E. Quattrini[5], P. Calligari[1], F. Picco[1], G. Deandrea[1], and A. Balestrazzi[2]

[1] C.R.A. – Istituto di Sperimentazione per la Pioppicoltura, 15033 Casale Monferrato, Italy.
[*]Email: zelasco@populus.it
[2] Dipartimento di Genetica e di Microbiologia, Università degli Studi di Pavia, 27100 Pavia, Italy.
[3] C.R.A. – Istituto Sperimentale per le Colture Foraggere, 26900 Lodi, Italy.
[4] Istituto Agrario di San Michele all'Adige, Centro Sperimentale, 38010 San Michele all'Adige, Italy.
[5]C.E.T.A.S. – Università degli Studi di Milano, 26900 Tavazzano (LO), Italy.

There is a global need for the sustainable production of food, fuel and fibre, due to the expanding earth population on limited and degrading land resources. The gene transfer technology offers the potential to reduce some inputs, required in agricultural systems, having negative side effects on health and environment. Crops engineered for insect pest resistance and tolerance to herbicides require substantially simplified management practices, compared to conventional varieties, and allow reductions in yield losses (Nickson, 2005). In addition, the use of genetically modified (GM) crops as production platforms for industrial chemicals and pharmaceuticals will reduce the reliance on chemical industry and animal-based production

Z. Xu et al. (eds.), Biotechnology and Sustainable Agriculture 2006 and Beyond, 413–418.

systems. In this context, the cultivation of woody plants, e.g., *Populus* spp., by the Short Rotation Forestry (SRF) approach, could represent a sustainable system able to produce a renewable feedstock for bioenergy and bioproducts associated with environmental and rural benefits. In order to fulfil all these requirements, the risks currently posed by the use of GM crops themselves need to be carefully evaluated, particularly those related to the interactions between transgenic plants and the environment.

In this work we tested, under greenhouse conditions, two classes of GM white poplars (*Populus alba* L.) expressing the *bar* gene conferring tolerance to the non-selective herbicide glufosinate ammonium, and the *StSy* gene encoding stilbene synthase (Confalonieri et al., 2000; Giorcelli et al., 2004). This enzyme, a specific component of the phenylpropanoid branch pathway, is required in various plant species for the biosynthesis of stilbenes in response to pathogen attacks as part of the plant defence response. In addition, several stilbenoids, including 3,5,4'-trihydroxystilbene (resveratrol), are considered to have beneficial effects on human health (Frémont, 2000).

The research activity was carried out over a 3-year period and several goals were fixed. Agronomical studies were conducted to assess the susceptibility of *StSy* GM poplars to different leaf diseases and to evaluate the tolerance to glufosinate ammonium of *bar* GM poplars. Biochemical analyses were performed in order to assess the long-term stability of resveratrol-like compounds produced in different tissues (leaves, stems and roots) of *StSy* GM poplars and to develop the possible use of these lines as a potential source for the production of pharmacological compounds. The stability of transgene expression in planta was also monitored over different seasons in apical and basal leaves of *StSy* and *bar* poplars. Finally, the persistence of recombinant DNA sequences in soil was investigated by PCR and Southern blot hybridization, in order to acquire information on the possible occurrence of horizontal gene transfer (HGT) events involving GM poplars and soil microorganisms.

The GM lines 5E and 6EA22 used in this study carried a single copy of the *bar* gene from *Streptomyces hygroscopicus* (Thompson et al., 1987) under the control of the constitutive 35SCaMV promoter. The GM lines 5ECA1 and 12ECA1 carried a single copy of the *StSy* gene from *Vitis vinifera* L. (Sparvoli et al., 1994) under the control of the same promoter. All transgenic lines carried the *nptII* marker gene associated with the NosP promoter. The plant material was transferred from sterile culture into peat pots containing a mixture of perlite and peat (1:4 ratio) and then acclimatized for 4 weeks in a growth chamber (22°C, 16/8 h photoperiod, photon flux density of 150 μmol m^{-2} s^{-1}). All rooted plantlets were transplanted into pots (30 cm diameter, 34 cm height) containing soil collected from agricultural fields. The plants were maintained in greenhouse under light, temperature and moisture regime similar to field conditions. All the plants, except for those used to study transgene stability, were coppiced during winter 2004 and 2005.

To assess the level of herbicide tolerance, the *bar* lines were sprayed (August 2004 and June 2005) with a dose equivalent to 5 L ha^{-1} glufosinate ammonium (Basta, Bayer Crop Science), i.e., the normal field dosage, and with a double dose (10 L ha^{-1}); untreated controls were also considered. The plants were observed for toxicity symptoms on leaves, stems and shoot tips every 2–3 days for 2 weeks after treatments. Results from these trials demonstrated that all the *bar*-transgenic plants showed a damage intensity significantly lower compared to untransformed plants, even when the double herbicide concentration was used.

The susceptibility to leaf diseases in the *StSy* transgenic and untransformed lines was assessed using artificial plant inoculations in greenhouse with conidial suspensions of *Pollaccia ramulosa* (Desm.) Ondřej (June), *Marssonina castagnei* (Desm. & Mont.) Magnus (September) and *Melampsora* spp. (September), respectively. Trials were performed in 2004 and 2005. Preliminary results did no evidence significant differences between GM and control lines. Further optimization of the reported methodology will be required.

The resveratrol-like compounds produced by leaves, stems and roots of *StSy*-poplars were evaluated by HPLC-DAD analysis as described by Giorcelli et al. (2004). Plant tissues were harvested in August and September 2004, subsequently in May, September and November 2005, and in March and May 2006. The analyses were carried out in triplicate. Differently from those of grapevine, the resveratrol synthesized by GM poplars were completely glucosylated in the position 3, producing mainly *trans*-resveratrol 3-glucoside (synonym: *trans*-piceid, *trans*-polydatin). The isomer *trans*-piceid was detected in the highest amount, especially in leaves where it was found in the range 46–364 mg/kg fresh weight during the full vegetative growth. *Trans*-piceid in the roots ranged 112–330 mg/kg fresh weight, with maximal values just after dormancy. The completion in 2006 of a 3-years follow-up, will allow us to formulate an explaining hypothesis for the observed seasonal variation of the *trans*-piceid concentration in the different tissues.

The piceid is a natural hydrophilic analogous of resveratrol, and as such is of pharmacological interest. To verify the possible use of the biomass produced from the *StSy* transformed lines as a potential source for the production of *trans*-piceid and *cis*-piceid, a lab scale preparation was performed. A 700 g sample of leaves collected from *StSy*-transgenic poplars was extracted in methanol and the extract was purified by flash-cromatography using two different resins (ENV+ and Toyopearl HW-40S). Pure *trans*- and *cis*-piceid were isolated by preparative HPLC using a Discovery C18 (250 × 50 mm, 10 μm) reversed-phase column and crystallised. Each step of the extraction and purification protocol was set and optimized, allowing to make a balance of the process. The leaves extract yielded 0.131% in weight of piceid, with a ratio *trans*/*cis* of ca. 2. The overall recovery was 70%, while the purity of the *trans*-piceid was of 88.9%, which is still not completely satisfactory for pharmacological studies. Further studies are in progress to improve the purification of the *trans*-piceid and to investigate its biological activity.

The level of transgene stability in planta was evaluated in the *bar*- and *StSy*-lines during the full vegetative growth (August 2004 and 2005) and at the onset of dormancy (November 2004 and October 2005). Total RNAs were extracted from apical and basal leaves and a northern blot hybridization analysis was carried out with gene-specific probes. In addition, the expression pattern of the endogenous *poUbi* gene was evaluated in the same tissue samples. Significant fluctuations in the steady-state level of the *bar* and *StSy* transcripts were found, suggesting that the activity of the constitutive 35SCaMV promoter was influenced by seasonal factors and organ position. To date, information concerning the molecular processes involved in the control of tree dormancy is lacking. Interestingly, species-specific DNA methylation mechanisms responsible for gene silencing have been identified (Mishiba et al., 2005). In white poplar, DNA methylation might be involved in the transgene silencing event. Investigation on the methylation status of the 35SCaMV-*bar* and 35SCaMV-*StSy* regions is currently in progress.

Soil samples were collected from the *bar* and *StSy* plots to monitor the persistence of recombinant DNA sequences. Molecular analyses failed to reveal the presence of NosP-*nptII*, 35SCaMV-*bar* and 35SCaMV-*StSy* sequences in the soil environment before GM poplar cultivation. Sampling was performed in May and August 2004 and 1 year later (August 2005) in the presence of transgenic plants. A progressive enrichment of pools carrying the expected amplification products took place during the cultivation of GM poplars. To assess the possible effects on soil microbial population, due to the persistence of the NosP-*nptII* sequence, the Km^R fraction of soil bacteria was determined. The study was carried out using cultivation-based plating techniques. An evident limitation of this methodology is that only culturable bacteria can be isolated and characterized. The total bacterial growth ranged from 10^6 to 10^9 CFU g^{-1} soil. The plate counts on the lowest kanamycin dose (10 μg m L^{-1}) ranged from 10^5 to 10^6 CFU g^{-1} soil while the bacterial fraction selected on the highest kanamycin dose (100 μg m L^{-1}) did not exceed 10^2 CFU g^{-1} soil. Approximately 4000 Km^R colonies were subjected to colony hybridization experiments with a specific *nptII* probe. No hybridization signals were detected, suggesting that the selected Km^R phenotype was dependent on endogenous mechanisms for antbiotic exclusion/degradation.

The reported data will contribute to a better understanding of the agronomical response and environmental impact of different classes of GM white poplars cultivated on a large scale, as a result of detailed analyses carried out at biochemical, molecular and cellular level.

References

Confalonieri M, Belenghi B, Balestrazzi A, Negri S, Facciotto G, Schenone G, Delledonne M (2000) Transformation of elite white poplar (*Populus alba* L.) cv. 'Villafranca' and evaluation of herbicide resistance. Plant Cell Rep 19: 978–982.

Frémont L (2000) Biological effects of resveratrol. Life Sci. 66: 663–673.

Giorcelli A, Sparvoli F, Mattivi F, Tava A, Balestrazzi A, Vrhovsek U, Calligari P, Bollini R, Confalonieri M (2004) Expression of the stilbene synthase (StSy) gene from grapevine in transgenic white poplar results in high accumulation of the antioxidant resveratrol glucosides. Transgenic Res 13: 203–214.

Mishiba K, Nishihara M, Nakatsuka T, Abe Y, Hiranu H, Yokoi T, Kikuchi A, Yamamura S (2005) Consistent transcriptional silencing of 35S-driven transgenes in gentian. Plant J 44: 541–556.

Nickson TE (2005) Crop biotechnology – the state of play. In: Poppy GM, Wilkinson MJ (eds) Gene Flow from GM Plants. Blackwell, Oxford, UK, pp 12–42.

Sparvoli F, Martin C, Scienza A, Gavazzi G, Tonelli C (1994) Cloning and molecular analysis of structural genes involved in flavonoid and stilbene biosynthesis in grape (*Vitis vinifera* L.). Plant Mol Biol 24: 743–755.

Thompson CJ, Movva NR, Tizard R, Crameri R, Davies JE, Lauwereys M, Botterman J (1987) Characterization of the herbicide-resistance gene *bar* from *Streptomyces hygroscopicus*. EMBO J 6: 2519–2523.

Molecular Breeding of Low-Phytic-Acid Grains in Rice by Using the Promoter of 18 kDa Oleosin

Mio Kuwano[1], Makoto Suzuki[1], Fumio Takaiwa[2], and Kaoru T. Yoshida[1,*]

[1]Graduate School of Agricultural and Life Sciences, University of Tokyo. 1-1-1 Yayoi, Bunkyo-ku, Tokyo 113-8657, Japan.
[2]National Institute of Agrobiological Sciences, 2-1-2 Kannondai, Tsukuba, Ibaraki 305-8602, Japan.
[*]Email: ayosida@mail.ecc.u-tokyo.ac.jp

In most plant seeds, phosphorus is stored primarily as *myo*-inositol 1,2,3,4,5,6-hexa*kis*phosphate ($InsP_6$; phytic acid). Monogastric animals are unable to utilize phytic acid phosphorus efficiently because they lack the digestive enzyme phytase, which is required to cleave phosphorus from phytic acid. Mineral phosphate is commonly added to animal diets as a supplement to prevent phosphorus deficiency. This causes high phosphorus concentrations in waste matter, and leads to the pollution of water bodies. Reducing the phytic acid content of seeds is a major breeding target, because it would both increase the availability of mineral nutrients and decrease the environmental phosphorus load.

The first step in phytic acid biosynthesis and inositol metabolism is the formation of 1D-*myo*-inositol 3-phosphate ($Ins(3)P_1$) from D-glucose-6-phosphate, catalyzed by $Ins(3)P_1$ synthase. We previously reported the isolation of a cDNA corresponding to the rice $Ins(3)P_1$ synthase gene, *RINO1*, and suggested that *RINO1* plays a key role in phytic acid biosynthesis in developing rice seeds (Yoshida et al., 1999).

Z. Xu et al. (eds.), Biotechnology and Sustainable Agriculture 2006 and Beyond, 419–421.

Reducing the expression of *RINO1* using transgenic methods might lead to a valuable reduction in the phytic acid content of seeds. However, the complete repression of *RINO1* gene expression in tissues other than seeds is very detrimental to plant development because RINO1 catalyzes the first step in inositol metabolism, as well as phytic acid biosynthesis. The use of a promoter that is active only in developing seeds, the storage site of phytic acid, to manipulate *RINO1* expression was suggested to be crucial for creating transgenic rice that produces seeds with the lowest possible phytic acid content (Feng and Yoshida, 2004).

Recently, we attempted to create transgenic rice that produced seeds with a low-phytic-acid phenotype by manipulating *RINO1* expression using the antisense *RINO1* cDNA fused to the glutelin promoter. The promoter of the rice major storage protein glutelin, *GluB-1*, drives extremely high, endosperm-specific expression during rice seed maturation. We obtained a stable transgenic line in the T_4 generation that contained four copies of the transgene and differed little from non-transgenic rice morphologically. Most of the T_5 seeds in this line contained more Pi than did non-transgenic seeds, with Pi constituting an average of 16% of the total phosphorus versus 3.1% in the non-transgenic seeds. The increase in Pi levels was achieved without significant changes in the total phosphorus levels, compared to non-transgenic rice. Ion chromatography analysis suggested that the increase in available Pi was accompanied by a molar-equivalent decrease in phytic acid P. In this line, the RINO1 protein levels in the T_5 seeds were markedly reduced at stages later than 14 days after flowering, but were not very different during the earlier stages.

In this study, we hoped to achieve a greater effect by using the rice seed protein 18-kDa oleosin, *Ole18*, promoter to direct *RINO1* gene expression in transgenic rice plants, because the *Ole18* promoter has the potential to drive high expression in the embryo and aleurone layer, which is the same spatial pattern of expression as *RINO1*.

The antisense *RINO1* cDNA was fused to the *Ole18* promoter. This construct was inserted into a binary vector and the resulting plasmid was transferred into *Agrobacterium tumefaciens* strain EHA105 by electroporation. Transgenic rice plants (*Oryza sativa* cv.

Kitaake) were produced using *Agrobacterium*-mediated transformation. *Agrobacterium* harboring the plasmid was used to infect scutellum-derived calli of rice. The regenerated seedlings were selected on media containing hygromycin, transplanted, and grown in a greenhouse.

Available Pi was extracted and analyzed using the colorimetric assay from one half of each seed and the total phosphorus was determined from the other half; then, the percentage of Pi in each seed was calculated. The number of chimeric transgene copies in the transgenic plants was determined using genomic Southern blot analysis. The amount of RINO1 protein in transgenic seeds was determined using immunoblot analysis.

Some of the transgenic lines using the *Ole18* promoter contained higher levels of Pi in seeds than did non-transgenic rice. Therefore, we selected these lines for further study. Most of the T_3 seeds of the transgenic lines that contained more than one copy of the transgene showed a great increase in the amount of Pi compared to non-transgenic rice. In some of these lines, the average Pi level was much higher than in the transgenic lines using the *GluB-1* promoter. No significant change in total phosphorus levels compared to non-transgenic rice was observed. The RINO1 protein levels of the T_2 seeds were reduced to less than 30% of those in non-transgenic seeds throughout the ripening stages. The decreased level of RINO1 protein suggests that the expression of the antisense *RINO1* transcripts efficiently represses *RINO1* gene expression, and that increased levels of seed Pi in the transgenic line are related to this reduction in RINO1 protein. These results indicate the feasibility of using molecular breeding to alter the storage form of phosphorus in seeds by manipulating *RINO1* gene expression under the control of the *Ole18* promoter.

References

Feng X, Yoshida KT (2004) Plant Biotechnol 21: 183–189.

Yoshida KT, Wada T, Koyama H, Mizobuchi-Fukuoka R, Naito S (1999) Plant Physiol 119: 65–72.

Regeneration, Micropropagation and Industrialization

A Highly Efficient Protocol for Micropropagation of North American Ginseng

Sijun Zhou* and Daniel C.W. Brown

Southern Crop Protection and Food Research Centre, Agriculture and Agri-Food Canada, 1391 Sandford St., London, Ontario, N5V 4T3, Canada.
*Email: zhous@agr.gc.ca

North American ginseng (*Panax quinquefolius* L.) is a native herb of Canada and the United States and is cultivated for its highly valued root used for medicinal purposes. It has a long production cycle as seeds are usually produced after a 3-year cultivation and must stratify for additional 12–18 months before germination. A clonal propagation method based on *in vitro* procedures would contribute to its genetic improvement by reducing the generation cycle time, would allow for the reduction in confounding genotype effects in field evaluations through the use of clonal material and would be useful for germplasm preservation.

The tissue culture of Korean ginseng (*Panax ginseng* C.A. Meyer) is well studied with reports on callus cultures (e.g., Choi et al., 2003), protoplast culture (Arya et al., 1991), shoot organogenesis (e.g., Bonfill et al., 2002), somatic embryogenesis (e.g., Choi et al., 1999 and cited literature). These techniques have largely not been used successfully with American ginseng. There are only a few reports on the *in vitro* cultivation of North American ginseng. These cases report on the formation of somatic embryos on callus culture of embryonic, seedling, root or leaf explants (Brown et al., 2001; Zhou et al., 2006 and cited literature). North American ginseng is still considered to be recalcitrant *in vitro* because of the difficulty in obtaining plants with a well-developed root system.

Z. Xu et al. (eds.), Biotechnology and Sustainable Agriculture 2006 and Beyond, 425–428.

Objectives

A laboratory-scale micropropagation system, based on a nine-stage protocol, has been developed for *P. quinquefolius* (Brown et al., 2001). However, the efficiency for plant recovery was low because of the difficulty in obtaining plants with a well-developed root system. We set out to optimize the micropropagation procedure so that tissue-cultured plantlets can be efficiently produced, recovered and transplanted into the field.

Methodology

Explants including cotyledons from stratified seeds and somatic embryo-derived tissues from abnormal regenerating shoots (tissue recycling) were used to induce somatic embryos. Cotyledons excised aseptically from zygotic embryos were placed abaxial side down on MS (Murashige and Skoog, 1962) medium containing 30–90 g L^{-1} sucrose. To make the somatic embryos separate or 'single', the cotyledons were pre-treated for plasmolysis with 1.0 M sucrose solution at 4°C for 48–168 h. In other cases, explants from tissue recycling were placed on MS medium containing 1.5 mg L^{-1} 2,4-dichlorophenoxyacetic acid (TR_1), or 1.0 mg L^{-1} 2,4-D and 1.0 mg L^{-1} α-naphthaleneacetic acid (TR_2).

Somatic embryos developed to the cotyledonary stage were transferred from induction media onto full or half-strength MS, B5 (Gamborg et al., 1968) or SH (Schenk and Hildebrandt, 1972) maturation media containing 3%, 6%, 9% sucrose, with or without 1% charcoal, and cultured for 1–2 months. Well-developed somatic embryos were transferred from maturation media onto MS or SH medium plus Gibberellic acid (GA_3) germination medium. The effect of GA_3 on germination was investigated by placing mature somatic embryos on SH media containing 2, 5, 10, 20, 30, or 40 mg L^{-1} GA_3.

When shoots were 0.5–1.0 cm in height, the plantlets were transferred from germination media to half-strength MS or half-strength SH (containing 0.5% activated charcoal) elongation medium. The effect of basal medium on development of plantlets was investigated by placing surface-sterilized stratified zygotic embryos on MS, B_5 and SH media in full, half and one third-strength respectively.

After 4 weeks on the elongation media, well-developed plants with a taproot or adventitious roots were potted in 65 × 65 × 60 mm plastic pots containing moistened Promix BX. The potted plants were cultivated and hardened in a growth chamber held at 20/16°C (day/night) under low light conditions. The acclimated plants were transplanted into field during summer season or kept in the chamber for another 3–4 months. The roots with dormant buds were transplanted into field or placed at 4°C for overwintering.

Results and discussion

A highly efficient system for plant propagation of North American ginseng (*Panax quinquefolius*) *via* somatic embryogenesis starting from cotyledon explants was established with simplified six-stage protocol: somatic embryo induction, maturation, germination (and tissue recycling), development of plantlets, acclimation into soil, and transplantation to the field. In the somatic embryo induction stage, somatic embryos can be induced from lateral edges and the adaxial surface of cotyledons cultured on growth regulator-free MS medium with high sucrose content and will result in a high frequency response (60%). The usual high frequency of abnormal and fused embryos can be partially controlled by pre-treating the cotyledon explants with 1.0 M sucrose for 48–96 h at 4°C. The pretreatment also results in an increase of embryogenesis response frequency (75%) and the number of embryos per explant from 10 to 21. A maturation stage is important for further development of somatic embryos and subsequent good germination. The optimal maturation medium is: half-strength MS or SH medium containing 3% sucrose and 1% activated charcoal. In the germination stage, gibberellic acid (GA_3) is necessary for breaking quiescence/dormancy of embryos. The optimal level of GA_3 is 10–20 mg L^{-1}. About 54% somatic embryos can produce normal plants. Elongated cotyledons and petioles from abnormal shoots after germination are good explants for somatic embryo production and result in high quality and number of somatic embryos. The frequency of embryogenesis on TR_2 medium is up to 90%, with up to hundreds of embryos on one explant. In the plant development stage, activated charcoal is beneficial to the plantlet especially for root development. The elongation medium, ½ SH plus 0.5% charcoal, results in 85% of germinated embryos developing into plants with a well-developed root system. Of over 600 potted plants, 400 plants have been transplanted into the field in May to July of 2004 and 2005. The growth chamber and field establishment rates, 5 weeks after transplantation, were 95.6% and 93.7%, respectively. The overwintering survival rate of transplanted plants in 2006 was 80.7%, similar to that of the seedling control (83.3%). The acclimatized plants are growing well and show the development of a dormant bud and a thickened root structure typical of seed-germinated plants.

References

Arya S, Liu JR, Eriksson T (1991) Plant regeneration from protoplasts of Panax ginseng (C.A. Meyer) through somatic embryogenesis. Plant Cell Rep 10: 277–281.

Bonfill M, Cusidó RM, Palazón J, Piñol MT, Morales C (2002) Influence of auxins on organogenesis and ginsenoside production in Panax ginseng calluses. Plant Cell Tissue Organ Cult 68: 73–78.

Brown DCW, Amyot L, Rintoul T (2001) Micropropagation of North American ginseng (Panax quinquefolius L.)—from test-tube to the field. In: Punja ZK (ed) Proceedings of the International Ginseng Workshop.

Choi YE, Yang DC, Yoon ES, Choi KT (1999) High-efficiency plant production via direct somatic single embryogenesis from preplasmolysed cotyledons of Panax ginseng and possible dormancy of somatic embryos. Plant Cell Rep 18: 493–499.

Choi YE, Jeong JH, Shin CK (2003) Hormone-independent embryo- genic callus production from ginseng cotyledons using high concentrations of NH4NO3 and progress towards bioreactor production. Plant Cell Tissue Organ Cult 72: 229–235.

Gamborg OL, Miller RA, Ojima K (1968) Nutrient requirements of suspension culture of soybean root cells. Exp Cell Res 50: 151–158.

Murashige T, Skoog F (1962) A reversed medium for rapid growth and bioassays with tobacco tissue cultures. Physiol Plant 15: 473–497.

Schenk RU, Hildebrandt AC (1972) Medium and techniques for induction and growth of monocotyledonous and dicotyledonous plant cell cultures. Can J Bot 50: 199–204.

Zhou S, Brown DCW (2006) High efficiency plant production of North American ginseng via somatic embryogenesis from cotyledon explants. Plant Cell Rep 25: 166–173.

The Use of Tissue Culture for Successful and Sustainable Cane

Satjaporn Chantawong[1], Upsorn Pliansinchai[1], Chalermpol Kirdmanee[2], and Pipat Weerathaworn[1]

[1]Mitr Phol Sugarcane Research Center Co., Ltd. Phukieo, Chaiyaphum, Thailand 36110.
[2]National Center for Genetic Engineering and Biotechnology National Science. 113, Thailand Science Park, Pahonyothin, Klong luang, Pathumthani 12120, Thailand.

Mitr Phol Sugar Group in 1996, one of the carriers, is to adapt tissue culture technique to support sustainable cane production in order to secure cane supply for the multi-usage e.g., sugar, ethanol and co-generation. Rapid multiplication efficiency has been done to produce disease-free cane setts of the new high potential cane varieties which produce high sugar and high yield through developed techniques described as following:

Plantlet production efficiency

Shoot induction

Previously, two media were used to induce shoot multiplication. Media suitable for each growth stage were developed following concept of some previous work (Sreenivasan and Sreenivasan, 1992; Ramanand and Singh, 2005). N1 and N2 media for meristem induction, N3 media for shoot induction, N4 media for shoot multiplication, N5 media for enhancing shoot elongation and vigor. This developed technique is able to increase number of explants/bud from

Z. Xu et al. (eds.), Biotechnology and Sustainable Agriculture 2006 and Beyond, 429–432.

400 to 1000 plantlets and reduce multiplication cycle time from 6 to 5 months.

Root induction

Previously, root induction medium partly contained sucrose 30 g L^{-1} (Zeng et al., 1983).However, our experience showed that this medium produced slow-growth seedlings and weak root system. Gonzalez et al. (1985) showed that sucrose enhanced adventitious root. Influence of sucrose concentration on number of roots and survival percentage after transplant were studied, varied from 0 to 100 g L^{-1}. Increasing sucrose up to 100 g L^{-1} can improve root numbers by 40.3%, seedling height by 29.7 and survival percentage rise up to 100%. In addition, it can reduce time of root induction from 30 to 15 days.

Efficiency of shoot and root induction was improved by these developed technique. Result in, vigorous and perfect seedlings obtained and production cycle reduce from 11–12 to 9 months.

Greenhouse transferring efficiency

It is considered through our experience that greenhouse transferring efficiency might be improved by using of some equipment instead of labor. It was shown that transferring 10,000 plantlets from the lab to polybags facilitated by soil mixed equipment finished in 17 days. Whereas, by two labors finished in a month. This can reduce labor cost by approximately 50%. Soil mixture formula were studied and the developed formula was proved to be as good as the previous formula in terms of plant growth and also reduce cost by 50%.

Hardening process

After transferring plantlets to greenhouse for 3 months, they are ready to be transplanted in the field. Previously, percent survival in the field was 80–95% and normally plants recovered in 2 weeks. Considering hardening concept 1) reduce plant succulence 2) reduce transpiration 3) induce root primordial 4) enhance regrowth. The previous transplanting method (transplanting immediately without cutting leaves and supplying water immediately) was compared with the developed method: withheld watering for 2–3 days before transplanting then cut 2/3 leaf blade of (reduce 2/3 transpiration area) and supply water immediately. The plantlets transplanted by developed methods survived up to 100% and recovered immediately. The plantlets transplanted by previous method survived 96 + 1% and recovered in 2 weeks. As a result the developed methods have been used as the normal practice.

Field quality control scheme

Plot selection

The multiplication plots have to be in isolate area or far from disease and pest area to prevent reinfection.

Plot preparation:

The plots have to be well-prepared and soil has been improved for high fertility, have enough water supply and well-drainage

Off-type inspection

Inspect for off-type variety cane monthly from 3 to 6 months and off type have to be remove.

Pest and diseases inspection and control

Inspect for diseases and pests monthly throughout the crop and eradicate whenever diseases and pest found.

Because of this strong quality control scheme, seed cane from these tissue culture multiplication plot can be obtained both in plant and first ratoon cane which can reduce cost of seed cane production by 50%.

Mitrphol Sugar Group is the only one sugar private company in Thailand which has invested for R&D. We have been successful in application of developed tissue culture technique to produce quality seed cane of new high potential varieties supply for our contract farmers. We can produce and distribute new cane varieties to replace previous cane varieties which have less potential and have been attacked by major diseases and pests. By this technique we can meet the cane sett demand in a short time and the new good varieties will be continuously distribute to the whole area result in the sustainable cane production and cane supply security. This might be because of the advantages of these technique compare to conventional multiplication: 1) need much less seed cane to start (one bud can produce 1,000 plants, less land and labor consume) 2) shorter time to produce (2.5 time less) 3) cane will be more vigor and diseases and pest free 4) can produce seed cane by 53 time more.

References

Gonzalez Morejon AA, Ojito MC, Castellanos Morales E (1985) Increase *in vitro* root formation in plantlets from tissue culture of an interspecific hybrid sugarcane. Centro Agrocola 12(3): 136–139.

Ramanand MLAL, Singh SB (2005) Comparative performance of micropropagation and conventionally raised crops of sugarcane. Sugar Technol 7(2 and 3): 93–95.

Sreenivasan TV, Sreenivasan J (1992) Micropropagation of sugarcane varieties for increasing cene yield. SISSTA Sugar J 4(18): 61–64.

Zeng DH, Lin MY, Han YN, Lin JG, Li HD, Guo CR (1983) Sugarcane tissue culture techniques. Fujian Agric Sci Technol 4: 11–14.

Development of Subclonal Variants from Interspecific Hybrids of Sugarcane

S. Rajeswari[1,*], Shini Sekar[2], and M. Krishnamurthi[2]

[1]Parry Sugarcane Research and Development Centre, Pugalur – 639113, Tamil Nadu, India.
[2]R&D Centre, E.I.D. Parry(I)Limited, Bangalore – 560049 , Karnataka, India.
[*]Email: rajeswaris@parry.murugappa.com

Introduction

Plant tissue culture has been useful in improving sugarcane by creation of variations and applying appropriate selection procedures for commercialization (Krishnamurthi and Tlaskal, 1974; Krishnamurthi, 1989). Using a tissue culture cycle it is possible to obtain increased genetic variability. The present study was taken up to develop subclonal variants from interspecific hybrid donors.

Materials and methods

Plant material:

Young shoots from 4 to 5 months old field grown interspecific hybrids (*S.officinarum* x *S.spontaneum*) were used as the source of explants.

The spindles were drawn from nine hybrid donor clones and cultured aseptically, on callus induction medium containing 2 mg.L^{-1} of 2,4-D and 20 mg.L^{-1} of caffeine. The inoculated bottles were kept at

Z. Xu et al. (eds.), Biotechnology and Sustainable Agriculture 2006 and Beyond, 433–436.

26°C ± 1°C in dark for 4 days after which they were subjected to 12 h of light and 12 h of darkness.

When the callus was abundant, it was transferred to the regeneration medium containing 10 mg.L^{-1} of kinetin and 20 mg.L^{-1} of caffeine. The cultures were kept at 26°C ± 1°C in 12 h light and 12 h darkness. The light intensity was 2000–3000 lux. The plantlets were allowed to grow for 1 month for further proliferation and regeneration.

Once the regenerated plantlets reached about 5 cm height, they were removed from the differentiating medium and sown in the rooting medium containing 1 mg.L^{-1} of IBA. The plantlets were incubated for 30 days at 26°C ± 1°C.

The plants with roots were removed from the medium and washed gently prior to planting in polybags.

The 120 plantlets from nine hybrid donors were planted in the field for multiplication. Each plantlet was considered as a subclone of hybrids and labelled accordingly.

The 120 subclones were planted in a randomized block design with three replicates, with a plot size of 1 row × 3 m along with donor clones. The biometric parameters viz., stalk thickness (in cm), stalk length(in cm) and internode length(in cm) were collected at the age of 10th month. The yield data was recorded based on single clump weight. The biochemical parameters viz., brix, pol percent cane, purity, fibre percent and Pure Obtainable Cane Sugar (POCS) percent were estimated at 12 months age.

Student 't' test was performed to find out the difference between donor clones and subclonal populations for biometric and biochemical parameters.

Results and discussion

Profuse callus was obtained from unfurled leaves. These when transferred to a medium without 2,4-D, in the presence of kinetin and external source of cool bright lights attenuated the induction of differentiation. Within 15–20 days shoots are formed. Root initiation took place within 8–15 days. It took 20 days to get well-developed roots.

The biometric and biochemical parameters and field data was collected at harvest at the age of 12 months. The characters varied substantially from the donors. The variations for one character are presented in Table 1.

Table 1. Range of variations for Pure Obtainable Cane Sugar (POCS%)

S.No	Interspecific hybrid donor clones	Donor clone value	Subclonal populations						No.of subclones
			0.01–2.5	2.6–5.0	5.1–7.5	7.6–10	10.1–12.5	12.6–14.3	
1	S.officinarum cv Badila x *S.spontaneum:* Donor	8.6	0	0	0	0	2	0	2
2	CoC 85-061x *S.spontaneum*	7.1	1	1	2	2	3	1	10
3	Co 86-032 x *S.spontaneum*	4.7	1	1	5	5	1	0	13
4	Co 88-011x *S.spontaneum*	7	0	2	22	10	11	0	45
5	CoG 94-077x *S.spontaneum*	5.2	3	0	3	4	3	0	13
6	PI 96-4024x *S.spontaneum*	2.1	2	1	0	0	0	0	3
7	PI 96-4024x *S.spontaneum*	11.3	0	1	1	2	0	0	4
8	PI 96-4066x *S.spontaneum*	6.4	1	0	3	11	6	0	21
9	PI 96-4069x *S.spontaneum*	5.7	0	0	3	6	0	0	9
								Total	120

Twenty variants were selected for parental line and energy cane, which had 16% fibre and 12% sucrose content. Thirteen subclones were selected with commercial potential based on optimum fibre level (13%), 200 cm stalk length, 10 cm internode length and POCS percent of 10. Thus tissue culture system supplements the breeding programme in developing parental lines and commercial clones through the modification in the genetic architecture of sugarcane.

Acknowledgement

The authors wish to thank E.I.D. Parry (I) Ltd, for permitting to publish this data.

References

Krishnamurthi M (1989) Development of subclonal populations in sugarcane and their genetic and field evaluation for commercial use. Ph.D Thesis, University of South Pacific, Fiji Islands, p. 400.

Krishnamurthi M, Tlaskal J (1974) Fiji disease resistant *Saccharum officinarum* var. Pindar subclones from tissue culture. Proc Intl Soc Sugarcane Technol 15(1): 130–137.

Forcing and In Vitro Establishment of Softwood Shoots from Large Stem Segments of Woody Plants

Faheem Aftab[1] and John E. Preece

Department of Plant, Soil and Agricultural Systems, Southern Illinois University, Carbondale, IL 62901-4415, USA.
[1]Current Address: Department of Botany, University of the Punjab, Q. A. Campus, Lahore-54590, Pakistan.
Email: faheem@botany.pu.edu.pk

Abstract

Softwood shoots can be forced from various lengths of large stem segments that have been removed from intact trees and shrubs. Many woody species have large numbers of latent buds and thus forcing may result in the production of many shoots. If the plants were originally propagated by seed, shoots that grow from lower sections may retain juvenile traits, such as ease of propagation. Forced leafy softwood shoots may either be rooted as stem cuttings or may be used as an explant source for *in vitro* studies and micropropagation. This is relatively a newer technique and its potential for micropropagation has not been determined in many woody plants. Contamination of the explants taken from the softwood-forced shoots is perhaps a major limitation associated with forcing under mist or fog. This paper focuses on several aspects influencing softwood shoot formation taking few representative examples, e.g., *Acer saccharinum* (silver maple), *Fraxinus pennsylvanica* (green ash), *Carya illinoensis* (pecan) and *Tectona grandis* (teak). The results

Z. Xu et al. (eds.), *Biotechnology and Sustainable Agriculture 2006 and Beyond*, 437–444.

suggest that this technique holds a promise for micropropagation of many recalcitrant woody species.

Introduction

In comparison with herbaceous plants, a limiting factor to initiate *in vitro* work on woody plants is the lack of suitable explants. In many instances, it is not possible to collect desirable explants directly from the field-grown plant sources. Softwood shoot forcing provides an opportunity to produce actively growing explants over several months of the year. This technique is not of a widespread use in commercial micropropagation laboratories, but appears to have great potential and may become adopted by labs in the near future (Preece and Read, 2003). Softwood shoots can be forced from various lengths of large stem segments that have been removed from intact trees (Henry and Preece, 1997a, b). In addition, it allows for the forcing of shoots from older, more juvenile portions of a woody plant (Preece and Read, 2003). Many tree species have large numbers of latent buds and thus forcing may result in the production of many shoots. These may be rooted as stem cuttings using techniques common in the nursery industry (Henry and Preece, 1997a) or may be used as an explant source for in vitro studies and micropropagation (Van Sambeek et al., 1997; Preece, 2003). Moreover, shoots can be produced for much of the year, greatly expanding softwood cutting propagation or explant initiation time for many temperate woody species (Van Sambeek et al., 2002). The available literature mainly focuses on temperate plant species (Henry and Preece, 1997 a; Preece et al., 2001). However, its potential use in tropical trees remains to be worked out.

Acer saccharinum (silver maple), *Carya illinoensis* (Pecan), *Fraxinus pennsylvanica* (green ash) and *Tectona grandis* (Teak) are woody species of great economic significance. Forcing and *in vitro* establishment in silver maple and green ash has already been reported (Aftab et al., 2005). Such studies however are lacking in pecan and teak.

The present paper focuses at the results with our research on pecan and teak in the light of our previous experience with silver maple and green ash. The purpose of this paper is to highlight the potential advantages of this technique for rapid propagation of many temperate or tropical plants of economic significance.

Materials and methods

Trees and stems of green ash, pecan, silver maple and teak were cut down or branches were cut and further cut to 40 cm lengths that were 4–9 cm in diameter (logs). All lateral shoots from the cut stem segments were removed before being placed horizontally in 52 × 25 × 6.5 cm (l × w × h) black polyethylene flats or other suitable flats with drainage holes. The lower halves of the stem segments were then embedded into the respective media. Four greenhouse media were selected in our experiments for forcing softwood shoots from green ash and silver maple (Aftab et al., 2005). These media were perlite, vermiculite, one perlite: one vermiculite (by volume), and a control without any medium (empty flats). In our work involving Pecan, we have used vermiculite as a medium for softwood shoot forcing. The cut pecan logs were placed horizontally on the medium in each flat. The log diameter in this treatment was variable and randomly placed over the medium. On an average, each flat had four logs. These flats were placed in the laboratory on June 07, 2004 under continuous light conditions. The medium in the flats was watered daily with care taken to avoid water contact with the new, emerging shoots. A second run for softwood shoot forcing from Pecan collected on the Southern Illinois University Carbondale campus was initiated on August 18, 2004. In addition, we have recently used autoclaved sand to force epicormic buds from teak at the University of the Punjab campus. As with other experiments, stem segments in the laboratory were watered daily with care taken to avoid water contact with newly emerging softwood shoots. Such shoots (≥4 cm long) could easily be harvested with scalpels.

Harvested pecan and teak shoots were surface disinfested for 15 min in 0.6% NaClO with 0.1% Tween 20, then rinsed, following the procedure of Aftab et al. (2005). After the surface disinfestation, the shoots were removed from the rinse water and nodal explants trimmed to 1.5–2.0 cm and shoot apices were cut to 2.0–3.0 cm long.

The explants were cultured in appropriate in vitro growth media as required for each species. The details for silver maple and green ash are described by Aftab et al. (2005). In pecan, we compared DKW (Driver and Kuniyuki, 1984) walnut medium, LP (Long et al., 1995) and WPM (Lloyd and McCown, 1981) each supplemented with 10 μM BA and 1 μM IBA with or without 1 μM TDZ leading to a 3 × 2 factorial arrangement of three media and two plant growth regulator (PGR) treatments. All media were supplemented with 30 g L^{-1} sucrose, 6.5 g L^{-1} agar and adjusted to pH 5.8. In teak, MS (Murashige and Skoog, 1962) medium was supplemented with various levels of BA or TDZ for the purpose of micropropagation. All cultures were incubated under a 16 h photoperiod at 22 ± 1°C except for teak that was incubated at 24 ± 1°C.

Results and discussion

Although our previous experience (Aftab et al., 2005) had shown that forcing epicormic buds in silver maple and green ash was generally better in terms of number of harvestable shoots under mist or fog as compared to lab conditions, the explants derived from such forced shoots were found to have high levels of microbial contamination when transferred to *in vitro* conditions. Thus in case of silver maple we could even get 6.7 shoots per stem segment under mist in the perlite/vermiculite mix. Green ash on the other hand produced a mean of 1.2 shoots. Microbial contamination, however, limited our success with subsequent *in vitro* cultures if the shoots were forced under mist or fog (Aftab et al., 2005). This was the reason that in our experiments pecan logs were forced only under lab conditions. Our results with pecan demonstrate that softwood shoot forcing is a good possibility to raise tissues to be manipulated under *in vitro* conditions. During the months of June to July, we have observed that pecan logs produce a maximum of five softwood shoots with an mean of three harvestable shoots (4 cm or above) per log (40 cm long) when placed horizontally using vermiculite as a sole greenhouse medium under the lab conditions (data not shown). Such logs were drenched daily with water using a garden watering can. Softwood shoot forcing to our knowledge has never been reported earlier in pecan but seems to hold promise during certain times of the year. Our studies showed that bud break was not possible at all during August to November, 2004, but was rapid during June and July 2004. Seasonal variations in shoot forcing of woody plants are well documented (Preece, 2003) and seem to hold for pecan as well. On the basis of this observation, we recommend that forcing epicormic shoots in pecan should be done during June to July (or perhaps earlier) so as to achieve a reasonable number of pecan's explants ready to be transferred to *in vitro* conditions. Once under *in vitro* conditions, such explants may be maintained and multiplied in the lab/greenhouse until the next spring when the acclimatized plants may be transplanted to field conditions.

Establishment of pecan under *in vitro* conditions is quite a challenge because of microbial contamination. Pecan shoots that forced from the logs under laboratory conditions were 6–8 cm long within a month, yielding three to four nodal segments and one shoot apex per forced-shoot for *in vitro* transfer to micropropagation media. In a separate experiment conducted on July 20, 2004, three NaClO concentrations, i.e., 0.6%, 1.2% or 1.8 % were tested for the surface disinfestation of nodal and shoot apex explants in pecan. Although the rate of contamination dropped to 68.19% (31.81% survival) at 1.8% NaClO, such a high concentration for the purpose of disinfestation appeared to be lethal as more than 80% of the noncontaminated cultures either turned brown (within the first few days of culture) or showed no sign of growth even after several weeks of culture initiation. A very high contamination rate in our studies was a major problem associated with pecan tissue cultures and hence did not allow a proper statistical analysis for a comparison of different media for micropropagation. However, in pecan tissue culture, we have found that all the six tested media, i.e., DKW, LP or WPM containing 10 μM BA and 1 μM IBA either with or without 1 μM TDZ could support (to a limited extent) bud growth and shoot enlargement. LP medium seems to be promising amongst all the tested media although the role of TDZ in terms of growth and development of pecan explants could not be ascertained even after two runs of the experiment.

Teak is a tropical hardwood tree species. Its response, on the other hand, seems diametrically different from the other three species discussed so far. Presently work is in progress but our preliminary data have shown an average of five shoots per log under greenhouse conditions if drenched daily with water using a garden watering can. An interesting aspect is our success with epicormic shoot forcing using autoclaved sand instead of a greenhouse medium. Use of autoclaved sand has shown better results in comparison with nonautoclaved sand and microbial contamination of the medium seems to be considerably reduced in our experimental conditions. However, this merits further study. To our knowledge, micropropagation in teak using epicormic bud activation has never been attempted before. We are quite optimistic though that it may potentially be a useful method to propagate teak.

TDZ (Thidiazuron, N-phenyl- N′-1, 2, 4-thiadiazol-5-yl urea) is one of several substituted ureas that has been investigated for cytokinin activity. Its use in plant tissue culture of recalcitrant woody plants has shown promise for micropropagation as well as callus induction and regeneration studies (Huetteman and Preece, 1993; Wilhelm, 1999; Preece et al., 2001; Preece, 2003; Ledbetter and Preece, 2004). We have observed that it is an excellent growth regulator for the initiation of callus and its subsequent maintenance over several months in teak. Thus, MS medium supplemented with 100 nM TDZ was found quite satisfactory in our studies for the initiation of embryogenic callus from shoot tip explants collected from a 70-year-old teak tree. This is an interesting finding since mature tree species seldom provide suitable explant material for *in vitro* studies. Lack of suitable explant material from adult source is a major limitation that makes it important to raise tissue through alternate methods such as epicormic buds. A relatively much higher TDZ level (1 μM) however was found optimum for callus maintenance over successive subcultures (Akram and Aftab, unpublished). Shoot multiplication in teak in our studies was possible in MS medium supplemented with 8 μM TDZ using nodal explants.

Shoot forcing as well as forcing epicormic buds is well documented for several temperate plants (Preece and Read, 2003). Since this is a relatively newer technique, its potential has not been determined so far in tropical trees. Our previous work with temperate woody plants as well as the on going research with tropical plants such as teak strongly suggests its applicability with wide range of woody plant species where adult source tree usually does not provide a suitable explant material.

Based on this information it may be recommended that this technique holds good potential to produce an ample quantity of explant material that may further be propagated either under *in vitro* or *ex vitro* conditions.

References

Aftab F, Mansouri K, Preece J E (2005) The influence of environment, media, and Zerotol on forcing and in vitro establishment of softwood shoots from large stem segments of Acer saccharinum and Fraxinus pennsylvanica Marsh. Propagation Ornamental Plants 5: 111–116.

Driver JA, Kuniyuki AH (1984) In vitro micropropagation of paradox walnut rootstock. HortScience 19: 507–509.

Henry P H, Preece JE (1997a) Production and rooting of shoots generated from dormant stem sections of *Acer* species. HortScience 32: 1274–1275.

Henry PH, Preece JE (1997b) Production of shoots from dormant Acer as influenced by length and caliper of stem sections. J Environ Hortic 15: 153–156.

Huetteman CA, Preece JE (1993) Thidiazuron: A potent cytokinin for woody plant tissue culture. Plant Cell Tissue Organ Cult 33: 105–119.

Ledbetter DI, Preece JI (2004) Thidiazuron stimulates adventitious shoot production from Hydrangea quercifolia Barte. Leaf explants. Sci Hortic 101: 121–126.

Lloyd G, McCown BH (1981) Commercially feasible micropropagation of mountain laurel (*Kalmia latifolia*) by use of shoot tip culture. Comb Proc Int Plant Prop Soc 30: 421–427.

Long LM, Preece JE, Van Sambeek JW (1995) Adventitious regeneration of *Juglans nigra* L (eastern black walnut). Plant Cell Rep 14: 799–803.

Murashige T, Skoog F (1962) A revised medium for rapid growth and bioassays with tobacco tissue cultures. Physiol Plant 15: 473–497.

Preece JE (2003). A century of progress with vegetative plant propagation. HortScience 38: 1015–1025.

Preece JE, Ledbetter DI, Zaczek JJ (2001) Rooting softwood cuttings collected from forced large stems of Oakleaf Hydrangea and American chestnut. Comb Proc Int Plant Prop Soc 51: 267– 270.

Preece JE, Read PE (2003) Novel methods in micropropagation. Proceedings of the 1st International Symposium on Acclimatization and Establishment of Micropropagated Plants. Acta Hortic 616: 71–76.

Van Sambeek JW, Lambus LJ, Preece JE (1997) Production of epicormic sprouts on branch segments of adult black walnut for in vitro culture. 88th Annual Report Northern Nut Growers Association, pp. 93–104.

Van Sambeek JW, Preece JE, Goggeshall MV (2002) Forcing epicormic sprouts on branch segments of adult hardwoods for softwood cuttings. Comb Proc Int Plant Prop Soc 52: 417–424.

Wilhelm E (1999) Micropropagation of juvenile sycamore maple via adventitious shoot formation by use of thidiazuron. Plant Cell Tissue Organ Cult 57: 57–60.

Protoplasts, Haploids and Embryo Culture in Crop Improvement

Introgression Breeding Program in *Lolium/Festuca* Complex Using Androgenesis

Toshihiko Yamada[1,*], Yang-dong Guo[2], Yuuko Mizukami[3], Ken-ichi Tamura[4], and Kazuhiro Tase[4]

[1]Field Science Center for Northern Biosphere, Hokkaido University, Kita 11 Nishi 10, Kita-ku, Sapporo, 060-0811, Japan.
[2]China Agricultural University, Beijing, 100094, China.
[3]Aichi Agricultural Research Center, Nagakute, Aichi, 480-1193, Japan.
[4]National Agricultural Research Center for Hokkaido Region, Hitsujigaoka, Toyohira-ku, Sapporo, 062-8555, Japan.
[*]Email: yamada@fsc.hokudai.ac.jp

Introduction

The genus *Festuca* L. and it is the closely allied genus *Lolium* L. have long fascinated agronomists, evolutionists and plant breeders. Intergeneric hybrids between closely related *Lolium* and *Festuca* species are being used to broaden the gene pool and provide plant breeders with options to combine high quality traits with broad adaptation to a range of environmental constraints. *Lolium* and *Festuca* species share valuable and complementary agronomic characters: for example *L. perenne*, offers good regrowth and nutritive value and is a good species for grazing whilst, *F. pratensis*, is more persistent and winter-hardy. Some amphidiploid cultivars of *Lolium/Festuca* have been developed. A major problem for amphidiploid breeding is the high level of homologous pairing between the different genomes that lead to genetic instability and a loss of hybridity in later generations. To overcome these problems and reduce transfer of deleterious *Festuca* traits, selective introgression of genes for desirable traits from *Festuca* into *Lolium* has become a favored methodology.

Z. Xu et al. (eds.), Biotechnology and Sustainable Agriculture 2006 and Beyond, 447–450.

Androgenesis was found to be an effective procedure for selecting *Lolium-Festuca* genotypes comprising gene combinations rarely or never recovered by conventional backcross breeding programs. Successful androgenesis in *Festulolium* hybrids, *L. perenne* x *F. pratensis* was established and a large numbers of dihaploids were developed (Guo et al., 2005). The object of this study is to initiate an introgression breeding program for improvement of winter hardiness in *L. perenne*/*Festuca* complex.

Materials and methods

Genotypes of *Lolium perenne* x *Festuca pratensis* (*Festulolium* hybrid), 'Prior', 'Bx350' and 'Bx351' were investigated in this study. PG-96 (Guo et al., 1999) with 2 mg L^{-1} 2,4-D, 0.5 mg L^{-1} kinetin was used as embryo (calli) induction media. Calli were transferred to the solid medium 190-2 supplemented with 0.1 mg L^{-1} 2,4-D, 1.5 mg L^{-1} kinetin for green plants regeneration. The ploidy level of androgenic progenies was analyzed by Partec CAII flow cytometry (Münster, Germany) with DAPI staining. Androgenic-derived plants with natural hardening during autumn were analyzed for freezing tolerance. Two dihapoids showing both good freezing tolerance and fertility potential were backcrossed with *L. perenne* 'Riikka'. Thereafter, BC_1 plants with superior freezing tolerance were backcrossed with *L. perenne*. For evaluation of freezing tolerance, crown tissues from each genotype were cooled by a programmed freezer to −17°C (dihaploids), −14°C (BC_1), and −15°C (BC_2), respectively. Co-dominant PCR markers distinguishable *Lolium*/*Festuca* have developed from several EST and SSR primers information. These markers were analyzed for genotyping among dihaploid, BC_1 and BC_2.

Results

Successful androgenesis in *L. perenne* x *F. pratensis* was accomplished using PG-96 medium for embryo/callus induction, and a large number dihaploids were produced. The PG-96 medium was composed of relatively complex organic acids and vitamin compounds (Guo et al., 1999). Androgenic progeny showed a large variation in freezing tolerance, 7% of progeny exceeding that of freezing hardy *F. pratensis* despite containing chromosomes of *L. perenne*, a more freezing-sensitive species and more than 60% of flowering progeny produced dehiscent anthers with pollen stainability ranging from 5% to 85%. Dihaploids with both freezing tolerance and fertility potential were backcrossed with diploid *L. perenne*. High freezing tolerant plants were found in the BC_1 progeny (Figure 1). Further backcrossing with *L. perenne* has been carried out and in the BC_2 some plants showed same level of freezing tolerance as *F. pratensis* (Figure 1). Co-dominant PCR markers distinguishable *Lolium/Festuca* were designed from *Lolium perenne* EST (unpublished) and *Festuca arundinaces* EST-SSR (Saha et al., 2004) etc. DNA markers analysis revealed that frequency of *Festuca* genotypes was detected 50% in a dihaploid plant and 27% in a BC_1 progeny plant. *F. pratensis*-specific markers were observed significantly with high frequency in superior freezing tolerant plants in BC_2 backcrossed progeny. These co-dominant PCR markers seem to be related with freezing tolerance of *F. pratensis*.

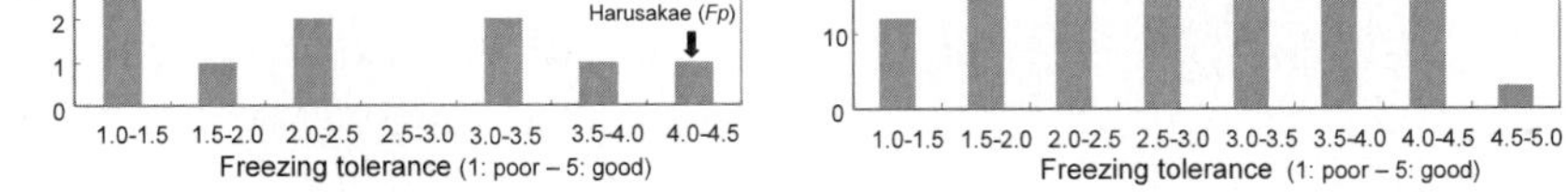

Figure 1. Freezing tolerance test in BC1 and BC2 backcrossed progeny derived from Androgenic Festulolium. (FL. Festulolium, Lp, Lollium perenne; Fp, Festuca pratensis)

Conclusion

High frequency androgenesis in *Festulolium* was established. These valuable materials from androgenesis will be used in an introgression breeding program to accelerate the breeding process and provide a novel robust forage grass for cultivation in marginal areas such as the eastern part of Hokkaido region of Japan. Also *Festuca*-specific co-dominant PCR markers could be useful for introgression breeding program.

References

Guo YD, Mizukami Y, Yamada T (2005) Genetic characterization of androgenic progeny derived from *Lolium perenne* x *Festuca pratensis* cultivars. New Phytologist 166: 455–464.

Guo YD, Sewón P, Pulli S (1999) Improved embryogenesis from anther culture and plant regeneration in timothy. Plant Cell Tissue Organ Cult 57: 85–93.

Saha MC, Mian MAR, Eujayl I, Wang L, May GD (2004) Tall fescue EST-SSR markers with transferability across several grass species. Theor Appl Genet 109: 783–791.

Recent Advances in Anther Culture of Coconut (*Cocos nucifera* L.)

P.I.P. Perera[1,*], V. Hocher[2], J.L. Verdeil[3], D.M.D Yakandawala[4], and L.K. Weerakoon[1]

[1] Coconut Research Institute, Tissue Culture Division, 61150 Lunuwila, Sri Lanka.
[2] Institute for Research and Development (IRD), UMR 1098 BEPC, IRD, BP 64501 – 911 Av. Agropolis, 34394 Montpellier Cedex 1 – France.
[3] CIRAD, TA40/02 Avenue Agropolis, 34398 Montpellier Cedex 5 – France.
[4] Faculty of Science, Department of Botany, University of Peradeniya, Sri Lanka.
[*]Tel: (94) 31 2255 300, Fax: (94) 31 2257395,
Email: prasanthi1970@yahoo.com

Introduction

The coconut palm (*Cocos nucifera* L) is one of the most important plantation crops, grown in about 90 countries in the world. It is given different terms, 'Tree of life', 'Tree from Heavens', 'Tree of Abundance' and 'Natures Super Market' considering the numerous values to the mankind, especially lives in the tropical region. The coconut palm is an aerially un-branched palm with a tall, slender stem and a massive crown. It is growing to a height of 12–24 m with 50–60 years of a life cycle. Coconut has a diploid chromosome complement of 2n=2x=32. It is a monocotyledon, belongs to the order *Arecales* (*Palmae*) and family *Arecaceae* (*Palmaceae*) and the sole species of the genus *Cocos*. Two main ecotypes are exists in this species, the Tall palms, *C. nucifera* typica (a cross-pollinating crop) and the Dwarf palms, *C. nucifera* nana (a self-pollinating

Z. Xu et al. (eds.), *Biotechnology and Sustainable Agriculture 2006 and Beyond,* 451–455.

crop). The Tall coconut palm is propagated only by seeds. It produces the heterogeneous seedlings by segregation and a higher variation exists within the progeny.

There is an urgent need to increase the productivity of the coconut plantations giving a higher income to the land owners while conserving the biodiversity. Therefore, genetic improvement of coconut palm is a priority research area to increase the productivity. It can be achieved either by conventional breeding methods or biotechnology.

However, production of new varieties of coconut by conventional methods is difficult due to several factors, including those common to all perennial crops such as inherent heterozygosity, long juvenile phase, time and cost limitations, inbreeding depression, long life time and most importantly unavailability of proper clonal propagation method. In general, conventional breeding techniques will not produce the major improvements sought without long term selection and lengthy back crossing procedures that will normally exceed more than 60 years. Development of inbred lines from a heterozygous tall coconut plant by conventional way is also a long, difficult and expensive because of the length of vegetative cycle and low propagation rate (Monfort, 1985). Therefore, to obtain a pure line from tall coconuts, it would take at least 60 years. The cost and land requirement for such an exercise is unjustifiable. There is a great potential of using modern technologies to support for the acceleration of the conventional breeding programs.

Haploid culture is a technique that can be used to improve the production and occurrence of the frequency of haploids. Doubled haploids (DH) are the pure lines produced by doubling the chromosome number of the haploids in which 100% homozygosity can be achieved within single step. Using homozygous lines it can be produced pure F1 hybrids. Recessive alleles in the parents are easily uncovered and express in DH plants and therefore, mutations can be easily detected. They provide a wide spectrum for the selection in breeding programs, as the gametic genotypes are fully expressed at plant level. Therefore, the value of the production DHs in a crop like coconut is unlimited.

The haploids are induced by triggering the normal developmental process of the gametes and turning in to the sporophytic pathway. The most common technique that is used for DHs production is anther culture. The technique has been successful with number of crop species after pioneer discovery done by Guha and Maheshwari (1964). Several groups of researches have been undertaken the studies on haploid plant production of coconut *via* anther culture, however only one callus (cited by Kovoor, 1981) and several direct embryos (Thanh-Tuyen and De Guzman, 1983; Monfort, 1985) at extremely low frequency (without any further development) have been recorded. Therefore it is very important to develop an anther culture protocol for mass production of DHs for coconut.

Materials and methods

Anthers excised from male flowers from 15 to 20 years old coconut palms (cultivar Sri Lanka Tall) were used as the explants. The collection of rachillae from the coconut inflorescence at desirable stages of microspore development was based on the age of the spadix, in the number of weeks before split opening (WBS). Anthers collected from inflorescences of 4 maturity stages (4, 3, 2 and 1 week before split opening of the spadix; designated 1–4 WBS) were used.

The rachilla was collected and they were subjected to heat (38°C) or cold (4°C) pre-treatments for 1, 3, 6 and 14 days. Then the anther were excised and surface sterilised prior to culture in androgenesis induction medium, modified Eeuwens Y3 liquid medium, supplemented with 100 μM 2,4-D, 9% (w/v) sucrose and 0.1% (w/v) activated charcoal (BDH acid washed). Each treatment was consisted of five replicates each containing 12 anthers per Petri dish, within 25 ml of liquid culture medium. All the treatments were compared with a control set of treatment without temperature stress. The petri dishes were sealed with Para film and the cultures were maintained under dark at 28°C for 8 months without sub-culturing. The number of calli and embryoids in each treatment was record. The anther derived calli/embryoids in all above cultures were then sub-cultured into the somatic embryo maturation medium (Y3 medium without any hormones) for 4 weeks followed by germination medium (modified Eeuwens Y3 medium supplemented with 5 µM BAP and 3.5 µM GA3). Calli and embryoids bearing anthers were sampled for histological analysis. Well developed calli and embryoids were used for ploidy analysis by flow cytometer.

Results and discussion

Anther response was observed only in one treatment where the anthers were pretreated at 38°C for 6 days. The pollen developmental stage is known to be critical for androgenesis and thus it is very important to determine the most suitable developmental stage of anthers for culture initiation. It is believed that the early, during or immediately after the first pollen mitosis is ideal for the induction of coconut androgenesis. Out of the four developmental stages tested, anthers at three WBS stage responded for the androgenesis. Anther response could be observed after 3 months of inoculation. Both embryoids and calli were produced under the same culture conditions. By repeating the optimum treatment combination three times, the anther derived structures were obtained at an average frequency of 21%. The embryoids/calli formed in this medium were sub cultured to maturation medium (Y3 medium without any hormones) followed by germination medium (Y3 medium containing 5 µM BAP) and the development of the embryos was observed.

The histological analysis revealed that the callus/embryoids have developed from the inner part of the anthers which suggests that they have originated from microspores. The flow cytometric analysis revealed that all the tested calli/embryoids were haploid (n=x=16) It indicates that the origin of the calli/embryoids are from the pollen grains of the anthers. This is the first time that successful production of embryogenic callus (haploid origin) from coconut anthers is reported.

Acknowledgements

We gratefully acknowledge the financial assistance provided by the Institut de Recherche pour le Developpement (IRD), Montpellier, France and Coconut Research Institute, Sri Lanka. We are grateful to C. Duperray (INSERM, Montpellier, France) for the technical assistance in cytometry.

References

Guha S, Maheshwari SC (1964) In vitro production of embryos from anthers of Datura. Nature 204: 497.

Kovoor (1981) Palm Tissue Culture: State of art and its application to the coconut. FAO Plant Production and Protection Paper 30. FAO, Rome.

Monfort S (1985) Androgenesis of coconut: Embryos from anther culture. Z Pflanzenzuchtg 94: 251–254.

Thanh-Tuyen NT, De Guzman EV (1983) Formation of pollen embryos in cultured anthers of coconut (*Cocos nucifera* L.). Plant Sci Lett 29: 81–88.

Acknowledgements

[illegible]

References

[illegible]

Protoplast Isolation and Cultivation from Embryogenic Suspension Cultures and Leaves of Grapevines (*Vitis* L.)

Jiang Lu[1], Xia Xu[1], and Judy Grosser[2]

[1]Center for Viticulture and Small Fruit Research, 6505 Mahan Drive, Florida A&M University, Tallahassee, FL 32317, USA.
[2]Citrus Research and Education Center, 700 Experiment Station Road, University of Florida, Lake Alfred, FL 33850, USA.

Abstract

Establishment of a protoplast-to-plant system in grapevines requires high quality protoplast isolation, capability of cell division and regeneration to plantlets. In this study, protoplasts were obtained from three somatic embryogenic suspension cultures and leaf tissues of four grapevines with a protoplast isolation mix of 1% cellulase R-10, 0.2% pectolyase Y 23, 1% macerozyme R-10 with 24 mM $CaCl_2.2H_2O$, 0.092 mM NaH_2PO_4, 0.7 M mannitol, and 6.15 mM 2-(N-morpholino) ethane sulfonic acid, plus 0.1% polyvinylpyrrolidone (PVP). The viability of the protoplasts was confirmed by fluorescein diacetate (FDA) staining, and subsequent cell wall formation, cell division, and callus development. Knowledge gained and methodology developed in this study are very helpful for the selection of parent tissue sources in the subsequent somatic hybridization.

Z. Xu et al. (eds.), Biotechnology and Sustainable Agriculture 2006 and Beyond, 457–460.

Introduction

Grapes (*Vitis* L.) are consisted of many species and varieties. Genetic improvement of grape cultivars involves in gene flow among these species/varieties. Conventional breeding has made great contributions to crop improvement, including fruit crops like grapes. However, certain difficulties exist among some grape interspecific hybridization due to genetic incompatibility. Exploration of other breeding alternatives is therefore necessary to expand the utilization of the rich grape germplasm.

Protoplast technology or protoplast fusion, which includes isolation, culture, and subsequent regeneration of protoplasts to whole plants, has made remarkable progress and been proved to be a practical approach in overcoming the genetic incompatibility between species (Grosser and Gmitter 1990; Davey et al., 2005). Major challenges faced in grape protoplast technology include the need to refine isolation procedures, culture media, and the ability to manipulate the conditions for regeneration and growth of plants (Papadakis et al., 2001). In this research, we report an efficient protocol for preparing viable protoplasts from somatic embryogenic suspension cultures and leaves of grapevines. Methods for culturing purified and fused products by polyethylene glycol (PEG) were also presented.

Materials and Methods

Somatic embryogenic suspension cultures of 'Autumn Royal Seedless' and 'Chardonnay' (*V. vinifera*), 'Tara' (*V. rotundifolia*), and leaves from *in vitro* grown plants of 'Cabernet Sauvignon' (*V. vinifera*), 'Orlando Seedless' (*Vitis* hybrid), 'Alachua' and 'Supreme' (*V. rotundifolia*) were incubated in a protoplast isolation mix for overnight digestion, then purification in a double layer sucrose/mannitol gradient (Grosser and Gmitter, 1990). Fusion was carried out by using 40% polyethylene glycol. Purified and fused products were cultured on liquid and solid media of MS (Murashige and Skoog, 1962), EME (Grosser and Gmittee, 1990), and NN (Nitsch and Nitsch, 1969) with 4.5 μM 2,4-dichlorophenoxy acetic acid and 1μM 6-benzyladenine. Double density of protoplasts were placed on solid media.

Yield was measured by the number of protoplasts produced per gram of fresh weight plant tissue. Viability of protoplasts was determined by using FDA staining. Plate efficiency was calculated by the percentage of cell cluster formation.

Results and discussion

Incubation with the protoplast isolation mix for overnight worked well for grape cell wall digestion and protoplast isolation. Depending on species and physiological stage of the plant tissue, an average of 10^7–10^8 protoplasts per gram of fresh tissue were produced. More than 90% of the protoplasts were viable as indicated by FDA staining and subsequent culture. Healthy, fully expanded young leaves, and vigorously growing SE suspensions harvested 5–10 days after subculture gave the best results.

Under microscope, high quality intact spherical leaf protoplasts were filled with green chloroplasts that were easily distinguished from protoplasts derived from suspension cultures, which also provided an excellent marker for identifying the fused products. Compared to solid culture, liquid culture was preferred because it was easy to monitor the progress of the culture and had high percentage of cell wall formation, cell division, and plate efficiency rates. More than 30% of the fused protoplasts aggregated into clusters when cultured on MS liquid medium, which was also higher than those unfused protoplasts of their parents. This could be a result of the hybrid vigor and ploidy doubling. Directly cultured on solid medium was not recommended at this point because it required a higher density for culture and was harder to monitor.

Individual protoplast could form a cell wall and later initiate either a callus through sustained division or a somatic embryo from *in vitro* culture. Cell wall formation from embryonic protoplasts and fused products began 3–5 days after liquid culture. Division was appeared 5–7 days after cultivation and formed clusters thereafter. Callus could not be seen with naked eyes until 3 weeks after the cultivation. However, no callus formation was observed on leaf protoplast cultures.

Grape protoplasts have been successfully isolated, purified, and cultured. Notable cell wall formation, cell division, and callus formation from grape protoplasts have been achieved in fused protoplasts. As we prepare this report, experiment results are continuously progressing. Methodology developed from this study has laid out a solid foundation for a successful somatic hybridization of grapevines.

References

Davey MR, Anthony P, Power JB, Lowe KC (2005) Research review paper: Plant protoplasts: Status and biotechnological perspectives. Biotechnol Adv 23: 131–171.

Grosser W, Gmitter Jr FG (1990) Protoplast fusion and citrus improvement. Plant Breed Rev 8: 339–374.

Murashige T, Skoog F (1962) A revised medium for rapid growth and bio-assays with tobacco tissue cultures. Physiol Plant 15: 473–497.

Nitsch JP, Nitsch C (1969) Haploid plants from pollen grains. Science 163: 85–87.

Papadakis A, Reustle G, Roubelakis-Angellakis KA (2001). Protoplast technology in grapevine. In: Roubelakis-Angelakid KA (eds) Molecular Biology & Biotechnology of the Grapevine. Kluwer Academic Publishers, Netherlands, pp. 353–392.

Protoplast Technology and Citrus Improvement

W.W. Guo[1,*], X.D. Cai[1], Y.J. Cheng[1], J. W. Grosser[2], and X.X. Deng[1]

[1]National Key Laboratory of Crop Genetic Improvement, Huazhong Agricultural University, Wuhan 430070, China.
[*]Email: guoww@mail.hzau.edu.cn
[2]Citrus Research and Education Center, IFAS, University of Florida, 700 Experiment Station Road, Lake Alfred, FL 33850, USA.

Introduction

Citrus is one most important fruit crop in China and worldwide. Protoplast fusion has been an effective and successful technique for citrus improvement by circumventing reproductive barriers such as nucellar polyembryony, pollen/ovule sterility, sexual incompatibility and long juvenility encountered in conventional breeding. Protoplasts isolated from embryogenic callus line, as source material are convenient and available at any time for genetic transformation of seedless citrus cultivars since most commercial cultivars are seedless, and routinely used epicotyl seedling segments are only available for seedy cultivars. Strategies and recent progress by protoplast fusion and protoplast transformation in our citrus improvement program will be provided.

Protoplast transformation and transgenic plants regenerated

Valencia orange (*Citrus sinensis* Osbeck) is the leading commercial citrus species in the world for processed juice products, but

Z. Xu et al. (eds.), Biotechnology and Sustainable Agriculture 2006 and Beyond, 461–464.

thermostable pectin methylesterase (TSPME) in it reduced the juice quality. To eliminate or greatly reduce TSPME activity in Valencia orange, an antisense gene cassette containing gene-specific sequence from TSPME cDNA and the enhanced GFP as selectable marker was transformed into protoplasts of early mature Valencia strain B6-68 mediated by PEG transformation, in an effort to obtain TSPME down-regulated transgenic Valencia lines. Transformed line was identified via expression of GFP, physically separated from non-transformed tissue, and cultured on somatic embryogenesis induction medium. One transgenic embryoid expressing GFP was recovered, shoots were regenerated and their growth was expedited by *in vitro* grafting. PCR analysis revealed that the PME gene was present in it. Southern blot analysis confirmed the presence of EGFP gene (Guo et al., 2005). Transgenic plants containing only GFP gene from other cultivars such as Hamlin orange, Murcott tangor were also regenerated by protoplast transformation (unpublished data). These transgenic plants are growing normally in the greenhouse for further evaluation and utilization.

Protoplast fusion and somatic hybrids regenerated

In the past decade, somatic hybrids from more than 30 interspecific and intergeneric fusion combinations have been produced in our program in China. Typical somatic hybrids include: 1) red tangerine + trifoliate orange, and red tangerine + citrange for rootstock improvement; 2) Bonnaza navel + grapefruit, Bendizao tangerine + Kinnow mandarin as tetraploid pollen donor for triploid scion breeding; 3) intertribal somatic hybrids between Page tangelo + *Murraya paniculata*, Bonnaza navel + *Clausena lansium*. Some hybrids such as Page tangelo + trifoliate orange (4X), Jincheng orange + HR (Hamlin orange + rough lemon) (6X) have flowered and their fertility is high. Jincheng + HR is also the first report on hexaploid flowering in *Citrus*. Most of these somatic hybrids have been molecularly characterized and could serve as valuable germplasm for citrus improvement as well as for plant cell genetics basic studies.

Targeted cybridization to transfer sterile cytoplasm from Satsuma to seedy cultivars for potential seedlessness

CMS (cytoplasmic male sterility) is controlled by mitochondria genome in higher plants. In the past decade, numerous citrus diploid leaf parent type cybrids with mitochondrial genome derived from their corresponding embryogenic callus parent were unexpectedly regenerated by standard symmetric fusion (Grosser et al., 2000). This phenomenon was extensive, and mainly on interspecific fusion combinations. Utilizing this phenomenon, fusions of Satsuma mandarin cv Guoqing No. 1 (G1), a cytoplasmic male sterile cultivar, with other seedy types (e.g., HB pummelo, Sunburst tangerine, and Lee hybrid (Clementine × Murcott) respectively), were designed and cybrid plants were successfully obtained (Guo et al., 2004). Recently, more cybrids of G1 Satsuma with seedy but elite Chinese local cultivars, e.g., Bingtang orange, Shatian pummelo were regenerated and molecularly characterized (unpublished data).

By fusion of embryogenic callus protoplasts with gfp transgenic mesophyll protoplasts, and facilitated by gfp expression and visualization in hybrid cells, somatic hybrid vigor or regeneration advantage was revealed and evidenced (Guo and Grosser, 2005). Currently, using the fusion model of embryogenic callus protoplasts + gfp transgenic mesophyll protoplasts, studies on somatic hybrid vigor and mechanism for cybrid regeneration via symmetric fusion are being conducted. Hope this study will further facilitate targeted cybridization to produce more cybrids between male sterile Satsuma and seedy citrus cultivars for potential seedlessness.

References

Grosser JW, Ollitrault P, Olivares-Fuster O (2000) Somatic hybridization in citrus: An effective tool to facilitate variety improvement. In Vitro Cell Dev Biol – Plant 36: 434–449.

Guo WW, Duan YX, Olivares-Fuster O, Wu ZC, Arias CR, Burns JK, Grosser JW (2005) Protoplast transformation and regeneration of transgenic Valencia sweet orange plants containing a juice quality related pectin methylesterase gene. Plant Cell Rep 24: 482–486.

Guo WW, Grosser JW (2005) Somatic hybrid vigor in *Citrus*: Direct evidence from protoplast fusion of an embryogenic callus line with a transgenic mesophyll parent expressing the GFP gene. Plant Sci 168: 1541–1545.
Guo WW, Prasad D, Cheng YJ, Serrano P, Deng XX, Grosser JW (2004) Targeted cybridization in *Citrus*: Transfer of Satsuma cytoplasm to seedy cultivars for potential seedlessness. Plant Cell Rep 22: 752–758.

Microspore Embryogenesis in the Apiaceae

A.M.R. Ferrie and T.D. Bethune

National Research Council/Plant Biotechnology Institute,
110 Gymnasium Place, Saskatoon, SK, S7N 0W9, Canada.

Introduction

The Apiaceae include a number of economically important vegetable, herb, and spice crops. These crops are used as food, flavourings for food and beverages, medicinals for humans and animals, and perfumes and cosmetics. The Apiaceae are grown worldwide but are most common in the temperate regions of the northern hemisphere. Increased yield, higher oil content, and better quality oil are the main objectives of improvement programs.

Doubled haploid technology is being used around the world to develop cultivars and uniform lines in many species, however little work has been conducted on the development of such systems for the Apiaceae species. Calli/embryoids derived from anther culture of celery (*Apium graveolens* L.) (Dohya et al., 1997), *Bupleurum falcatum* L. (Shon and Yoshida, 1997), carrot (*Daucus carota*), fennel (*Foeniculum vulgare* Mill.), and mitsuba (*Cryptotaenia japonica*) (Matsubara et al., 1995) have been reported. Plantlets were regenerated from celery, carrot, and *B. falcatum*. These same studies evaluated microspore culture. Heart-shaped embryos were observed in celery but there was no further development (Dohya et al., 1997). The objectives of this study were to evaluate the Apiaceae for microspore culture response, and to compare the agronomic performance of resulting doubled haploid plants against the corresponding parental lines.

Z. Xu et al. (eds.), Biotechnology and Sustainable Agriculture 2006 and Beyond, 465–468.

Materials and methods

Donor plant seed of 15 Apiaceae species: ajowan (*Carum coptimum* L.), angelica (*Angelica archangelica* L.), anise (*Pimpinella anisum* L.), caraway (*Carum carvi* L.), carrot, celery, coriander (*Coriandrum sativum* L.), cumin (*Cumunum cyminum* Linn.), dill (*Anethum graveolens* L.), fennel, gota kola (*Centella asiatica* L.), lovage (*Levisticum officinale* Koch.), mitsuba, parsley (*Petroselinum crispum* Mill.), and parsnip (*Pastinaca sativa* L.) was obtained from Richter Seeds, Ontario, Canada. Dill seed was obtained from the University of Saskatchewan, Saskatoon, Canada. Plants were grown in the greenhouse or growth cabinets.

These Apiaceae species were evaluated for microspore culture response using the *B. napus* microspore culture protocol (Ferrie, 2003). Embryos derived from the microspore culture experiments were cultured on B5 medium (Gamborg et al., 1968) with no growth regulators. Plantlets with a well established root and shoot system were treated with 0.34% colchicine and planted into pots and grown in the greenhouse. Plants were self-pollinated to encourage maximum seed production. Seed was harvested from each plant individually.

Although doubled haploidy protocols have not been optimized for all Apiaceae species, doubled haploid plants and seed were produced from the experiments described previously. These lines were evaluated under field conditions. Plant height, yield and oil quality were evaluated.

Results and discussion

Of the 15 species evaluated, microspore swelling, divisions, and embryo development was observed in eight of them (angelica, anise, caraway, carrot, dill, fennel, lovage, and parsnip). These species were selected for further optimization and development of the doubled haploidy protocol. Factors evaluated included, genotype, donor plant growth conditions, pretreatments, developmental stage of the microspore, media constituents, and culture conditions. Embryogenesis went through the same developmental stages as *Brassica* embryo development (i.e., globular, heart, torpedo, cotyledonary). Although the protocol is not optimized at this time, embryos did develop from the different treatments evaluated. These embryos were regenerated to produce haploid plants. As no spontaneous doubling was observed, colchicine was used to double the chromosome number. Doubled haploid plants and resulting seeds were produced from anise, caraway, carrot, dill, fennel, lovage, and parsnip.

Under field evaluation, differences in height were observed between the parental lines and the DH lines. Of the nine dill lines evaluated, eight were shorter than the parental cultivar by at least 20 cm. Within each DH line, the plants were more uniform in height and morphology when compared to the parental line. This would be expected, as uniformity is a key feature of the doubled haploidy process. The shorter plant stature observed for many of the DH dill lines might be beneficial as this species is prone to harvest loss due to lodging. Seed yield from the DH dill lines showed some differences. The three most productive doubled haploid lines averaged 61% higher yield than the parental lines. Of the nine DH dill lines evaluated in the field, seven of them had a seed yield similar to or greater than the parental line.

Conclusions

The Apiaceae species have been considered recalcitrant when it comes to androgenesis, especially microspore culture. Very little research has been published in this area. Fifteen Apiaceae species were evaluated for microspore culture response. Of these 15 species, eight showed some indication of embryogenesis (i.e., swelling, microspore division, or embryo development). Doubled haploid (DH) plants have been produced in anise, caraway, carrot, dill, fennel, lovage, and parsnip. Seed from these DH plants have been generated in anise, carrot, dill, and fennel.

Field trials were conducted in 2003, 2004, 2005, and 2006. The DH lines differed from each other and the parental line in terms of germination, height, maturity, oil content, and oil quality. Efforts to identify DH lines with superior agronomic or quality characteristics are ongoing.

References

Dohya N, Matsubara S, Murakami K (1997) Callus formation and regeneration of adventitious embryos from celery microspores by anther and isolated microspore cultures. J Japanese Soc Hortic Sci 65: 747–752.

Ferrie AMR (2003) Microspore culture of Brassica species. In Maluszynski M, Kasha KJ, Forster BP, Szarejko I (eds) Doubled Haploid Production in Crop Plants pp. 205–215. Dordrecht: Kluwer Academic Publishers.

Gamborg OL, Miller RA, Ojima K, Nishio T, Dore C (1968) Callus formation and regeneration of adventitious embryos from carrot, fennel and mitsuba microspores by anther and isolated microspore cultures. Exp Cell Res 50: 151–158.

Matsubara S, Dohya N, Murakami K (1995) Acta Hortic 392: 129–137.

Shon T-K, Yoshida T (1997) Induction of haploid plantlets by anther culture of Bupleurum falcatum L. Jpn. Japanese J Crop Sci 66: 137–138.

Technology Transfer to Developing Countries

Integration of Cold Signal Transduction Pathway Related to ABA 8'-hydroxylase in Arabidopsis

Xin Zhou[1], Fei He[2], Fengxia Liu[1], Xue Zheng[1], Chao Di[1], Shaoxia Zhou[1], Hongya Gu[2], Wenying Xu[3], and Zhen Su[1]

[1]State Key Laboratory of Plant Physiology and Biotechnology, China Agricultural University, China.
[2]College of Biological Sciences, Peking University, China.
[3]Institute of Genetics and Developmental Biology, Chinese Academy of Sciences, Datun Road, Beijing 100101, China.

Abscisic acid (ABA) is one of plant hormones, playing essential roles in developmental regulation and stress signal transduction pathways, such as seed dormancy, seed germination, and environmental stress responses, etc. In recent years, the ABA 8'-hydroxylase was considered to play key role for ABA degradation. Arabidopsis P450 CYP707A subfamily genes were identified by reverse genetic approach to encode ABA 8'-hydroxylases (Kushiro et al., 2004; Saito et al., 2004). CYP707A subfamily includes four members in Arabidopsis, named CYP707A1 (At4g19230), CYP707A2 (At2g29090), CYP707A3 (At5g45340), and CYP707A4 (At3g19270). CYP707A genes were up-regulated by ABA, salinity, drought and osmotic stresses (Saito et al., 2004). However, there are no reports available about the interaction between CYP707A transcripts and cold stress.

Z. Xu et al. (eds.), Biotechnology and Sustainable Agriculture 2006 and Beyond, 471–474.

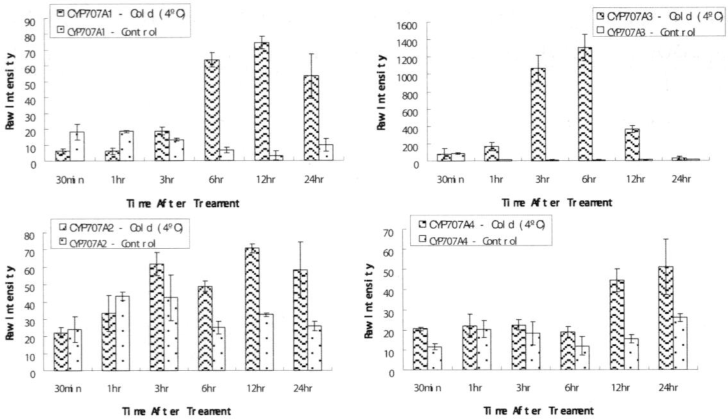

Figure 1. Expression level of CYP707A subfamily genes by cold stress in Arabidopsis Columbia wild type (col-0). (Original data from AtGenExpress.)

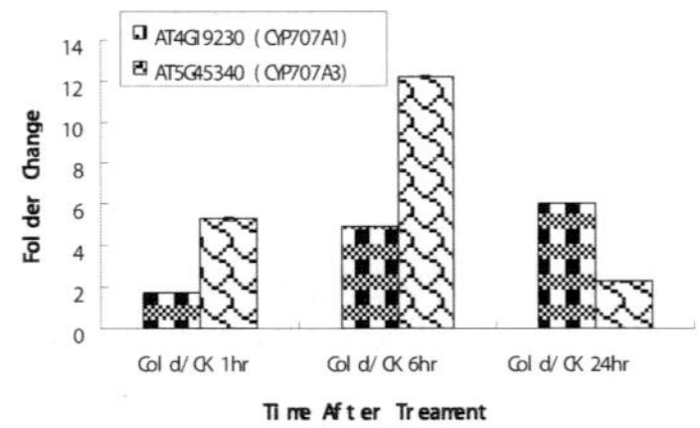

Figure 2. Real time PCR of CYP707A1 and CYP707A3 under cold stress in Arabidopsis Columbia wild type (col-0). Each bar shows the ratio between cold sample and control sample with triplicate reactions.

We did Arabidopsis GeneChip data mining studying the regulation of CYP707A genes in relation to cold respond. CYP707A subfamily genes showed differential expression patterns under cold stress (Figure 1). Especially, CYP707A3 expressed significantly higher than other genes in early cold treatment. The expression level of CYP707A3 increased under cold stress, starting at 1 h, and reached peak at 3–6 h of cold treatment, then dropped to control level at 24 h. Quantitative Real-Time PCR confirmed the expression specificity of CYP707A3 gene under cold stress (Figure 2). We also analyzed the late-stage cold-induced expression level of CYP707A gene in five cold-tolerant wild Arabidopsis ecotypes. The CYP707A3 was highly up-regulated at 24 h of cold treatment in these cold-tolerant ecotypes but not in Columbia ecotype (data not shown). In the meanwhile, we did data mining for CYP707A genes' expression in the time series of ABA treatment experiment (from 30 min to 3 h). These genes all showed dramatically induced in 3 h's exogenous ABA treatment (data from AtGenExpress).

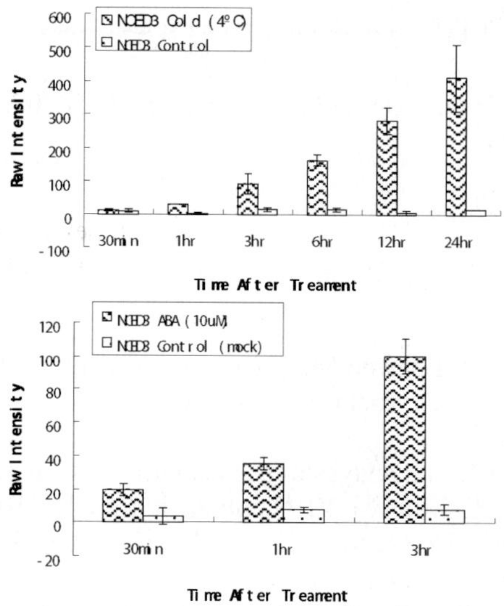

Figure 3. A: Expression level of NCED3 gene by cold stress in shoot samples. **B**: Expression level of NCED3 gene by ABA treatment. (Original data from AtGenExpress.)

NCED3 gene, one of the key genes in ABA biosynthesis pathway, was significantly up-regulated by cold stress (Figure 3A) and ABA treatment (Figure 3B).

Therefore, we propose that CYP707A3 and NCED3 expression may be regulated both by cold and ABA during cold stress (Figure 4). Our study might provide a new cold signal transduction pathway in Arabidopsis. Further simulation and reverse genetics approaches will be conducted for integrating ABA 8'-hydroxylase into cold signal transduction pathway.

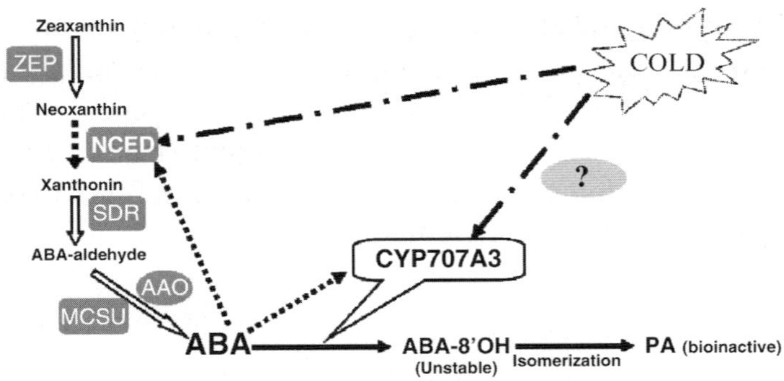

Figure 4. Hypothesis about how cold stress affects CYP707A3 and NCED3 expression in Arabidopsis.

References

http://arabidopsis.info/

Krochko JE, Abrams, GD, Loewen MK, Abrams SR, Cutler AJ (1998) (+)-abscisic acid 8'-hydroxylase is a cytochrome P450 monooxygenase. Plant Physiol 118: 849–860.

Kushiro T, Okamoto M, Nakabayashi K, Yamagishi K, Kitamura S, Asami T, Hirai N, Koshiba T, Kamiya Y, Nambara E (2004) The Arabidopsis cytochrome P450 CYP707A encodes ABA 8'-hydroxylases: key enzymes in ABA catabolism. EMBO J 23: 1647–1656.

Lee B, Henderson DA, Zhu JK (2005) The Arabidopsis cold-responsive transcriptome and its regulation by ICE1. Plant Cell 17: 3155–3175.

Saito S, Hirai N, Matsumoto C, Ohigashi H, Ohta D, Sakata K, Mizutani M (2004) Arabidopsis CYP707As encode (+)-abscisic acid 8'-hydroxylase, a key enzyme in the oxidative catabolism of abscisic acid. Plant Physiol 134: 1439–1449.

Rice Genetics and Genomics Information as the Ultimate Tool for Rice Improvement

Takuji Sasaki, Jianzhong Wu, and Takashi Matsumoto

National Institute of Agrobiological Sciences, 1-2, Kannondai 2-chome, Tsukuba, Ibaraki 305-8602, Japan.

Abstract

With the completion of the genome sequence, rice has become a very valuable crop not only in agriculture but also in basic and applied research. Accompanying this historical achievement, fundamental knowledge on the biology of the rice plant has been elucidated and various strategies have been developed to further clarify the molecular basis of its growth and development. These include microarray of DNA sequences representing genic and/or intergenic information for inheritance, non-coding RNA identification regulating gene expression, and proteome information including interaction required for hormonal signal transduction. The availability of an accurate, complete, map-based sequence provides a real opportunity to identify all the genes corresponding to phenotypes and thereby promote ground breaking research using conventional genetic tools. This is clearly shown by the increase in the number of rice genes controlling phenotypes that have been identified so far based on genomics information. Useful applications in physiological studies have also led to the clarification of gibberellin signal transduction pathway or discovery of novel genes involved in flowering among others. Another important point for basic research that will greatly facilitated by the standard genome sequence information of *Oryza sativa* is the clarification of biological diversity among the different

Z. Xu et al. (eds.), Biotechnology and Sustainable Agriculture 2006 and Beyond, 475–485.

species in the genus Oryza. Comparative analysis of the two major subspecies of cultivated rice, indica and japonica, and that of relative wild species will not only clarify the evolution of rice but will also facilitate the discovery of genetically important SNPs related to linkage disequilibrium. These basic tools and knowledge will provide a great opportunity to transfer the highly desirable alleles to actual breeding programs for improvement of current rice cultivars to feed the still increasing world population.

Introduction

Genome analysis of many species can provide the platform on how to understand the diversity of life. It also paves the way for a paradigm shift that may revolutionize various approaches in genetics and breeding similar to the formulation of Mendel's laws of genetics in 1865 and the discovery of the double-helical structure of DNA in 1953. Since rice is not only the most important staple in the world but is also one of the most widely studied plant species, a paradigm shift in rice research must be widely accepted by various groups from ordinary consumers and farmers, to basic plant scientists including geneticists, molecular biologists and breeders. An urgent need to supply sufficient amount of food to the increasing world population particularly in Asian and African countries is now widely recognized by the scientific community. At the same time, there is also a need to establish a model cereal crop that could be used in understanding other cereal crops that comprise the overall food source for the entire humanity. Extensive analysis of the rice genome is probably one of the most important contribution of the rice research community to address these needs. Therefore, the genomic information and tools developed in the long course of rice genome research may now be used to address these major issues so that the large investment in various projects on rice genome analysis can be converted to public scientific and economic benefits.

The history of rice breeding must have started in accordance with the beginning of agriculture. The early farmers must have compared characteristics of rice plants and their seeds, selected more favorable seeds and continuously sowed them for further harvest. At that time, farmers must have relied mainly on natural variation as source of plants with good agronomic potential. Through the natural system of ecological diversification, which involved continuous cycles of hybridization, differentiation and selection, favorable agronomic characters of rice must have been maintained for thousands of years. Since there is no clear evidence of a definitive strategy which led to cross pollination between different rice plants, early cultivars must have been accidentally generated by chance through cross pollination that might have occurred when two different rice groups were grown at proximate distance and flowered at almost the same time. In Japan and also presumably in many other areas and countries, such accidental crosses between different plant types might have led to production of rice plants with better qualities which were then selected by local farmers and replanted in succeeding planting seasons. Many so-called local varieties must have been generated in such manner and later used as resources for more organized breeding strategies based on established theories of inheritance formulated about 100 years ago. As a result, we now have more than 120,000 rice varieties around the world well-adapted to various culture conditions. However, increase of population especially in Asia where rice is the main staple requires further development of new breeding strategies to further increase rice production under much restricted environment such as decreasing lands for cultivation, much reduced water supply and various stresses that continuously reduce production. A large-scale program in rice genome research, which was launched in early 1990s in several countries and which led to the sequencing of the entire genome is expected to provide the stimulus for more innovative rice breeding strategy. This review will focus on what has been elucidated so far by extensive rice genome analysis in the last 15 years and how the results will be used for rice improvement.

Public access to rice genome information

Molecular analysis of the rice genome has made significant progress since the first rice molecular genetic map was published in 1989 (McCouch et al., 1988). The ultimate goal of deciphering the compete genome sequence was eventually attained in 2004 (International Rice Genome Sequencing Project, 2005). In the process, many molecular tools have been generated as follows:

1. RFLP markers and PCR-applicable SSR and SNP markers
2. BAC / PAC libraries and BAC clones with end-sequence information
3. cDNAs with full-length or partial sequences
4. Map-based genome sequence of a japonica cultivar and whole-genome shotgun sequence of an indica cultivar

These tools serve as the foundation for structural and functional analysis of the rice genome. All informations are publicly available in well established databases such as INE (http://rgp.dna.affrc.go.jp/E/giot/INE.html) and Oryzabase (http://shigen.lab.nig.ac.jp/rice/oryzabase/top/top.jsp). Upon completion of the genome sequence, the Rice Annotation Project Database (RAP-DB, http://rapdb.lab.nig.ac.jp/) has been released to provide a comprehensive and highly reliable analysis of predicted genes. It primarily maps rice full-length cDNA sequences to the Nipponbare genome sequence and contains the manually curated annotation of gene models. So far, about 30,000 rice genes are unambiguously attributed to the genome sequence. Similar analysis of the complete genome sequence in the TIGR Rice Genome Database (http://www.tigr.org/tdb/e2k1/osa1/) provides automated annotation that uses several prediction programs to obtain all plausible genes in the genome. The rice genome information is integrated with information in other cereal crops in the Gramene database (http://www.gramene.org/). Gramene also aims to support analysis of wild rice genomes to understand the domestication of rice and to facilitate the discovery of useful alleles for improvement of modern varieties to overcome biotic and/or abiotic stresses, increase yield and better eating qualities.

Genomics databases must be supported by controlled vocabularies for the description of the genes, molecular function, biological process, cellular component of gene products and phenotypes to facilitate uniform queries across various systems. In the case of rice, the Gene Symbolization, Nomenclature and Linkage subcommittee of the Rice Genetics Cooperative (http://www.shigen.nig.ac.jp/rice/

rgn/vol22/vol22_1.html) established the general rules and standardized symbols for known genes. On account of the completion of the genome sequence, it has also recently provided the guidelines for a universal system for naming of rice genes. The adoption of a common gene nomenclature for rice will facilitate structural, functional and evolutionary comparisons of genes and genetic variation among cereal crops, plants and other organisms.

Structural analysis of the rice genome

The fundamental characteristics of the rice genome such as gene models, gene duplication, segmental duplication, transposable element content, and cross talks between nuclear and organelle genomes have been analyzed in detail based on the high-quality rice genome sequence. In addition, the structure of the centromeres of chromosomes four and eight has been completely elucidated (Nagaki et al., 2004; Wu et al., 2004; Zhang et al., 2004) representing the first two centromeres so far clarified in higher plants. The genome information that has been accumulated so far provide an overview on the diversity in the genus Oryza and the need to explore the genome structure of various rice species (AA to JJ genomes) as well as to discover new alleles, to identify genes involved in agriculturally important traits, and to understand evolutionary history of the rice genome.

The availability of a standard genome sequence facilitates comparison among rice plants with high accuracy and within a short period. This is because of combination of the gold standard Nipponbare genome sequence and development of more sophisticated sequencer generating hundred mega-base nucleotides of short sequences (100–200 nt) within a short period of time and at a much reduced cost (Margulies et al., 2005). If these short sequences are overlaid on the standard sequence, it is expected that regions without repetitive sequences or insertions and/or deletions could be promptly aligned to generate a correct assembly of corresponding sequences. This was clearly shown by sequencing ten contiguous BAC clones of *O.rufipogon* assigned to the sequential Hd3a and FT1 regions (Margulies et al., 2005). This assignment was aided by using extensive

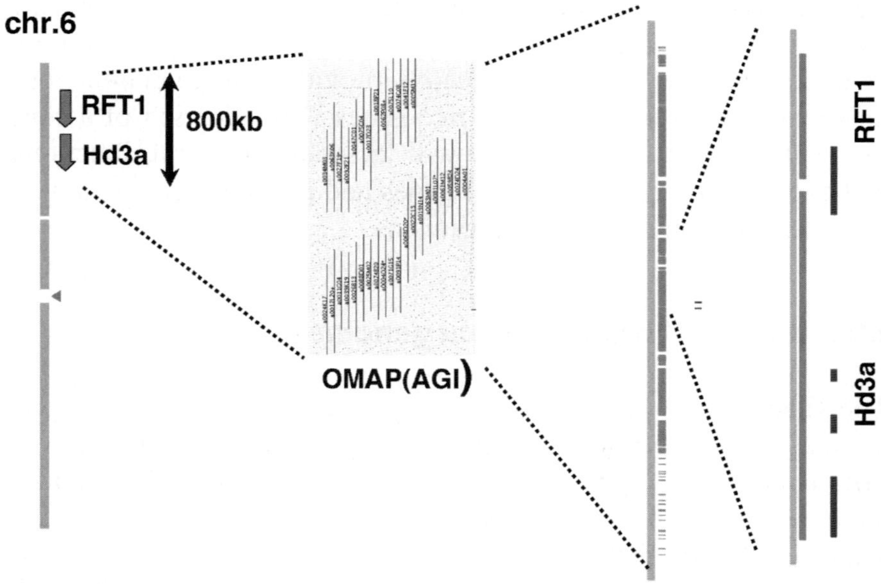

Figure 1. Contiguous *O. rufipogon* BAC clones assigned to the Nipponbare chromosome 6 region containing the RFT1 and Hd3 genes. The position of the two genes in Nipponbare chromosome 6 is shown on the left. The BAC contigs of *O. rufipogon* (AGI) in the corresponding region are vertically shown in the middle. In the right, sequence contigs from pyrosequencing method (red rectangles) are aligned to the Nipponbare sequence (light green bar). The annotated exons of Hd3a and RFT1 gene are indicated as blue bars.

fingerprint data generated by OMAP (http://www.omap.org/nsf.html). As shown in Figure 1, the genomic region of *O.rufipogon* corresponding to *Hd3a* and *FT1* was completely sequenced without any gaps. However, overlaying of short sequences was not successfully attained outside the FT region. These two opposing cases might indicate the existence of conserved genic and diverged intergenic sequences between *O.sativa* and *O.rufipogon*. In the case of *O.sativa* ssp. *indica*, both the genic and intergenic BAC-end sequences could be successfully overlayed to the japonica genome sequence thereby facilitating the construction of contigs (International Rice Genome Sequencing Project, 2005). These two cases suggest how the *O.sativa* standard genome sequence can be efficiently used to detect SNPs and to reconstruct the genome sequence of other types of rice plants.

Functional analysis of the rice genome

The term 'functional analysis' covers a wide range of meaning and in accordance with the progress of technology in genome science, a much deeper understanding of the mechanism of phenotype expression has become available. Identification and functional analysis of genes by map-based cloning is now efficiently and successfully carried out in rice using various tools obtained so far such as SSRs or SNPs, which could be used to detect polymorphisms and to narrow down the target genomic region (Kojima et al., 2002). Once a fine genetic map of the target region is established, screening of corresponding DNA fragments in BAC or fosmid clones can be accurately performed. However it is still necessary to continue the fine genetic mapping by adding new markers designed using genome sequences of selected DNA fragments to narrow down the candidate region to less than 50 kb or less because of the relatively small size of the rice gene. Occasionally the difference of gene expression profile corresponding to the phenotype can be used to identify the target gene. However, this approach is not always effective because the difference of gene expression does not necessarily correspond to the difference in phenotype. It is generally recognized that genes accompanying the phenotype are also variably expressed from time to time.

Forward genetic method relies on artificial disruption of the genome sequence and the information of the flanking sequence of the disruption point. Combining this sequence information with the corresponding phenotype can lead to the identification of genic or intergenic sequence responsible to the phenotype. Disruption tools that are now widely used in rice include T-DNA, *Ac/Ds* and *Tos17.* There are many disrupted rice lines generated by several researchers in several countries. An international collaboration has been established to facilitate sharing of resources and to efficiently perform functional analysis of the rice genome. This strategy could be very effective if various regions of the whole genome could be disrupted evenly. However, most of the available disruption tools such as T-DNA, *Ac/Ds*, or *Tos17* show preference of insertion into the genic region. This should not be a problem if the phenotype is only regulated by expression of genic region of the sequence. It is now widely known that intergenic regions such as promoter, *cis*-element or distantly separated sequence can invariably affect phenotype expression (Konishi et al., 2006).

One of the main focus for current studies on rice functional analysis concerns the elucidation of molecular mechanisms involved in the expression of quantitative trait loci (QTL). So far, QTL has been considered a statistical trait with no clear definition on what type of gene or sequence contributed to the trait. However, by combination of statistical genetics and molecular genetics, the molecular nature of QTL has been identified and represents actual interaction of genic or intergenic effect involved in QTL. This is clearly shown in case of heading date of rice controlled by photoperiod response (Yano et al., 2000). However, QTL is a complex trait and it is well known that detection of a specific trait is not guaranteed with any parental combination due to variation in nucleotide sequence of each corresponding locus of the QTL. A study to prove this hypothesis is now under way to correlate nucleotide sequence variation and degree of expression (Kliebenstein et al., 2006). Existence of QTL is only detectable by forward genetics and also the isolation of each gene involved in the QTL is available by map-based cloning using chromosomal segment substitution lines (CSSL). There are many reverse genetics resources useful for assisting the acceleration of elucidation of gene involved in QTL. However these resources must be systematically made available in comprehensive databases so that the relationship between disrupted sequence and phenotype can be clearly elucidated.

Transfer of novel molecular information to breeding

Breeding aims to create new varieties with more preferable or novel characters. For this purpose, both cross-aided and vector-aided methods have been used. The cross-aided method involves hybridization between different rice types and has been used for a long time. The vector-aided method involves introduction of non-rice genes or DNA by using *Agrobacterium tumefasciens*. This non-conventional method can overcome breeding barriers most commonly encountered due to lack of crossability resulting from chromosomal and genic differences. As a result, several agronomically important genes including herbicide tolerance and insecticide tolerance have been incorporated in rice. For example, the genetically engineered Bt-rice which is resistant to naturally occurring soil bacterium has been released and its successful commercialization has been achieved in Iran since 2005 (http://www.isaaa.org/kc/CBTNews/press_release/briefs34/ESummary/iran.htm). The so-called golden rice containing the pro-vitamin A synthesis genes is expected to reduce malnutrition in poor Asian and African countries which results in blindness among children suffering from vitamin A deficiency (Hoa et al., 2003). Transformation of rice itself does not require information of genome structure, but if targeted genome sequence is preferred for effective expression of an introduced alien gene, such study can be greatly facilitated by the genome sequence information. In fact, homologous recombination of rice has been successful due to efficient utilization of the genome sequence (Terada et al., 2002).

The utility of genome information for cross-aided breeding is realized in marker-assisted selection system. This system could be very effective if markers closely linked to the target trait could be developed and the gene corresponding to the trait can be identified based on the position of markers. Even if it is not possible, trials to find out polymorphisms in the sequence as close as possible to the target trait can be performed to get a reliable marker. In addition to establishing markers, cross-aided method requires a vast amount of supply of variable gene resources correlated to their degree of expression. Comparative genomics could provide the answer for analyzing the corresponding alleles in wild rice species.

Gene expression is highly influenced by the environment. Drastic changes in the environment such as global warming and deficiency of water undoubtedly influence growth of rice plants and finally rice yield. Recent advances in rice genomics offer new opportunities to enhance the efficiency of rice breeding strategies. The benefits of extensive genome research could be of great use to humanity if they could be used for practical rice cultivation in paddy fields that would lead to sustainable production. The challenge now is to bridge the lead to sustainable production. The challenge now is to bridge the gap that separates this new paradigm to actual breeding targets in order to generate rice varieties with higher-yielding potential, more nutritive value, and more resistance to biological and abiological stresses.

References

Hoa TT, Al-Babili S, Schaub P, Potrykus I, Beyer P (2003) Golden indica and japonica rice lines amenable to deregulation. Plant Physiol 133: 161–169.

International Rice Genome Sequencing Project (2005) The map-based sequence of the rice genome. Nature 436: 793–800.

Kliebenstein DJ, West MA, van Leeuwen H, Louder O, Doerge RW, St. Claire DA (2006) Identification of QTLs controlling gene expression networks defined *a priori*. BMC Bioinformatics 7: 308–325.

Kojima S, Takahashi Y, Kobayashi Y, Monna L, Sasaki T, Araki T, Yano M (2002) *Hd3a*, a rice ortholog of the *Arabidopsis FT* gene, promotes transition to flowering downstream of *Hd1* under short-day condition. Plant Cell Physiol 43: 1096–1105.

Konishi S, Izawa T, Lin SY, Ebana K, Fukuta Y, Sasaki T, Yano M (2006) An SNP caused loss of seed shattering during rice domestication. Science 312: 1392–1396.

Margulies M, Egholm M, Altman WE, Attaya S, Bader JS, Bemben LA, Berka J, Braverman MS, Chen YJ, Chen Z, Dewell SB, Du L, Fierro JM, Godwin BC, Helgesen S, Ho CH, Irzyk GP, Jando SC, Alenquer ML, Jarvie TP, Jirage KB, Kim JB, Lanza JR, Lefkowitz SM, Lei M, Li J, Lohman KL, Lu H, Makhijani VB, McDade KE, McKenna MP, Myers EW, Nickerson E, Nobile JR, Tomasz A, Volkmer GA, Wang SH, Wang Y, Weiner MP, Yu P, Begley RF, Rothberg JM (2005) Genome sequencing in microfabricated high-density picolitre reactors. Nature 437: 376–380.

McCouch SR, Kochert G, Yu ZH, Wang ZY, Khush GS, Coffman WR, Tanksley SD (1988) Molecular mapping of rice chromosomes. Theor Appl Genet 76: 815–829.

Nagaki K, Cheng Z, Ouyang S, Talbert PB, Kim M, Jones KM, Henikoff S, Buell CR, Jiang J (2004) Sequencing of a rice centromere uncovers active genes. Nat Genet 36: 138–145.

Terada R, Urawa H, Inagaki Y, Tsugane K, Iida S (2002) Efficient gene targeting by homologous recombination in rice. Nat Biotechnol 20: 1030–1034.

Wu J, Yamagata H, Hayashi-Tsugane M, Hijishita S, Fujisawa M, Shibata M, Ito Y, Nakamura M, Sakaguchi M, Yoshihara R, Kobayashi H, Ito K, Karasawa W, Yamamoto M, Saji S, Katagiri S, Kanamori H, Namiki N, Katayose Y, Matsumoto T, Sasaki T (2004) Composition and structure of the centromeric region of rice chromosome 8. Plant Cell 16: 967–976.

Yano M, Katayose Y, Ashikari M, Yamanouchi U, Monna L, Fuse T, Baba T, Yamamoto K, Umehara Y, Nagamura Y, Sasaki T (2000) Hd1, a major photoperiod sensitivity quantitative trait locus in rice, is closely related to the Arabidopsis flowering time gene *CONSTANS*. Plant Cell 12: 2473–2484.
Zhang Y, Huang Y, Zhang L, Li Y, Lu T, Lu Y, Feng Q, Zhang Q, Cheng Z, Xue Y, Wing RA, Han B (2004) Structural features of the rice chromosome 4 centromere. Nucleic Acids Res 32: 2023–2030.

The Development of Agricultural Biotechnology Capacities in Palestine through the UNESCO Biotechnology Educational and Training Center at Bethlehem University

Naim Iraki*, Omar Dar-Issa, Basma Sandouka, Michael Sansour, Rami Arafeh, Amani Abu-Sa'da, and Nida' Salah

UNESCO Biotechnology Educational and Training Center, Bethlehem University, P. O. Box 9, Bethlehem, Palestinian Authority, Israel.
*Email: niraki@bethlehem.edu

Introduction

Attempting to promote the development of agricultural biotechnology and plant molecular biology in developing countries of the world, Prof. Indra Vasil (University of Florida, USA) the former chairman of the UNESCO Biotechnology Action Council (UNESCO BAC) established, about 10 years ago, five biotechnology educational and training centers (BETCEN) in regions of developing countries in the world including China, Mexico, Hungary, South Africa, and Palestine. The mission of these centers is to promote the development of biotechnology research and build, through educational and training activities a well trained man power capable of conduction of modern research in plant molecular biology and agricultural biotechnology. During the first 3 years of activity, the UNESCO BETCEN in Palestine was granted about $200,000 from the UNESCO BAC to finance educational and training activities and to support small-scale research projects in Palestine and Arab countries. In addition, the center received, in subsequent years, research grants of about $1.2 million from American and German funding

Z. Xu et al. (eds.), Biotechnology and Sustainable Agriculture 2006 and Beyond, 487–490.

agencies to finance five research projects in plant biotechnology in collaboration with Israeli, Arab, and German research institutes. In this paper, we briefly report about the educational and research activities held in this center during the last decade, as well as about the outcomes of some of the completed research projects and their current application in the Palestinian agriculture.

Educational activities and research fellowships

Our educational activity involved the following; first, courses and workshops on Plant Tissue Culture Techniques, delivered by the late Prof. Abed Watad from the Agricultural Research Organization (ARO), Israel, and Prof. Hassan Abu-Qauod from An-Najah National University, Palestine. Second, workshops on Plant Viral Diseases and Detection of Plant Viruses by Electron Microscopy, was given by Prof. Abed Gera and Prof. Sara Spiegle, both from ARO, Israel. Third, an intensive course on Biological Control and a workshop on The Use of Entomopathogenic Nematodes as a Biocontrol Agent. These two educational activities were given by Prof. Ralf-Udo Ehlers and his team from the University of Kiel in Germany. In addition, our team held several introductory courses on biotechnology and biological control addressed to high school, and undergraduate students in the West Bank. The total number of scientists from Palestine, Jordan, Egypt, and Morocco participated in all courses and workshops exceeded 300. In addition to this educational activity, our UNESCO BETCEN awarded sixteen research fellowships, each of $5000, to students in Palestine, Lebanon, Morocco, Tunisia, and Egypt to support research on plant biotechnology.

Research work and the development of biotechnology capacities

In 1998, our laboratory started an intensive research work involving multilateral scientific collaboration within the frame of several research projects financed by American and German funding agencies. The first project involved development of a regional viral indexing and certification program for plant propagation material in the Middle East. The work was collaborated with laboratories from seven countries in the Middle East including; Palestine, Israel, Jordan, Lebanon, Egypt, Tunisia, and Morocco. The work of our team focused on development of serological and molecular techniques for the detection of viruses in grapevine, citrus, and potatoes. The project was financed by the U.S. Agency for International Development (USAID) for a period of 5 years. The second project dealt with studying the molecular basis for pathogenicity of *Clavibacter michiganensis* subsp. *michiganensis*, *Erwinia herbicola* pv. *gypsophilae* and *E.herbicola* pv. *betae* and its application for diagnosis. The main part of our work in this project focused on application of diagnosis methods based on PCR amplification of parts of the bacterial pathogenicity genes using specific primers. The work was collaborated with one laboratory at the University of Bielefeld, Germany, and another two laboratories in Israel. The third project, also financed by the DFG, Germany, studied the evolutionary divergence, reproductive biology and conservation of the royal irises *Iris* section *Oncocyclus*. The research work was in collaboration with the University of Mainz in Germany and the Hebrew University of Jerusalem in Israel. The work of our team involved evaluation of levels of genetic diversity within populations of two endangered species of irises in the West Bank and Israel ; *Iris haynei* and *I. atrofusca* using random amplified polymorphic DNA (RAPD) for the assessment of the taxonomic status of these two taxa in order to recommend appropriate conservation measures.

The knowledge and expertise acquired by our team through work on the above topics, constituted a solid basis for the establishment of a well-equipped diagnostics laboratory for the detection of bacterial and viral phytopathogens. This laboratory is the sole phytopathogen diagnostics facility available in Palestine. The Palestinian Ministry of Agriculture utilizes the services of this laboratory for a nationwide annual screening of grapevine, citrus, and stone fruit seedlings for the occurrence of viral and bacterial pathogens. The annual survey is important for implementing the required quarantine measures.

The fourth research project conducted in our laboratory involved research on insect-pathogenic nematodes and their use for the control of white grubs (*Maladera matrida*) in the Middle East. We have isolated and identified ten strains of nematodes by means of molecular methods involving restriction digestion and DNA sequence homology of part of the nematode genome. One of our isolates is possibly a novel species. In addition, we applied two of our isolates for the control of white grubs in fields of sweet potato in Gaza. This project was financed by the DFG, Germany and collaborated with ARO, Israel, and University of Kiel, Germany. The fifth project deals with development of tomatoes with combined resistance to Tomato Yellow Leaf Curl virus (TYLCV) using both virus-derived resistance and molecular marker-assisted breeding. The funding and collaboration are as those mentioned for the first project. We have succeeded to deve lop a transgenic tomato plant resistant to TYLCV.

AUTHOR INDEX

Abu-Sa'da, Amani 485
Aftab, Faheem 437
Ahn, Yeh-Jin 227
Ainley, W. Michael 111
Akbergenov, Rashid 201
Arafeh, Rami 485
Arana, Fernando 299
Arslan, Neşet 381
Ayaydin, Ferhan 149
Azria, D. 175

Badea, Ana 349
Balestrazzi, A. 413
Barrell, Philippa 77
Baskar, Ramaurthy 77
Beerhues, Ludger 345
Bell, Jeremey 333
Berger, Fred 131
Bethune, T.D. 465
Bhatnagar, S. 397
Bonadei, M. 413
Borchetia, S. 235
Borthakur, A. 235, 242, 253
Bouton, Joseph 333
Boyce, A.N. 327
Brand, Lukas 77
Brown, Daniel C. 103, 125, 425
Buh, M. 191

Cai, Charles Q. 111
Cai, X.D. 461
Calligari, P. 413
Carbonera, D. 413
Carman, John G. 303
Chan, Rebecca M.L. 283
Chandran, S. 327
Chantawong, Satjaporn 429
Charls, S.M. 363
Chaudhury, Abed 131
Chen, F.C. 261
Chen, Grace Q. 227
Chen, Hong-Hwa 167
Chen, Wen-Huei 167
Chen, Z. Jeffrey 153
Chen, Zhen 299
Cheng, Wang 231
Cheng, Xiaofei 333
Cheng, Y.J. 461
Chew, O. 175
Choi, Young-Im 265
Cidade, Luciana C. 223
Cifuentes, Carolina 353
Cleveland, L.L. 361
Çöçü, Satı 381
Collingwood, Trevor N. 111
Confalonieri, M. 413
Costa, Marcio G.C. 223
Cserháti, Mátyás 149, 163
Curtis, Mark 77

Dar-Issa, Omar 485
Das, Sudripta 235
Deandrea, G. 413
Deng, X.X. 461
Deng, Xing Wang 81
Dennis, E.S. 131, 205
Dhankher, Om Parkash 49
Di, Chao 471
Dixon, Richard A. 231
Dobra, J. 171
Dobrev, P. 171
Dodo, H.W. 261
Dolferus, Rudy 205
Dolgov, S.V. 279
Drossard, Jürgen 13
Dudits, Dénes 149, 163

Dunwell, Jim M. 35
Dutta, R.K. 235
Dwivedi, Krishna 305

Eason, Jocelyn 157
Edwards, Keith 289
Elliott, Estella G. 305
Emons, Anne Mie C. 353
Eudes, F. 349

Fagoaga, Carmen 211
Falkenburg, Dieter 399
Fehér, Attila 149
Feldmann, Ken 319
Ferrie, A.M.R. 465
Fischer, Rainer 13
Flores, Ricardo 211
Fregene, Martin 409
Fujioka, Shozo 319

Garrison, Robbi J. 111
Gaudet, D. 349
Gaudinova, A. 171
Ge, Xiaochun 199
Ge, Yaxin 333
Geelen, Danny 121
Gennaro, M. 413
Gezan, Salvador 289
Gheyselinck, Jacqueline 77
Giersberg, Martin 399
Giorcelli, A. 413
Gooding, Mike 289
Goossens, Alain 135
Graf, R. 349
Gregory, Philip D. 111
Grosser, Judy 457, 461
Grossniklaus, Ueli 77
Gruden, K. 191
Gruissem, Wilhelm 201
Gu, Hongya 471
Gunawardana, D. 175
Guo, W.W. 461
Guo, Yang-dong 447
Gürbüz, Bilal 381

Haisel, D. 171
Han, Kun Jun 333
Hanhineva, Kati 115
Haque, Md. Shahidul 10
Harris, Sam 319
Hasegawa, Kana 209
Hattori, K. 107
Hayashi, Kokoro 209
He, Bing 97
He, Fei 471
Heaton, Andrew C.P. 49
Hellwig, Stephan 13
Hensel, Goetz 143
Hinchliffe, Doug 153
Hocher, V. 451
Hohn, Thomas 201
Hong, Yan 397
Hong, Yuan Qing Deborah 397
Hopkins, Andrew 333
Horváth, Gábor V. 149, 163
Hsiao, Yu-Yun 167
Huang, Jian 63
Huang, Tian-Hsiang 167
Huang, Yinghua 215

Iba, Koh 209
Inzé, Dirk 135
İpek, Arif 381
Iraki, Naim 485

Jamison, Michelle S. 305
Jaskowiak, M.A. 363
Jensen, Susanne 409
John, U.P. 175
Johnston, Amal 77
Jones, Todd 141
Jongsma, M.A. 191
Jørgensen, Kirsten 409

Karenlampi, Sirpa 115
Karplus, Valerie J. 81
Kavas, M. 315
Kawasaki, Tsutomu 209
Kim, Do-Young 265

Kim, Hee Jin 153
Kim, Jeong-Soon 305
Kim, Tehryung 49
Kirdmanee, Chalermpol 429
Kis, Petra 163
Klein, Patricia 305
Köhler, Claudia 309
Kojima, Chojiro 209
Kok-Jacon, Geraldine 405
Kokko, Harri 115
Konan, K.N. 261
Krishnamurthi, M. 433
Kuiper, M.J. 175
Kumlehn, Jochen 143, 399
Kuoh, Chang-Sheng 167
Kuwano, Mio 419
Kwok, Shing 319

Lacey, Jeff 305
Lai, Fang-Ming 141
Laroche, A. 349
Lee, Youngsook 265
Leu, Yann-Lii 167
Li, Deng-Di 295
Li, Guo-Jing 199
Li, Xue-Bao 295
Li, Yan 299
Ling, Lee Ai 327
Liu, Benye 345
Liu, Chang-Jun 231
Liu, Fengxia 471
Liu, Qiaoquan 283
López, Carmelo 211
Lu, Jiang 457
Luo, Ming 131
Luo, Mingzhu 131
Lupotto, Elisabetta 389

Ma, Xuefeng 333
Mackin, L. 175
Malbeck, J. 171
Mankin, Luke 141
Marthe, Conny 143
Martin, R. 171
Martinoia, Enrico 265
Matsumoto, Takashi 477
Mattivi, F. 413
Meagher, Richard B. 49
Mei, Kangfeng 141
Mendes, Amanda F.S. 223
Mendoza, Alfonso Hermoso de 211
Miki, Brian 125
Mills, Clare 289
Mirici, Semra 381
Miskolczi, Pál 149
Mitchell, Rowan 267
Mizukami, Yuuko 447
Mohan Verma, Virendra 375
Mok, D.W.S. 171
Mok, M.C. 171
Møller, Birger Lindberg 409
Moore, Gloria A. 223
Morant, Marc 409
Moreno, Pedro 211
Motyka, V. 171

Nair, Helen 327
Navarro, Luis 211
Nazarian Firouzabadi, Farhad 405
Noel, Joseph P. 231
Noh, Eun-Woon 265
Novakova, M. 171

Ogihara, Yasunari 323
Oksman-Caldentey, Kirsi-Marja 135
Öktem, H.A. 315
Oliver, Sandra N. 315
Oresic, Matej 205
Otoni, Wagner C. 223
Özcan, Sebahattin 381

Parmaksız, İskender 381
Pathirana, Ranjith 157
Pattanayak, Jayasree 305
Peacock, Jim W. 131
Peña, Leandro 211
Pence, V.C. 363
Pennell, Roger 321

Perera, P.I.P. 451
Petolino, Joseph F. 111
Picco, F. 413
Pinontoan, Reinhard 209
Pischke, Melissa 49
Plair, B.L. 363
Pliansinchai, Upsorn 429
Polotnianka, R.M. 175
Pompe-Novak, M. 191
Pooggin, Mikhail M. 201
Poole, Rebecca 289
Pospisilova, J. 171
Preece, John E. 437
Pua, Eng-Chong 89

Qin, Ji 405
Quattrini, E. 413

Radhakrishnan, Parthiban 319
Rajeswari, S. 433
Ravnikar, M. 191
Razali, Zuliana 327
Ribarev, E. 175
Riehl, Marcus 399
Rischer, Heiko 135
Rubin-Wilson, Beth C. 111

Saalbach, Isolde 399
Sadasivaiah, S. 349
Salah, Nida' 485
Salazar, Andres 319
Sancak, Cengiz 381
Sandouka, Basma 485
SanfaHon, HjiPne 103
Sansour, Michael 485
Sarıhan, Ercüment O. 381
Sasaki, Takuji 475
Sass, László 163
Schestibratov, K.A. 279
Schillberg, Stefan 13
Schulenberg, Lisa L. 111
Sekar, Shini 433
Sevimay, Cafer S. 381
Shenton, Christina 289
Shewry, Peter 289
Shi, Hai-Yan 295
Shim, Donghwan 265
Shimamoto, Ko 209
Sivakumaran, A. 175
Skeggs, Peter 289
Slapar, N. 191
Smith, Aaron P. 49
Soares-Filho, Walter S. 223
Song, Won-Yong 265
Sørensen, Charlotte 409
Spangenberg, G.C. 175
Stelly, David 153
Stile, Maria Rosaria 389
Storchova, H. 171
Štrukelj, B. 191
Su, Zhen 471
Sun, Samuel S.M. 283
Suurs, Luc 405
Suzuki, A. 107
Suzuki, Makoto 107, 419
Svircev, Antonet 103
Sze, Sing-Hoi 153

Tabata, Ryo 209
Takaiwa, Fumio 419
Talbot, J.P. 175
Tamura, Ken-ichi 447
Tandon, Pramod 385
Tase, Kazuhiro 447
Thaxton, Peggy 153
Tian, Lining 103, 125
Török, Katalin 163
Trieu, Anthony 319
Triplett, Barbara A. 153
Tsai, Wen-Chieh 167
Tsuneyoshi, T. 107
Turóczy, Zoltán 163
Twyman, Richard M. 13

Ulrich, Tom 305
Umezawa, Toshiaki 341
Uranbey, Serkan 381
Urwin, P.E. 181

Valkov, Vladimir 143
Vanderschuren, Hervé 201
Vang, Louisa 227
Vankova, R. 171
Verdeil, J.L. 451
Vincken, Jean-Paul 405
Viquez, O.M. 261
Visser, Richard G.F. 405

Wan, Jianmin 319
Wan, Yongfang 289
Wang, Guang Yuan 397
Wang, Hei-Chia 167
Wang, Hong 295
Wang, Jiulin 319
Wang, Xiu-Lan 295
Wang, Zeng-Yu 333
Weerakoon, L.K. 451
Weerathaworn, Pipat 429
Welham, Sue 289
Wen, Rui 103
West, Phillip 157
Wheeler, Tim 289
Winget, G.D. 363
Wong, Hann Ling 209
Worden, Andrew 111
Wright, Elane 333
Wu, Chuan-Yin 319
Wu, Jianzhong 475
Wu, Keqiang 125
Wu, Tian-Shung 167

Xi, Yajun 333
Xia, Hui-Jun 299
Xia, Yiji 199
Xiao, Xirong 333
Xu, Hong-Yuan 97
Xu, Hong-Zhang 97
Xu, Wen-Liang 295
Xu, Wenying 471
Xu, Xia 457
Xu, Zhi-Hong 3
Xue, Yongbiao 63
Xue, Zheng 471

Yaeno, Takashi 209
Yakandawala, D.M.D. 451
Yamada, Toshihiko 447
Yang, Guang 299
Yoshida, Kaoru T. 419
Yücel, M. 315

Zelasco, S. 413
Zhang, Ji-Yi 333
Zhang, Ke 319
Zhang, Peng 201
Zhang, Shuocheng 103
Zhang, Zai-Bao 299
Zhao, Lan 63
Zhao, Xiaochun 205
Zhao, Zuo-yu 273
Zhou, Feng-Jue 97
Zhou, Shaoxia 471
Zhou, Sijun 425
Zhou, Xin 471
Zhu, Huifen 199
Zhu, Tong 289